Science, Faith, and Politics

Science, Faith, and Politics

FRANCIS BACON AND THE UTOPIAN
ROOTS OF THE MODERN AGE

A Commentary on Bacon's *Advancement of Learning*

Jerry Weinberger

CORNELL UNIVERSITY PRESS

ITHACA AND LONDON

International Standard Book Number 0-8014-1817-8
Library of Congress Catalog Card Number 85-47707
Printed in the United States of America
Librarians: Library of Congress cataloging information
appears on the last page of the book.
The paper in this book is acid-free and meets the guidelines for
permanence and durability of the Committee on Production Guidelines
for Book Longevity of the Council on Library Resources.

To Diane, Seth, and Davida

CONTENTS

Contents

PREFACE

In this book I hope to restore the now-forgotten eighteenth-century view that Francis Bacon was the greatest of all the "moderns"—the thinkers from Machiavelli to Hobbes who recommended turning the human intellect from the contemplation of God and nature to the scientific project for mastering nature and fortune. Even though Rousseau thought Bacon was perhaps the greatest philosopher to have lived, in recent times Bacon has been all but forgotten, with attempts to understand the principles of modern life and politics focusing on Machiavelli, Hobbes, and Locke. But it is not my intention just to set the record straight with regard to Bacon's greatness. For in our times we have come to doubt the goodness of the modern project, especially as it consists in the progress of modern science. With such doubts abounding, it is no wonder that studies of Rousseau proliferate and that Nietzsche and Heidegger have become more respectable and more studied. I hope to show that Bacon not only was a central founder of the modern project but also knew more about its limits and problems than its later critics. Like Nietzsche and Heidegger, Bacon understood the modern project to be the destiny of human reason. But unlike them, Bacon did not subvert modern rationalism as he exposed its defects. Rather, he thought it possible to understand the causes of human destiny, thereby to bring reason to bear on the problems of reason. In rediscovering the power of Bacon's thought, I have sought to unearth principles by which we can take our bearings in our modern, technological world.

I intend to show that although Bacon was a founder of the modern project, he did not take its point of view for granted. In fact, from the standpoint of an earlier wisdom, the teaching of classical

9

political philosophy, he understood both the necessity and the limits of the new course for human intellect and ambition. Consequently, when he made the case for the scientific mastery of nature, Bacon did more than merely supplement Machiavelli's realistic new political science. Like Machiavelli, Bacon thought that classical political thought was inadequate for coping with the problems of Christian politics. Unlike Machiavelli, he also thought that we could understand and prepare for the modern age only by properly understanding the wisdom of Plato and Aristotle, which Machiavelli had so boldly rejected. As far as the modern project rests on Baconian rather than on Machiavellian grounds, the turn from ancient to modern thought was not a categorical rejection of classical political philosophy. According to Bacon, only the ancient wisdom could disclose how far modern hopes and methods—and their unintended consequences—are truly new and how far they are more familiar than we suppose. And it is easier to grapple with familiar problems than it is to face what is altogether strange.

Why this book is a commentary on Bacon's *Advancement of Learning*, referring to his other works only as seems appropriate to the course of that treatise, I explain in my Introduction. But before I proceed, I should mention some matters of texts and format. The *Advancement of Learning* was an early work, published in English in 1605, seven years after the first edition of the *Essays* and fifteen years before the account of the new inductive method for studying nature, the *Novum organum*. In 1623, Bacon published an expansion and Latin translation of the *Advancement of Learning* titled *De dignitate et augmentis scientiarum*. The *De augmentis* consists of nine separate books: the first is a translation of the first book of the *Advancement of Learning* and the remaining eight are a translation and expansion of the second book of the *Advancement of Learning*, which contains the division of the sciences. For two reasons the present study focuses on the early *Advancement of Learning* rather than on the later *De augmentis*. First, the later additions do not alter significantly the account of the sciences judged to be lacking or sufficient, and the translation of book 1 of the *Advancement of Learning* is almost direct. Second, and more important, Bacon admitted that in preparing for the translation to be read in "all places" he had acted as his own *Index expurgatorius*: throughout the whole of the *De augmentis*, Bacon omitted all comments offensive to the Church of Rome.[1] As we will see in the text that follows, the problems of religion and Christianity, including Ro-

1. See Spedding's note, *Advancement* 277.

man Catholic Christianity, are central to Bacon's teaching. Therefore the *Advancement of Learning* is complete, and the *De augmentis* is
not.[2] For this reason my commentary deals with the *Advancement of
Learning* with an eye on the *De augmentis* rather than vice versa. In
the footnotes, I have identified the important differences between the
two versions.

My procedure will be to paraphrase each distinct portion of Bacon's text, using Bacon's versions of Latin quotations when appropriate, and then to analyze and characterize the argument so that the
reader can follow the argument without reading Bacon's text itself.
Each section paraphrased begins with bracketed page numbers identifying its location in Bacon's text and ends with a horizontal line
marking it off from the following commentary. For all purposes I
have used the authoritative edition of Bacon's works prepared by
Spedding, Ellis, and Heath between 1857 and 1874, published in both
London and Boston. The Boston edition was organized differently
from the London edition; I have used the London edition because it
was reprinted in facsimile by Friedrich Frommann Verlag in 1963
and is therefore likely to be the more readily available. The bracketed page numbers refer to the third volume of that edition, which
contains the *Advancement of Learning*. Throughout the paraphrase of
Bacon's text, I have for the most part used Spedding's translations
of Bacon's quotations from classical literature. In the footnotes I have
identified Spedding's translations, and where there is no such notice,
the translation is my own. In identifying and locating Bacon's classical sources, I have used Oxford Classical Texts unless I have indicated otherwise. Throughout I have used the standard scholarly
conventions for references to classical and other texts so that readers
should be able to find their places in most editions or translations.

I am pleased to thank in print those institutions and individuals
who have helped me with this book. At early stages of my work, the
Earhart Foundation of Ann Arbor, Michigan, provided much-needed
financial assistance. I appreciate its trust and support. Susan Plesko
expertly typed draft after draft.
My parents instilled in me a love of learning and hard work. For

---

2. Bacon's concern about the secrecy of political science may explain why he chose
to announce in the *Advertisement Touching an Holy War* that the *De augmentis* acquitted
his promise to present a division of the sciences rather than to announce it in the
obvious place, the *De augmentis* itself. Bacon certainly knew that the announcement
was important, for he instructed Rawley to have the *Advertisement Touching an Holy
War* translated even though it was an incomplete fragment. Spedding wonders about
this matter; see *BW* I, 415–20; *Advertisement* 3–7.

this legacy I will always be grateful. My wife and children, to whom this volume is dedicated, were as patient as they could have been with a husband and father who was too often absorbed in study and writing. I can only hope that the present work justifies their sacrifice.

Mark Blitz, Harvey Mansfield, Jr., and Arthur Melzer read all or parts of earlier drafts. They and anonymous reviewers provided much good advice about how the book could be made better. I have been fortunate to have had several wonderful teachers. Foremost among them is Harvey Mansfield, Jr.; his teaching, thought, and scholarship have been my standards, and to him I owe more than I can repay.

JERRY WEINBERGER

East Lansing, Michigan

ABBREVIATIONS

Advancement   *The Two Books of Francis Bacon of the Proficience and Advancement of Learning Divine and Human.* The *Advancement of Learning* is located in volume 3 of *BW*, 259–491.

Advertisement   *Advertisment Touching an Holy War.* The *Advertisment* is located in volume 7 of *BW*, 9–36.

Att.   Cicero. *Epistulae ad Atticum.* Text and trans. D. R. Shackelton Bailey. Cambridge: Cambridge University Press, 1967.

BW   *The Works of Francis Bacon,* ed. James Spedding, Robert Leslie Ellis, and Douglas Denon Heath. 14 vols. London, 1857–74.

D   *Discorsi sopra la prima deca di Tito Livio.* In *Tutte le opera di Niccolò Machiavelli,* a cura di Francesco Flora e di Carlo Cordié. Rome: Mondadori Editore, 1968.

De aug.   *De dignitate et augmentis scientiarum.* The *De augmentis* is located in volume 1 of *BW*, 423–837.

Ep.   Seneca. *Epistulae morales.*

Essays   *The Essays or Counsels, Civil and Moral* (1625). The *Essays* are located in volume 6 of *BW*, 371–517.

EW   *The English Works of Thomas Hobbes of Malmsbury.* Ed. Sir William Molesworth. 11 vols. London: John Bohn, 1839–45.

Fam.   Cicero. *Epistulae ad familiares.*

Historiae   *Scriptores historiae Augustae.* Ed. E. Hohl. 2 vols. Leipzig: Teubner, 1965.

Instauration   *The Great Instauration.* James Spedding's translation of Bacon's prolegomena to the whole of his intended project, *Instauratio magna.* The *Instauration* is located in volume 4 of *BW*, 3–33.

Nic. Eth.   Aristotle. *Nicomachean Ethics.*

P   *Il Principe.* In *Tutte le opere di Niccolò Machiavelli,* a cura di Francesco Flora e di Carlo Cordié. Rome: Mondadori Editore, 1968.

Science, Faith, and Politics

Introduction: Ancient Utopianism
and the Modern Age

This book is a commentary on an early seventeenth-century text, but my old-fashioned approach to an old treatise is meant to illuminate the most pressing problem of the modern age: what may loosely be called the "problem of technology." No one can doubt that such a problem exists and that it stalks our lives and our thoughts. The free and the unfree political systems of our day are bound together by identical hopes for technological abundance and by shared fears of nuclear war and environmental damage. More generally, the scientific promise of boundless liberty seems belied and threatened by scientific determinism and materialism. And the more science promises human self-reliance, the more we search for missing gods. We feel besieged by the very means that grant us power, and we are alternately proud and ashamed of our impious mastery over nature.

It might be easy to understand these problems if technology were simply applied natural science, so that humanistic knowledge could judge them both as any end orders means that it neither produces nor applies and so that humanistic knowledge could separate the human from the divine. But in fact modern humanism is grounded on technology: the humanistic perspective is no less scientific than the scientific project is humane and technological. Nietzsche and Heidegger have shown that modern natural science from its beginning assumed the world to be the whole of things that can be used, changed, and conquered by a knowing subject. Accordingly, the world was taken to be made up of tools and objects, with objects understood to be the eternal, objective reality underlying the changeable usefulness or "value" of every thing. For modern science, the changeless order of

17

nature is revealed by way of, and as the ultimate source of power for, human art. The changeless "objective" truths of nature appear, paradoxically, only as nature is the object of conquest rather than the mysterious source of guidance and awe. God is dead because scientific rationalism separates eternal being from becoming, with the effect that everything divine becomes objectively knowable, while the objectively knowable becomes the basis of the means for human will. We are estranged from an awesome god because science makes the divine all too human and familiar.[1] Reason's knowledge of the spontaneous order of nature is assimilated to the freedom of human art. And as we learn from Kant, *the* source of modern humanism, scientific reason shows the essence of man to be practical, or moral, freedom, because nature's subjugated objective order can be no guide or model for human purpose. Because they are morally free, human beings do not need their artful projects even though only reason understood as art discloses the changeless truth of freedom's difference from nature: only with the light of science can humanity be separated from the objective, natural human possibilities or from art as it freely attends to unfree, determined needs.

For these reasons the problem of technology is not just a matter of distinguishing humanistically between means and ends, and between man and God, so that we may restore them to rational balance. Rather in the modern age the project of reason causes us to doubt whether ends and means are related at all and whether reason separates man from God only to make the eternal or divine a mere means. Small wonder that we are today confused about means, ends, and the gods: to know ourselves as ends is possible only by freeing means and art from the limits of any eternal end, and as long as every end is human, we are wholly free from means and the need for art. In the modern scientific age, idealism that treats ends as human freedom, materialism and technology that unlock nature for her means, and rational theology that seeks a loving god who works no miracles each claim to be the exclusive source of answers to the questions of practical life. In fact they are all cut from the whole cloth of the scientific project, but as elements of the modern age, they contend with the humorless, dogmatic fervor of ones who cannot do without their enemies. They are necessary partners, but they can never dance in harmony.

1. See, e.g., Martin Heidegger, "Die Frage nach der Technik," in *Vorträge und Aufsätze,* pt. I (Pfullingen: Neske, 1967); *Einführung in die Metaphysik* (Tubingen: Niemeyer, 1953); "Nietzsches Wort 'Gott ist tot,'" in *Holzwege* (Frankfurt: Klosterman, 1952).

The most general lesson of this book is that the contending elements of the modern age are best understood as natural forms of political partisanship and that as such they are topics of political philosophy. The book's claim is that we learn this lesson from Francis Bacon, which is surprising, since Bacon predated German critical philosophy and since he was such a partisan, not to say founder, of the modern scientific project itself. My claim is that in his *Advancement of Learning* Bacon tells us more about the causes and problems of the modern age than any other thinker, at least in part because, while he was one of its founders and so not situated within it, he was not moved by the enthusiasm and partial ignorance of the practical founder.

As if these claims were not startling enough, they are demonstrated by a very special kind of commentary on Bacon's text, one that is different from most conventional forms of textual analysis or historical research. I turn to Bacon because he stands in and outside our modern age: he speaks to those who preceded him as well as to our own critical needs. My concern is historical only insofar as the modern age has a historical course, which includes the discovery of history as a social and cultural force. Bacon need not be studied in the terms of some narrow historical context, because he gives an account of our modern history's *nature*. He tells us what human beings must be in order to have experienced modern history. For Bacon, mankind can have a history, but mankind's natural possibilities determine the course of that history, not vice versa. I am simply willing to consider that Bacon might have been right and that any attempt to explain the truth or importance of his teaching as the reflection of some historical moment can itself be explained by his argument about what modern history is. The commentary to come must bear the weight of such claims, but before we can turn to our study, we need to know the broader context of ideas in which Bacon's teaching stands. Some remarks are necessary to explain why we should turn to Bacon's teaching at all. And we need to know what our mode of study will be and why it must be employed. In the course of these remarks we will glimpse the views developed in the commentary itself.

In this book I intend to show that Bacon was more than an enthusiastic partisan of the new program for human reason, the conquest of nature by a new natural science. I contend that Bacon understood the new project's limits and problems from a perspective beyond its point of view. But according to the most powerful understanding of

19

the modern age—bequeathed to us by Nietzsche and Heidegger—
Bacon could not have occupied such ground. According to Nietzsche
and Heidegger, Bacon could not have surveyed a new chapter in the
history of reason because he was already in the grip of what was truly
new. Bacon could not have seen what was unique and troublesome
about the modern age because he was already its child. For Nietzsche
and Heidegger, the real beginning of modernity must be traced to
the origins of rationalism in Greek philosophy, in particular to Socrates and Plato. The novelty of the modern scientific age pales before the development of rationalism and its guise, metaphysics. In
disclosing the forms of thought and action, metaphysics began with
"self-evident," that is, dogmatic, distinctions between subject and object, being and becoming, whole and part, nature and value, and ends
and means. The project of rationalism was to harmonize the elements of these distinctions by way of knowledge. Modern science is
the corrosive legacy of this original dogmatism: as the most successful form of knowledge it presupposes the distinctions, but at the same
time it discovers unbridgeable chasms between them. These chasms
lead to the practical paralysis—the "nihilism"—of the scientific age,
where idealism, materialism, technology, and rational theology are
locked together by shared, dogmatic presuppositions and are torn
asunder as exclusive and inadequate claimants to knowledge about
rational practice.

In describing the elements of the modern age as the result of now-
forgotten dogmatism, the account that I have just described is more
than a little powerful. But to accept it we must wean ourselves from
the most fundamental terms that make problems of our current
problems of thought and practice. Whatever our current predica-
ment, the dangers of such a leap should be well known to anyone
familiar with practical affairs. In going beyond the "nihilism" de-
cried by Nietzsche and Heidegger, we give up the moral categories
necessary for any intelligible talk of such humble practical necessities
as justice and moderation. One intention of this book is to take a step
toward showing that this account of the problem of technology is
wrong. But in doing so I will not contend directly with its spokes-
men. Rather I will show that some, but not enough, of their account
of our beginnings is correct.

It is true that at least Machiavelli, the first modern founder, was
ultimately less original and self-knowing than he appears. In this respect he reflected some prior dogmatism. But Machiavelli seems
dogmatic precisely when we contrast his thought with the open questioning of the classical political philosophy of Socrates, Plato, and

Aristotle, as we learn from Francis Bacon, the second founder, who recommended a project he *knew* to be dogmatic. From classical political philosophy Bacon learned that the ambition of technology is the ultimate source of dogmatism, which is either idealistic or materialistic. He learned that such ambition is rooted in mankind's political nature. From the ancients, Bacon learned the causes of the scientific pretension of reason to harmonize nature and human action either by way of masterful art or by way of absolute, demonstrable knowledge, and he learned how this pretension is inseparable from the human openness to the gods. And finally, from the ancients Bacon learned why in his age—and in ours—managing such dogmatic pretension required something quite different from the classical teaching about concrete political practice. From the standpoint of reason that knows both justice and moderation, it is possible to show the causes of reason's now-perplexed dogmas. To grasp these notions, and to make our way to Bacon himself, we have to understand just how classical political philosophy was not dogmatic. We have to say something about what it meant for classical political philosophy to be utopian.

We have admitted that the very first founder was both dogmatic and not wholly original. It will remain to be seen just how the Italian doctor was not wholly original, but we can glimpse his dogmatism if we compare his thought to that by comparison with which he regarded his own as wholly new. If one considers the founders of the modern scientific project, it is clear that they spoke for the powers of reason more as partisans than as open-minded questioners. They rejected scholastic dogmatism but only to clear the way for their own version of demonstrable certainty. And to justify doing so, they had to reject classical utopian thought. Classical utopian thought disclosed nature and political life in the light of one human possibility: the life of restful contemplation. In doing so, it revealed the limits of human attainments—for practical men as measured by contemplation and for knowledge as measured by its objects. Reason can disclose the ends that determine human excellence, but human nature has neither the spontaneity of nonhuman nature nor the permanence of its forms. Rather, human nature must make its way to its end by means of art, which is as changeable as it is always less than perfect. While ends can be disclosed, they cannot be immanent in practice, and therefore they cannot be demonstrated as ends with the certainty required for practice. For this reason they are the objects of contemplation and wonder; they have more of the nature of eternal problems than they do of the mariner's north star. Precisely

21

because nature differs from and yet is tied to art, theory disappoints the practice that so needs theory. When classical utopianism imagined what political life would be like if it were modeled not on contemplation and wonder but on the certain knowledge of the highest human excellence, it did so not for the sake of producing such a life but to show the crucial tension between theory and practice. Classical utopia disclosed the comedy of practical hopes for the sake of moderation. And it disclosed the limits of knowledge for the sake of a kind of piety. Knowing about the being of justice, *the* political virtue, could never lead to perfectly just practice because such knowledge could never be comprehensive or clear enough to satisfy the demands of practice. Unlike gods or the self-sufficient objects of contemplation, men are always subject to needs whose urgency outstrips their knowledge of justice.

As we will see in the course of our study, classical utopianism taught that these needs are caused by more than material scarcity or changeable circumstances. In fact, they are caused by the very need that practice has for knowledge and by the tension between them. They arise because human beings are characterized by reason *and* art.[2] As rational animals endowed with speech, human beings can distinguish between need and self-sufficiency, between dependent part and resplendent whole. And as artful creatures they strive to become self-sufficient and whole. As human beings are both artful and rational, they are open to the possibility of freedom, which is coeval with the human experience of need. But the ancients thought freedom dangerous: there is a reason, known to the ancients, why freedom, so dear to our modern age, is not the prescription of classical utopian thought. We need to see very briefly just why it is not.

The ancient utopian thinkers took human society to be an order of several productive arts pursued by persons who, unlike the brutes, act not out of mere necessity but for the sake of the good, for self-sufficiency.[3] As it is imaginable by the rational animal, perfect self-sufficiency is the godlike freedom from the need to produce. Of course if all human beings tried to exercise such freedom, they would surely perish. But if they did not strive for such self-sufficiency, if they attended only to the truly most immediate needs, they would be no different from the brutes who practice no arts. Therefore human society is an order of arts that produces what are in fact luxuries: things that remove us from the immediate demands of the body but that,

2. Cf. *Nic. Eth.* 1094a1–b11; *Politics* 1252a1–23, 1252b27–1253a18.
3. See n. 2 above; Plato, *Republic* 369b5–372e1.

22

although they appear to provide true self-sufficiency, are really re-
finements of the immediate bodily needs.

Were all men to know the real truth about luxury, that it provides
merely apparent self-sufficiency, they would act like the contemplative Socrates, who did not practice the art that he knew. But of course
then the very existence of society would be impossible. Because no
one desires what is merely apparently good or self-sufficient,[4] for men
to be more than mere brutes the arts must be pursued from the mistaken belief that they and their objects provide true self-sufficiency.
For human beings, society can exist only when they are deluded about
luxury. Every art is pursued as if self-sufficient wholeness consists in
the most of what only it can produce. And each art's rational desire
for the good is an aspiration to an impossible freedom from need
because, to be self-sufficient, it assumes the spontaneous service of
all the other arts. But then controversy arises: should military strat-
egy, which produces victory, serve or be served by medicine, which
produces health? Every art is not only an art but also a possible claim
to rule, and if there is to be any order to the arts, and hence any
production, one claim must subordinate other claims. But no art or
artful product—not even a poetic dialogue or philosophic speech—
is really perfectly free or self-sufficient. Therefore even the most
persuasive subordination of the arts to a ruling art will mask some
measure of controversial repression. For this reason, coercion must
always underlie persuasion.

Classical utopianism was actually realistic. It taught that because
reason and art are always linked, society as an order of productive
arts always produces political claims that incline toward unrealistic or
naive utopianism—and tyranny—because their very tendency is to
forget their dependence on other conflicting claims. Every claim can
forget its own political character—its need both to coerce and to persuade other reasonable claims—as if it were the perfectly free and
just principle of a noncoercive, spontaneous order of the arts. But
such a claim can only lead to the most tyrannical coercion because it
could accomplish its impossible goal of overcoming the need for
coercion and persuasion only by wholly and violently suppressing all
other claims. For classical utopian thought, the very existence of human society depends upon the artful pursuit of perfect freedom and
justice, but this is always the most dangerous delusion.

Such delusion is reflected in the concern practical human beings
have to comprehend perfect, primordial origins of social life. Origi-

4. *Republic* 505d5–10.

nal, golden times are depicted as free from scarcity, for which reason they are taken to be free from all coercion and injustice. But the
common myth of civil religion simply expresses the latent, utopian
pretense, or hope, of any productive art to provide perfect freedom
and justice, as if their opposites were caused by the mere scarcity of
what it alone knows how to produce. As Plato says by way of the
Stranger, even in an age of Cronos, where there is no scarcity, the
only creature for whom there would not be a problem about justice
would be the pig, not the rational animal.[5] It is not scarcity that causes
injustice and need but the human longing to be self-sufficient like
the gods, a longing that rational, artful human society cannot do
without. The common myth of civil religion is not the result of idle
curiosity. Rather it is a sign of the inherent, utopian danger of political society: men cannot forget about perfect, primodial beginnings
because they are always tempted to recreate them. The truth is that
such times, whether past, present, or future, could only be terrible,
as in fact real beginnings actually are. The myth may cover up this
harshness, causing men to forget it. And such forgetting may prevent us from imitating its license, but the image depicted in the myth
is really the very source of the worst injustice. Reason is always tempted
to clarify human beginnings by aspiring to the apparent power of art
or by granting to art the apparent certainty of reason. As it thus arises
from political life, philosophy is tempted to see the whole of nature
and man as if it could be fully comprehended either by art by itself
or by reason by itself. Philosophy is therefore tempted to claim to
know the whole to consist of only the moving bodies that can give
themselves to art or of only the intelligible, mathematical forms loved
by the mind.[6] Such views debunk convention in the light of nature,
for conventions are neither as fixed as eidetic or mathematical enti-
ties nor as malleable as matter in motion. But only well-formed conventions restrain the naive utopian political hopes inherent in every
order of productive arts. As idealism or materialism, philosophy
springs from and supports the rational pretensions of the arts, and
only political philosophy can expose its errors.

For classical utopian thought, the truth about political rule is that
it can never be perfectly free or just. Moreover, the desire to know
and provide such rule is as dangerous as it is necessary. For this reason, political life could be ordered only by the moral virtues that,

5. Cf. Plato, *Statesman* 226c4–6, 269c4–72d4; *Republic* 372a5–73a8.
6. See *Nic. Eth.* 1094a1–b11; *Metaphysics* A; Plato, *Republic* 504d4–e3; also see
Chapters 5 and 8 below.

though modeled on the likes of Socrates' contemplative freedom, could never be as complete. As the city is a political order of productive arts, the moral virtues were special modifications of these arts, which, we recall, Socrates did not practice. For the classical utopians, the moral virtues were a kind of oblivious attention to the arts that make up the city. For instance, while the art of acquisition is constrained and harsh, and open to the tyrannical claim to just self-sufficiency, the virtue of this art, liberality, would cultivate the practice of graceful spending, so as to make the spender not take acquisition too seriously, as if he did not once have to acquire needily or did not depend upon those who do. Of course such virtues depend not upon knowledge but rather upon a special kind of ignorance, an ignorance of original and continuing need and an ignorance of the virtues having been produced by a convention-producing art. Although the virtues serve the arts by moderating their necessary claims, this needy service would have to be forgotten lest the moderating grace of virtuous freedom be exposed as less than it must be taken to be. In sum, freedom and justice could be understood in the light of the moral virtues, not vice versa. Therefore, practical affairs could never be better ordered than by moderation, which requires at least some toleration for injustice and need. But even moderation could never be easy or likely, because it could be produced only by the prudent use of religious myths that cannot but refer to perfectly free and just beginnings. The very means for inducing the oblivion required for moral virtue suggest the harshest hopes of the productive arts. The best such myths can hide the harshness of beginnings, but their very focus on perfect times reveals the inescapable limits on the prudence by which they can be used. This conclusion is not hopeful, but to think otherwise would be to imagine human beings wholly different from their actual condition. It would be to repeat the artful delusion that misunderstands what it means for human beings to experience freedom and need and for justice to be *the* political virtue.

By comparison the founders of the modern scientific project are more like dogmatic partisans—or at least more like practical founders with axes to grind—than like questioning, ironic political philosophers. And they made it clear that their project required rejecting the tradition of classical utopian thought. Machiavelli, Bacon, and Hobbes all complained that classical utopian thought impeded the power of men to conquer nature and fortune and to discover demonstrable principles of justice. This rejection of classical utopianism was nowhere better stated than it was at first in Machiavelli's famous diatribe against Plato in the fifteenth chapter of *The Prince:* those who study imaginary republics and principalities rather than new ones, and those who abandon what is actually done for what should be done, will come to ruin. Bacon complained that the ideal of restful contem-, plation bred contempt for experience and the practical arts, so necessary for the "conquest of nature in action." Likewise Hobbes claimed boldly that, when founded on the principles of the new science of matter and its "clear and exact method," the realistic study of morals and politics would disclose an infallible means to safe, commodious living: a "true and certain rule of action by which we might know whether that which we undertake be just or unjust."[7] Beginning with Machiavelli, the founders of our modern age promised the human conquest of fortune and a freedom not just for some in relation to others but for humanity as a whole. Machiavelli's realism turned men from imaginary worlds but only so that men might conquer fortune and manage political affairs with perfect certainty. No such promises of certainty could be farther from the questions induced by a Platonic dialogue, in particular, the *Republic.* And to the ancient utopians, such promises could only be dogmatic, and dangerous, if they were ever thought to be possible.

The ancient utopians considered it necessary to manipulate myths so as to hide the dangerous link between man's practice of the productive arts and the rational longing for self-sufficiency. But the moderns argued that if religion could be made to reflect these facts realistically, so that all would know that all human possibilities, and especially the moral virtues, are actually comprised of the earthly passions and needy desires served by the practical arts, it would be possible to achieve perfect justice by producing a perfect economy of liberated desire and physical satisfaction. And it is but one step from such a materialistic economy to the idealistic view that human freedom transpires beyond the realm served by any artful project. As I noted earlier, when reason assimilates art, human freedom can be severed from every artful activity.

From the ancient utopian point of view, the realistic modern project merely reflects the dangerous and deluded hope that injustice and dependence spring from the scarcity of what only the productive arts

7. *Instauration* 7–8, 23–24; *Novum organum* 1.3, 2.1–4; *Advancement* 294–95, 475; Thomas Hobbes, *The Elements of Philosophy,* in *EW* I, 9 (1.7). Cf. Hobbes, *De cive,* Ep. Ded., *EW* II, i–viii; *EW* I, 8–10, IV, 1; *Leviathan,* chap. 26. See J. Weinberger, "Hobbes's Doctrine of Method," *American Political Science Review* 69 (December 1975):1336–53.

know how to make. From the ancients' point of view, the new project for reason repeats the rational pretension of every art. It seems to believe a new version of the dangerous myth of civil religion: by focusing on realistic rather than golden beginnings—on what came to be known after Hobbes as the "state of nature"—it would be possible to effect the golden dream of every art. Moreover, the course of this project, from Machiavelli's realism to Bacon's science of nature, and from this to Hobbes's science of perfect justice and thence to the idealistic doctrine of freedom, could have been predicted from the ancients' analysis of the dogmatic possibilities of reason and art. Because they spring from the openness of the artful, rational animal to the divine, the modern project could have been comprised of nothing more or less than the contending and contradictory elements of materialism, idealism, technology, and eventually rational theology.

From the classical point of view, not only was Machiavelli's realism dogmatic, but the course of the modern age from its dogmatic beginning to its present perplexity about means and ends was no mere accident. It is no wonder, then, that Machiavelli rejected the ancient teaching. But it is therefore remarkable that Bacon turned to the utopian form of writing, in his *New Atlantis*, toward the end of his career. Bacon's work as a whole is obviously dedicated to replacing classical with modern thought: in his *Novum organum*, he presented his famous new method for experimental natural science. And he defended his new practical project for learning, the mastery of nature, against the speculative, contemplative tradition in the *Advancement of Learning*, first published in 1605 and again, in Latin translation, just three years before his death. In the *Advancement*, Bacon repeated Machiavelli's argument against classical utopian thought. But if the mere existence of the *New Atlantis* seems anomalous, the character of the work is even more so. Although it depicts a perfectly just and harmonious world provided by the unlimited material products of modern science, it differs markedly from the enlightened, realistic world of secularized Christianity predicted by Machiavelli, Descartes, and Hobbes. As anyone who reads it can see, the Christianity described in the *New Atlantis* is as pious as it is secular, the society is not egalitarian or free, and its science is shrouded in secrecy, denying the possibility of full enlightenment to all or even most who live there. These obvious facts accord more with classical utopianism than they do with the modern scientific project. Surprising as it may be, their importance has not been noted often, which perhaps explains why Bacon's moral and political thought has been taken to be located in the *Essays* and in professional writings rather than in his dramatic argument for a wholly new project for human learning.[8]

When the *New Atlantis* is examined more closely, it becomes clear that it imitates Plato's utopia as much as it announces a new perfected world. In fact, it reflects on the modern project from the standpoint of the ancient utopian political philosophy. To see how it does so is to gain entry to Bacon's critical political philosophy. In particular, we learn how Bacon's treatise *The Advancement of Learning* went beyond Machiavelli to disclose the elements and causes of the problems that beset the project he began and Bacon advanced and recommended. If we understand this achievement we may see how reason might be brought to illuminate our own troubled times, when reason's very ambition obscures the reasonable balance of means, human ends, and the gods.

In order to understand the *New Atlantis*, we must look very briefly at Plato's *Republic, Timaeus,* and *Critias,* which it explicitly imitates. As *the* work of classical utopian thought, the *Republic* presents an account of the best city, the city that would embody perfect justice. Socrates fashions this city on the model of the arts as unerring and self-sufficient, a model suggested by the two most important interlocutors, Thrasymachos and Glaucon, both of whom wish they could be perfectly self-sufficient and therefore perfectly free and just.[9] This wish is nothing but the utopian political claim latent in the practice of the productive arts, and because it is such, the perfectly just city modeled on it proves to be impossible. As a perfect productive art, the best city's justice requires two separate principles that are taken to be the same: that each person perfect one productive art and that each person mind his own business. But these principles are not compatible if some art involves minding others' or everyone's business. And in fact the philosophers and the guardians mind the business of the artisans while the artisans practice the only art Socrates calls common: the money-making art.[10] The problem is that what is

---

8. *New Atlantis, BW* III, 130–31, 134–35, 146–51, 154, 165. Gough seems to see some of these facts, but he ignores their importance; see *New Atlantis,* ed. Alfred B. Gough (Oxford: Clarendon Press, 1915), xxxiii. Scholars almost uniformly take the *New Atlantis* to be nothing but a model for a future scientific academy, with the notable and important exception of Howard White, *Peace among the Willows: The Political Philosophy of Francis Bacon* (The Hague: Nijhoff, 1968).

9. *Republic* 336b1–37d2, 338c1–3, 340d1–41a4, 348c11–12, 358b1–62c8, 370a7–c5.

10. Ibid., 334b3–6, 345e5–52a3, 433a1–34d1, 519c8–21b10, 540e5–41b5.

determined to be one's own is nothing but the practice of some productive art. And because every art ultimately minds everyone's business, there is no standard to help us differentiate between what is and what is not one's own. Communism would solve this problem if whatever were one's own were exactly what also belongs to everyone else, but the communism introduced in the *Republic* proves to be impossible because there is no such thing as a human body that can share everything or can have nothing that is irreducibly its own.[11]

The artful desire for perfect freedom and justice produces a city that turns out in fact to be a collection of well-trained soldiers who protect the practitioners of the common art of money making, who are not said to practice the communism that supposedly moderates the soldiers. Despite the fact that avarice is a source of injustice,[12] the best city as a whole proves to be avaricious, which leads us to suspect that its sole activity might be conquest and acquisition. Now, although Socrates presents the best city as a kind of possible new beginning, he does not do so freely but rather as the result of persuasion and force.[13] And although this force is playful in the dialogue, it cannot but remind of the deadly force that was ultimately brought to bear on Socrates' freely chosen practice of questioning the human things and those who purport to know them.

From Socrates' presentation we learn that, although men who need the arts cannot but look to such beginnings, the best we can do is to cover up the truth about beginnings—obscuring the true harshness of the human need to acquire by way of art, which is apparent in real beginnings and latent in all political practice. Even in the best city, then, it is not the demonstrable knowledge of justice "in itself" that moderates the best citizens' activities. Rather it is a noble lie asserting that the inhabitants of the city are autochthonous.[14] Were it otherwise, then their first perfected art must have been the art of violent acquisition, and the most such a lie can do is to cause them to forget the harshness of original, artful acquisition, thereby to minimize the likeness of their present activities to it. It will not be surprising, then, if their present activity proves in fact to be something like unbridled, or tyrannical, acquisition.

We know what their activities are from the *Timaeus* and *Critias*, which continue the discussion in which Socrates "founds" the best city. In the *Timaeus*, some men present at the discussion of the *Republic*, but

11. Ibid., 449a1–57c2, 462c10–e2, 466c6–72b2.
12. Ibid., 416d3–17b9, 547b2–69c9.   13. Ibid., 327a1–28b3.
14. Ibid., 414b8–17b9.

not the *Republic's* interlocutors, meet the next day to consider the best city "in motion," or engaged in its most suitable activity.[15] This activity proves to be a war (not political philosophy) fought by the best city, now identified as early Athens, against the artful, prideful, and tyrannous Atlantians who threaten the peoples of Greece. Now, in the *Timaeus*, Critias relates the war very briefly, explaining that, after the Athenians defeated the invaders and liberated the Greeks who would not help themselves, both Atlantis and the Athenian warriors were destroyed by a flood that was a divine purge rather than a natural catastrophe.[16] But after this brief speech, Timaeus relates a cosmogony that takes up the rest of the entire dialogue. In the *Critias*, which begins where the *Timaeus* ends, Critias describes first the Athenians and then the Atlantians, but before he relates the details of their war, the dialogue breaks off abruptly.

We wonder why Critias' speech must be preceded by Timaeus' long, strange cosmogony. We can discern the answer by thinking about how the *Critias* is unfinished. The dialogue breaks off just as Zeus is about to make a speech to announce his punishment of the Atlantians, who have become lawless and ambitious.[17] We do not hear the speech, but we *do* know that the Athenians defeated the Atlantians and that they were both destroyed by divine purge. But why would Zeus, who rules by law, punish the innocent Athenians?[18] Answering this question is the point of Timaeus' cosmogony. In accounting for the origins of the visible whole, his speech attempts to convert the "noble lie" of the *Republic* into a truth: his cosmogony would explain how the citizens of the best city could be literally children of the earth.[19] If they can be shown in such a light, then justice-loving Athenians would not have to settle for being fooled by noble lies, because their origins may truly have been golden and so perfectly free and just. Timaeus, an eminent citizen of a foreign city, is described as an accomplished philosopher given to astronomy and cosmology, that is, to the formal, mathematical account of the moving whole.[20] But in the dialogue, he brings philosophy to the task of accounting for the *origins* of the whole, not just of earth and stars, but also of all the things on earth, including the best city. Timaeus brings natural philosophy and mathematics—not political philosophy—to the aid of the justice-loving Thrasymachos and Glaucon, who speak for the artful hope that such origins were the most beautiful and the best and that they might be recreated.[21] If injustice caused the divine purge, then the blame

---

15. *Timaeus* 19b3–20c3.   16. Ibid., 21a7–25d6.   17. *Critias* 121a8–c5.
18. Ibid., 121b7–8; *Timaeus* 25b5–c6.   19. *Republic* 414e1–6.
20. *Timaeus* 20a1–5, 27a2–b6.   21. Ibid., 29d7–30c1.

can be attributed to Zeus, not to the Athenians or to the original demiurge.

But Timaeus' cosmogony proves to be inadequate for its task. Not only does Plato tell us that it is only a likely tale, but even what is given in order to show the Athenians' justice proves to be shocking: as Timaeus tells the tale, the very fact of the Athenians' autochthony would mean that they love bloodshed more than justice. Attempting to give a philosophic account of the whole, Timaeus in his ignorance of the human things does not adequately explain the differences between animals and between animals and vegetables.[22] In the absence of such an explanation, it is certain that if the original Athenians, or any people, were children of the earth, then they must at first have been needy cannibals. And if they were not such children, they must have been unjust. The Athenians' war against the Atlantians could be as just as it could be only on the basis of the noble lie, not on the basis of Timaeus' cosmogony. The best they could have done would be to have forgotten the harshness of original beginnings so as not to be tempted by their unbridled license. But this would have made them at best moderate, not perfectly just or free, and unlike human law, Zeus might punish anything less than perfection.

For Plato, the artful desire for perfect justice prompts curiosity about perfect, free beginnings, but the truth about them reflects the fact that for human beings there are always needs that cannot be met justly: human society is always an order of productive arts, pursued for the good life or for self-sufficiency; for this reason every society is political, and every political order rests upon some controversial repression. On the one hand, knowing this truth can give license to the savage harshness latent in all artful acquisition. But on the other hand, striving for a perfect beginning is the necessary, justice-loving source of the worst injustice. Therefore the best that practical men can do is to listen to lies that obfuscate the truth about beginnings, like the noble lie and unlike Timaeus' tale, for the sake of moderation. For this reason the philosopher Timaeus, not the political philosopher Socrates, presents the truth-loving cosmogonical tale meant to satisfy Thrasymachos' and Glaucon's extreme justice-loving desires. Apart from political philosophy, natural philosophy, whether it be materialistic, idealistic, or both, really speaks for partisan enthusiasm. Political philosophy teaches that there is no demonstrable knowledge of perfect beginnings and that there is, consequently, no demonstrable possibility of perfectly just practice. To see the best

22. Ibid., 76e7–77c5; see Seth Benardete, "On Plato's *Timaeus* and Timaeus' Science Fiction," *Interpretation* 2 (Summer 1971):26–30, 59–63; *Republic* 469b5–71c3.

possibilities of justice and virtue, the perplexing tension between art and reason, and so between convention and nature, must be understood to be the eternal problem that it is. The belief that it might be resolved is the most dangerous and inescapable political hope.

The title and content of Bacon's *New Atlantis* are clearly meant to remind us of Plato's *Republic, Timaeus,* and *Critias.* Like them, it relates the war between Plato's citizens of the best regime and the ancient Atlantians. In Bacon's story, lost European sailors visit the astounding, scientific people of Bensalem, the new Atlantis, where they learn about the Bensalemites' science that masters nature, which they are invited to disclose to the entire world. In an account of their beginnings, the new Atlantians explain that long ago they defeated the old Atlantians, or rather defeated the part of the old Atlantian expedition that challenged them while the other part challenged the Greeks. Bacon boldly reveals a possibility Plato cautiously disguised: the old Atlantians met by the Greeks were exterminated to a man.[23] In contrast to the savagery of the Greeks' "just war," the early Bensalemites, led by their king Altabin, defeated *their* old Atlantians in a bloodless battle and then set them free. Divine revenge *or* natural catastrophe destroyed the old Atlantis, the other parts of the world, and the Greeks, but the Bensalemites avoided such a harsh fate because some time after the battle a new king named Solamona set them on the path of modern science.[24] Either the gods or good fortune spared the Bensalemites, for the gift of their science helped them to avoid a fate they did not deserve. Like the *Republic,* the *New Atlantis* relates a new beginning, for an entire world rather than merely for one city. But Bacon's new beginning not only is presented as possible but also appears to avoid the harshness of the beginning that Bacon boldly says haunts Socrates' best city and Socrates is constrained more cautiously to disguise.

However, the beginnings described in the *New Atlantis* may only appear to be pacific and just. Because we hear nothing of what transpired before the just defeat of the old Atlantis, we do not know whether some earlier harsh beginning might yet be reflected in the new one. After all, the world may have been destroyed by natural catastrophe rather than by divine revenge, so Bensalemite science may yet be helpless to avoid some deserved divine judgment. But after the advent of science, the Bensalemites witnessed a miraculous revelation that gave them the truths of Christianity.[25] If the salvation provided by science is rooted in and reflects necessary crimes, Chris-

23. *Advancement* 141–43.   24. Ibid., 142, 143, 144–47.   25. Ibid., 137–39.

tianity made compatible with science shows the way to recognize them
and to atone for them. Therefore men need not be deterred by the
likes of the law-ruling Zeus or by moderation from pursuing and enjoying a world in which science promises to free them from the urgent needs that seem to be the roots of injustice.

Such an argument would place Bacon squarely within the modern
project were it not for one important fact: the *New Atlantis* is unfinished, like the *Critias* (and so the *Republic* and *Timaeus*); and moreover, it is unfinished for the very same reason—it lacks a final speech
that would complete an account of perfectly just practice. According
to long-accepted scholarly opinion, begun by Rawley, Bacon's secretary, the *New Atlantis* is unfinished because it lacks an account of the
best regime, a subject Rawley says Bacon abandoned in order to pursue natural history.[26] The tradition is correct about what the *New Atlantis* lacks, but it misconstrues this lack, and so misses its enormous
importance, because it does not heed what Bacon himself actually said.
In the Latin translation of the *Advancement of Learning*, *De augmentis
scientiarum*, Bacon remarks that "if his leisure time shall hereafter
produce anything concerning political knowledge, the work will
perchance be either abortive or posthumous."[27] The *New Atlantis* is
both abortive and posthumous. Furthermore, in the *Advancement of
Learning* government is clearly identified as a separate science, but
about it Bacon says that it is a subject "secret and retired," because it
is both too hard to know and not fit to be uttered.[28] Although the
New Atlantis clearly promises perfect justice and commodious living,
Bacon's remarks plainly suggest that it contains a secret, excessively
difficult, and unspeakable political teaching.[29]

But why should it do so if science and Christianity can provide
perfect freedom and justice? Either the *New Atlantis* lacks a secret and
retired teaching about government, or it contains such a teaching that
would merely appear to be missing, but in either case the *New Atlantis* would teach the same thing as Plato's three dialogues: a demonstrable account of perfect applied justice is impossible because it is

---

26. Ibid., 127, 166. See F. H. Anderson, *The Philosophy of Francis Bacon* (Chicago:
University of Chicago Press, 1948), 24, 36, 40, 259; *New Atlantis*, ed. Gough, xxvii—
xxix.

27. *De aug.* 792 (Spedding's trans. in *BW* V, 78–79).

28. *Advancement* 473–76. In the *De augmentis,* Bacon says he passes over even the
"manifest and revealed" parts of government because of his deference to the king to
whom his treatise is offered. Cf. *De aug.* 745–49, 792; *Advancement* 474–76.

29. See J. Weinberger, "Science and Rule in Bacon's Utopia: An Introduction to
the Reading of the *New Atlantis,*" *American Political Science Review* 70 (September
1976):865–85.

too hard to know and dangerous because it is not fit to be uttered. It seems that, however much he recommended a world freed by a new productive science and justified by secularized Christianity, Bacon would have seen Hobbes's boast of a perfect, practical science of justice as reflecting the utopian delusion of the productive arts. And as we will see, Bacon did regard Machiavelli as dogmatic and not wholly original. Bacon seems to have agreed with Plato that it is dangerous to inquire into impossible perfect beginnings. But then we must ask why, if Bacon agreed with ancient utopian thought, he was so much more daring than Plato regarding first and new beginnings. Unlike Plato, Bacon never says that his account of new beginnings is a likely tale, and there is simply no doubt that Bacon, like Machiavelli and Hobbes, intended for all mankind to tread its way.

To answer these questions we must turn to the *Advancement of Learning*. By an indirect remark made in another incomplete, posthumous work, *An Advertisement Touching an Holy War*, Bacon tells us that his division of the sciences is complete.[30] In the *Advancement* Bacon presents his division or account of the sciences that are sufficient and that need to be improved, defending his new project for learning not only against the divines and other traditionally learned men but also against the "politiques." The *Advancement* addresses *all* the sciences, including government. But since government is a secret and retired science, the *Advancement* is a complete but secret and retired account of government and of the relationship between government, the mastery of nature, and the Holy Faith, whose God seems to know nothing of Zeus' law-loving revenge, even though His Son reminds us of harsh, realistic beginnings. Bacon knows of the means to present such a teaching, for in the treatise he notes as one method of writing the one used "by the discretion of the ancients, but disgraced since by the impostures of many vain persons, who have made it as a false light for their counterfeit merchandises." This is the "enigmatical and disclosed" method, whose purpose is to "remove the vulgar capacities from being admitted to the secrets of knowledges, and to reserve them to selected auditors, or wits of such sharpness as can pierce the veil."[31] Bacon uses this method in the *Advancement of Learning*.

On the one hand, Bacon's boldness places him at the beginning of the modern scientific project. But on the other hand, his caution suggests that he thought this project best understood from the per-

---

30. *Advertisement* 13–14.   31. *Advancement* 404–5.

spective of the older tradition of utopian political thought. While Bacon himself gave impetus to the project whose slogan is "a true and certain rule of action by which we might know whether that which we undertake be just or unjust," he acknowledged the thought that such a promise is both dangerous and impossible. While recommending the mastery of nature, he doubts whether the artful conquest of nature's penury can accomplish its true goal, which is to overcome mankind's obstreperous political nature. Bacon is situated at once within the modern project and beyond it: his understanding of the ancient utopian teaching encompassed that project's causes, limits, and problems. To the extent that he participated in its founding, Bacon found himself in a situation which, while intelligible only from the utopian perspective, was not experienced by the ancients themselves. Bacon had to recommend what he knew to be problematic; he therefore had no choice but to present his comprehensive teaching by way of the ancients' "enigmatical method." Why this method had to be used in *any* case he learned from the ancients, but why he had to recommend science enigmatically he learned both from the ancients and by himself.

Because Bacon employs the "enigmatical method," our study must take the form of a close textual commentary. Since our task is to "pierce the veil," we have to see that Bacon teaches as much by what he does not say as by what he does. But what he does not say is always clearly indicated by something he has said by way of contradiction or by way of an allusion that is inappropriate, given the purpose it is ostensibly to serve. The elegant fabric of his argument emerges from three interwoven strands: the text as it appears, the contradictions, and the false allusions as they fit together with themselves, the contradictions, and the text. The hardheaded reader may be inclined to doubt that Bacon employed such a curious method. But true hardheadedness consists in taking seriously Bacon's explicit statements on this matter, however much they may differ from contemporary, self-certain, and "scientific" practices.

Although I began my studies by tracing every clue and side road, I cannot continue to be so free with space and time. Therefore I have had to pick and choose. There is much that I have been unable to say. There is no danger in this constraint, however, because my intention is to return the reader to Bacon's text. Nothing can take its place, and it is a complete and available standard by which my choices can be judged. Bacon's argument develops slowly and fugally. While my aim has been to follow his path, I have tried to let the reader

35

know what lies ahead without spoiling the surprise and without too much repetition. Bacon's treatise is divided into two books, with no other formal divisions, headings, or chapters. The separate parts, chapters, and sections of this study are my own; although they conform in theme to the sections in Bacon's text, it remains to be seen whether their titles conform to Bacon's intention. However, some overview of the chapters to come may be helpful, since, to demonstrate the argument sketched briefly in this introduction, we must follow Bacon, who does not lay out his path in advance.

Chapter 1 treats Bacon's dedicatory epistle, in which he details the conditions that necessitate his treatise. Caused by Christian morality and politics, these conditions call for the new scientific project for human learning. But for reasons having to do with justice, the new project requires an apology, which precedes its praise and the description of its scope and occupies book 1 of Bacon's treatise and likewise Part 1 of my study. In the course of his apology, first to the "divines," then to the politicians, and then to the learned themselves, Bacon explains how and why his apology differs from Socrates'. Although Bacon must understand political philosophy in order to understand his age, he cannot bring it to bear on practical life as did Socrates and cannot use it to ground a political science such as Aristotle's.

Since his project and his apology are occasioned by problems of Christian politics, Bacon begins with an apology to the divines, which I treat in Chapter 2. In this chapter, Bacon explains that Christian charity is not unique but is a form of generic charity that is rooted in the openness of all political life to the divine. Generic charity is the temptation, coeval with political life, to think that art and reason can be identical. Bacon shows that generic charity, and not just Christian charity, calls forth natural and mathematical philosophy and subverts political moderation. Moreover, he suggests that the new science of nature is itself charitable. In Chapter 3, I examine Bacon's apology to the politicians. Here Bacon argues that Christian charity has undermined moral virtue, which moderates generic charity. Consequently, Christian charity has distorted the political life of the modern age. Christian charity came to rule the world because the Romans replaced classical moral virtue with their dedication to political honor. As the treatise proceeds, Bacon explains in detail just how it was able to do so. But already Bacon demonstrates that if we are to understand the modern age, we must understand the difference between moral virtue as it was described by classical political philosophy and honor as it was practiced by the Romans. Bacon thought, we learn in this chapter, that Machiavelli, his predecessor and the original founder of the modern age, did not understand the crucial difference.

In Chapter 4, which is divided into three parts, we work our way through Bacon's apology to the learned men themselves. Bacon has said that the scientific conquest meant to cure the politics of Christian charity is itself charitable. His task is to explain how this charity limits the new project for learning. To explain the point fully, Bacon must show that Machiavelli's new, realistic political science was not sufficient to subdue Christian politics and that it had to be supplemented by the new project for the scientific conquest of nature. In the three parts of Chapter 4, we find Bacon's argument that Machiavelli misunderstood the true power of Christian charity and belief because he did not know well enough how virtue, honor, and charity differ. He did not know the difference because he turned from the teaching of the ancient utopians and therefore did not know how much his own views were in the grip of Christian charity. In this chapter, it becomes clear why Bacon is forced to apologize for his project: the new science is modeled on the secret avarice of charity. Only a project modeled on charity can speak to Christian charity, and therefore the new knowledge cannot wholly overcome the difficulties of Machiavelli's political science. The new project is itself charitable. As a result its dangers are those that are coeval with political life. Only the utopian wisdom of the ancient utopians reveals this truth; it explains how charity is the ineradicable hope of artful men to an impossible and immoderate freedom from the constraints of political life.

In Chapter 5 we examine Bacon's postscript to his apology. Building on the preceding discussion, Bacon explains how and why innovation in the arts will determine modern times. He shows how such innovation requires management by principles beyond its scope. In identifying these principles, Bacon argues that, as the ancient utopians taught, contemplative theory is superior to practice, even though the force of modern history calls for the identity of theory, now understood as materialistic or mathematical natural philosophy, and practice.

In the sixth and final chapter of Part 1, we examine Bacon's two-part account of the gods as the ultimate source of history's force. Bacon first describes what any creator god must be, explaining that any such god is at once necessary for political life and the cause of man's need for human justice and moderation. Bacon then explains that, as the Roman concept of honor became the concept of human sin, the Christian version of such a god now governs the political fate of the modern age. Because of this fate, the true teaching of the ancient utopians can at best merely warn of the injustice of Machiavelli's political science, which cannot be expunged from the project for conquering nature. The ancient teaching can at best merely warn of the dangers in the new identity of theory and practice.

In book 2 of his treatise, Bacon presents the division of the sciences that follows from the newly exalted status of the practical arts. We examine the division of the sciences in Part 2 of our study. In Chapter 7 we discover Bacon's argument that the new identity of theory and practice distorts the political science taught by Plato and Aristotle. In the present age, the political philosopher cannot be the model for those who govern and who cause political life to move. This is the proximate reason why Machiavelli's political science, and its supplement, the new project for scientific conquest, can always be unjust and immoderate. In Chapter 8 we see Bacon's argument for the superiority of contemplation and wonder to any dogmatic philosophy of nature, whether it be materialistic or idealistic. Such dogmatic philosophy springs from belief in the identity of theory and practice, a belief that is ineradicable from political life and dominant in the present age. Chapter 9 works through Bacon's detailed account of how the new status of the practical arts subverts the old political science. Because the political philosopher is no longer a possible model, the statesman is no longer the best possible practical human being. And in the absence of the statesman, the ambitions of the poets and the sophists can no longer be restrained by a rhetoric that serves statesmanship. The old political science is no longer useful, but Bacon warns that it will one day be needed to disclose the limits of the modern project.

Bacon is now ready to present the deeper causes of the dangers of the new age. Therefore, in Chapter 10, we see Bacon's account of the private and public human good. The new project for learning can be Machiavellian, that is to say unjust and immoderate, because it distorts the relations between the private and public goods. And finally, in Chapter 11, we see Bacon's demonstration that the morality and politics of the new project for learning actually are Machiavellian because they blur the distinction between private and public propriety. My book, and Bacon's treatise, concludes by returning to the question of the gods. Bacon shows that the necessity of Machiavellian politics is the revenge of any creator god who serves man's political life. The divine gift of the arts to man is not an unmixed blessing, for as long as man is both the rational and the artful animal, human reason will never wholly overcome dogmatic and contentious belief.

PART I

BACON'S APOLOGY

[1]

Dedicatory Epistle:
On the Perfect Christian Prince

[261–264]

Bacon opens his treatise with a dedicatory epistle to King James I. "Under the law" there were daily sacrifices, which proceeded "upon ordinary observance," and freewill offerings, which proceeded "upon a devout cheerfulness." Likewise there belongs to kings from servants both tributes of duty and presents of affection. Regarding the former, Bacon hopes not to be wanting, according to his duty and the king's pleasure. Regarding the latter, he thinks it "more respective" to offer an oblation referring to the "propriety and excellency" of the king's individual person than to offer one referring to the business of his "crown and state." Being not presumptuous to discover what scripture declares to be inscrutable, Bacon observes the king with the "eye of duty and admiration." He leaves aside the "other parts of virtue and fortune," and he wonders at the king's intellectual virtues and faculties: his capacity, memory, apprehension, judgment, and order of elocution. Of all the men Bacon has known, the king most leads a person to agree with Plato "that all knowledge is but remembrance" and that the mind originally knows everything, which is restored after it is sequestered in the body. As the scripture said that the wisest king's heart was "as the sand of the sea," being both large and small, so the king is able to comprehend the greatest and the least matters. This is god given, because it would seem to be a natural impossibility for the "same instrument to make itself fit for great and small works." The king's gift of speech is like Augustus Caesar's as Tacitus described it, "flowing and prince like." Unlike servile speech that is labored, affected by art or precept, or framed

43

after some pattern, the king's speech flows as from a fountain and yet streams and branches itself into nature's order.

In the king's civil estate, his virtue emulates and is contented by his fortune: virtuous disposition with fortunate regiment, virtuous expectation of greater fortune with prosperous possession, virtuous marital fidelity with happy fruit, and a Christian desire for peace with well-intentioned neighbors. And just as in these civil matters, in intellectual matters there is "no less contention" between the king's natural gifts and the universality and perfection of his learning. Since Christ's time there has been no more learned king, a judgment supported by a consideration of the emperors of Rome, including Caesar the dictator and Marcus Antoninus, the "emperors of Grecia or of the West," and the lines of France, Spain, England, and Scotland. Many kings can appear learned or can support the learned, but it seems almost a miracle that a born king should have such learning.

The king combines divine and sacred literature as well as "profane and human," causing him to be invested with the "triplicity" ascribed to the ancient Hermes: the power and fortune of a king, the knowledge and illumination of a priest, and the learning and universality of a philosopher. Such propriety deserves to be expressed not only in present admiration or in the history of ages to come but also in "some solid work, fixed memorial, and immortal monument, bearing a character or signature both of the power of a king and the difference and perfection of such a king." Therefore Bacon presents as an oblation "some treatise tending to that end." The sum will consist of two parts: the first will concern the "excellency of learning and knowledge, and the excellency of the merit and true glory in the augmentation and propagation thereof," and the second part will account for the "particular acts and works" that have been undertaken or omitted for the advancement of learning. Bacon hopes by such an oblation to spur the king to conclude particulars for the advancement of learning, since he himself cannot "positively or affirmatively" advise the king or propound to him "framed particulars."

We begin by recalling the question posed in the Introduction. Why, when he agreed with Plato that the most dangerous problems of politics spring from the necessary hopes of the productive arts, did Bacon so powerfully abet these hopes? Socrates was forced to give his account of the perfectly just city by those who represented the hopes of the city's productive arts. His apology, demanded by the actual city, required an explanation for his criticism of the hopes and knowledge of these arts and for his philosophizing and refusal to practice his own productive art. Bacon counsels in favor of the productive arts, but as we see in the sequel, he is forced, like Socrates, to give an apology that turns out to be, in large measure, Socratic. We need to know what necessity caused Bacon's counsel and his apology, and in the dedicatory epistle, he begins with just the matter of how much his treatise is free and how much it is bound by public duty. At first he asserts that his treatise is simply a free gift. But this outward claim does not withstand close scrutiny.

Bacon asserts that an Old Testament distinction between dutiful and free offerings determines that his own free gift to the king should refer to the king's private business rather than to public matters. He implies that the Old Law and present duty identify the ordinary and the dutiful with public matters and the occasional and free with private matters. Furthermore, the observant eye of duty, or ordinary, public consideration, limits the private matters of a king to be addressed. According to Solomon's claim that a king's heart is inscrutable,[1] Bacon will mention only what the philosophers call the intellectual virtues. Now, to begin with, the limits Bacon mentions do not spring from the Old Testament, as he says. The Old Testament does distinguish between ordinary and occasional sacrifices, with the one being constrained by some necessity while the other is not.[2] But there is no corresponding distinction between public and private. This second distinction is Bacon's addition, founded not on the Old Law but rather on the New, from which we learn to distinguish between what belongs to Caesar and what belongs to God.[3] The prophet Samuel did not introduce a distinction between public and private matters, let alone between the political and the divine. Rather he acceded to popular demand that a king be appointed to rule Israel as did the kings of other nations. This action displeased both God and Samuel, who warned of a king's demands. But still, God, not Samuel, anointed a king to rule over His exclusively chosen people.[4] According to the Old Law, freedom and constraint were experienced within the political horizons of a people whose exclusive boundaries are determined by God's choice. Under the New Law, public and private are distinct as between exclusive or particular political constraints and private but universal freedom promised by God. At the outset, then, Bacon describes the freedom of his treatise by blurring the difference between the Old and the New Laws. Somehow the freedom of his trea-

---

1. Prov. 25.3.   2. Lev. 7.16, 23.1–33; Num. 28, 29.
3. Matt. 22.21; Mark 12.17; Luke 20.25.   4. 1 Sam. 8, 10, 12.

tise is determined by the possibility that although it pretends them to be different, the universal empire of Jesus separates public and private matters no better than could the divine but political and exclusive empire of the Jews.

It is not surprising, then, that while the difference between public and private matters requires Bacon to speak freely only of the king's intellectual virtues, he does not confine his remarks to these virtues. Like a gossip who mentions what he will not say, he compares them to the harmony of the king's fortune and the virtues relevant to his civil estate. Moreover, a mere glance at the treatise reveals that it touches the king's private affairs as well as his public affairs. And however free and private it might be, as a published treatise it is public and constrained. We wonder who the king is with respect to whom Bacon ignores the difference between public and private.

We learn that the king's intellectual virtue seems almost impossible because "in nature" it seems impossible for the same instrument to fit both great and small works. Bacon tells us that the king's mind is like a naturally impossible universal tool, so much so that it proves Plato's opinion that knowledge is but remembrance and that the mind by nature knows all things. Yet if the mind were a perfect tool, it would be no tool but rather a perfect reflection of the whole of nature. And it is certainly questionable whether such a mind, or nature reflected in such a mind, is possible. Presuming for the moment that it is, Bacon says that the king is the most learned king or temporal monarch "since Christ's time." But Bacon does not just compare the king's triple goodness (kingly power and fortune, priestly knowledge and illumination, philosopher's learning and universality) to qualities possessed by the Christian rulers. He also boldly mentions the pre-Christian dictator Caesar. That is, if there have been no more learned rulers since Christ's time, then the king's universal, toollike mind must be comparable to that of the universally ambitious Caesar, who Bacon later shows may have differed from the odious Cataline only by a difference in fortune.[5] If, as Bacon says, his treatise bears the signature of the Christian prince, then such a prince resembles a tyrant who cares nothing for the difference between private and public propriety. Bacon imitates this indifference in order to address a Christian prince, and at least at the beginning, the presence of such a prince is the force that constrains his speeches. Therefore even though the king's toollike mind is not inscrutable, Bacon does not

5. *Advancement* 440–41; see Chap. 10 at nn. 24, 43 below.

speak for himself but rather refers to Plato to indicate whether such a mind is possible and by whom it might be desired.

Bacon says that the king's mind should convince a person of Plato's argument in the *Meno*. Plato's *Meno* is perhaps the most important dialogue for discerning the reasons for Socrates' apology to the city of Athens. In the *Meno*, Socrates argues in the presence of his future accuser Anytus, defender of the city's artisans and politicians, that virtue cannot be taught and that the sophists are superior to any Athenian gentleman in teaching what is taken to be virtue.[6] In the dialogue, Anytus proves to be wiser than his rude, harsh threats against Socrates imply. Anytus refuses to defend the artisans or the politicians in the freer, private context of the dialogue. He refuses to suggest the most obvious source of instruction in virtue: the laws, which are the products of the politicians' and lawgiver's art.[7] Anytus seems to understand Socrates' subtle argument that the city has more in common with the sophist than it does with philosophy, for the conventional virtues praised by the city are the result of a productive art which the sophists pretend to teach along with all the other arts.[8] The sophists claim to possess the art of appearance or imitation, by means of which they pretend to a mastery of all of the several productive arts. Anytus, the spokesman for the productive artists and the politicians, who are themselves productive artisans,[9] divines the guilty truth of the city's virtue and the reason for Socrates' refusal to equate virtue with a teachable productive art. If the laws are not divine but are produced by human art that merely imitates nature, then both the virtues and the justice of the several regimes can come to light as the sophist and the tyrant understand them—as appearances that facilitate the rule of the strongest.

Bacon likens the king's mind to a universal, productive tool that encompasses all things great and small.[10] As such a perfect tool, the king's mind could reflect the whole of nature only by accomplishing nature's spontaneous self-sufficiency. According to Plato the one art that pretends to such perfect mastery is the sophist's tyrannical art of the seeming, and if the king reminds us of the *Meno*, then Bacon would seem to describe the king, who perfectly combines political power and wisdom, as the perfect sophist. But the king is also likened to the ambitious dictator Caesar. And the king is not just like Caesar; his speech is like Augustus Caesar's speech as described by

6. *Meno* 89d–95b; cf. *Phaedo* 72e1–77a5.   7. *Meno* 92d7–93a4.
8. *Republic* 492a–93d, 596d; *Sophist* 240d1–41c6, 260d5–61b4.
9. *Apology* 23e5.   10. 1 Kings 4.29.

Tacitus: "princelike, flowing as from a fountain and yet embracing nature's order." The king's speech reflects his mind. In the second book, Bacon informs us by way of Tacitus that Augustus' directness in declaring his desire for "no less than the tyranny" served to mask that very desire and so served his prosperity.[11] The present reference is to Tacitus' comparison of Caesar's, Augustus', Tiberius', Caligula's, and Claudius' (!) eloquence to the "borrowed eloquence" that Nero needed,[12] but the later reference suggests that Augustus' forthright eloquence, so different from Tiberius' dissimulation, was in fact a more artful form of dissimulation. Tacitus makes this suggestion in comparing ancient Rome to modern Rome, in which empire and wealth unleashed the ambition for tyranny.

Like Plato, Bacon knows that the artful conquest of nature in action is the same as the sophist's dangerous pretense to have mastered all of the productive arts. Also, like Plato, Bacon knows that the sophists speak not just for themselves but for the most dangerous tendency of the city's justice itself. Unlike Socrates, Bacon makes an apology, as we will shortly see, that is constrained not simply by the city as such but rather by the *Christian* prince who stands for the danger of the city as such. Socrates could defend his criticism of the productive arts because he could praise philosophy—or at least philosophy properly understood—and because such philosophy could disclose the grounds for moderating political ambition. Bacon can only apologize for his very encomium of the productive arts themselves. We wonder, then, if Bacon's response to the utopian pretense of the arts is explained by the difference between the city that compelled Socrates and the prince to whom Bacon must speak. This difference is of course Christianity. As Bacon suggests by his clever references to Tacitus, modern times differ from ancient times in that a new empire has risen. The empire Bacon confronts is Christian rather than Roman, but the two are somehow connected; where Socrates could weigh the merits of politics and philosophy, Bacon can only weigh the merits of two different kinds of Roman tyrants described by Tacitus. One of them, Caesar, was a bold tyrant who failed to be an emperor and whose tyrannical founding of an empire was thus requited. The other, Augustus, was a dissembling tyrant who became an emperor and who died a happy death.

Bacon's treatise is needed to educate a Christian prince who can-

11. *Advancement* 467. Here Bacon refers to Tacitus' comparison of Pompey to Gaius Marius and Lucius Sulla, all of whom aimed at the principate; Augustus of course dissembled his ambition (Tacitus, *Historiae* 2.38).

12. *Annales* 13.3.

not be addressed without confusing public and private propriety. In addition, however, Bacon's treatise is a response to Christian political life. Somehow the problem of Christian justice requires a treatise proposing a new task for learning. As Bacon's treatise develops, it becomes clear that this task is the unifying of theory and practice and the elevation of the practical arts to be the model of the mind's perfection. In fact, then, the king to whom the treatise is given is praised as a paragon of the new learning, which equates the perfection of the mind with the perfection of a productive tool. But when the king is changed by Bacon's gift, he will therefore be like the two kinds of tyrant described by Tacitus, the one who was justified and the one who was not.

Bacon's examples of the two tyrants (and indeed the whole of this dedication to the king) cannot but remind us of Machiavelli's political science, which purported to manage political fortune by justifying the tyrant, by reconciling the tyrant's unjust, lawless ambition with the public good. Already, then, we know that, for Bacon, Christian princes and peoples must be formed by more than Machiavelli's principles, which alone were not sufficient to justify the tyrant. But it seems that the new learning goes beyond Machiavelli only to confront anew the failure to justify the tyrant and to determine proper boundaries between public and private life. The perfected Christian prince will be both kinds of tyrant, one justified and the other not. Bacon will show that these two kinds are the poles between which the modern age will chart its course. And we learn this lesson first not from Machiavelli but from Plato's critique of the sophists. Although the two kinds of tyrant are always possible, they were not, as they are for the Christian era, the sole practical horizon for the classical utopian thinkers. But then, these thinkers knew nothing of the historical legacy of the Jews.

It is no wonder that the excellence of learning requires an apology that occasions a treatise different from the one Bacon describes as his gift to the king. Bacon first promises the king a tribute to learning's excellence and an account of the works so far undertaken to advance it. But because the first part requires an apology, the two offered parts become the middle of four distinct parts: the apology on behalf of learning, the tribute to learning's excellence, the account of works, and a new division of the sciences. The apology and the division are added to Bacon's first description of his gift, and we must wonder whether the charges addressed by the apology are reflected in the new face of the intellectual globe.

[2]

The Apology to the Divines:
The Charitable Sources of Pride

[264–268]

To clear the way for true testimony about the dignity of knowledge, Bacon must deliver it from the disgraces that knowledge has received. These arise from ignorance "severally disguised," and they appear sometimes in the zeal and jealousy of divines, the severity and arrogance of politiques, and the errors and imperfections of learned men themselves. The divines say that knowledge must be accepted with limitation and caution and that aspiring to knowledge caused the original sin and the fall. Referring to Paul, they say that knowledge is like the serpent that makes man swell (*scientia inflat*).[1] The divines refer to Solomon, who said that there is no end of making books, that knowledge is weariness of the flesh, that in spacious knowledge is "much contristation," and that increasing knowledge increases anxiety.[2] And they repeat Paul's warning that we should not be spoiled by vain philosophy.[3] The divines further say that experience shows how learned men have been arch heretics, how learned times incline toward atheism, and how contemplation of second causes "derogates" from our dependence on God, the first cause.

To disclose the divines' ignorance and error and the "misunderstanding in the grounds thereof," Bacon says that they appear not to have known that the fall was tempted by the proud knowledge of good and evil, with an "intent" in man to give law to himself and not to depend on God's commandments. The fall was not tempted by "the pure knowlege of nature and universality," which man used to

1. 1 Cor. 8.1.   2. Eccles. 12.12, 1.18.   3. Col. 2.8.

name the creatures in Paradise, according to their properties and as "they were brought before him." No quantity can cause the mind to swell, because nothing can fill or extend man's soul except God and the contemplation of God, as is shown by Solomon's remark that the eye is never satisfied with seeing nor the ear with hearing.[4] If there is no fullness, then the "continent is greater than the content," and so of knowledge, the mind, and the senses that are the mind's reporters, Solomon said, after his calendar of times and seasons, that God made all things beautiful and "has placed the world in man's heart, yet cannot man find out the work which God worketh from beginning to the end."[5] God made man's mind, in other words, as a glass capable of joyfully imaging the universal world, just as the eye enjoys receiving light. The mind is not only delighted to behold the variety and vicissitude of things and times but is also raised to discover the "ordinances and decrees" that are observed infallibly throughout them. Solomon implies that the "supreme or summary law of nature," or God's work from beginning to end, might not be discoverable. But its possible inaccessibility does not "derogate" from the mind's capacity, because it can be explained by such impediments as shortness of life, ill conjunction of labors, ill tradition of knowledge, and many other inconveniences to which man's life is subject. Solomon says in "another place" that man's spirit is God's lamp, with which he searches the "inwardness of all secrets."[6] Therefore, no "parcel of the world" is denied to man's invention and inquiry. If such is the mind's capacity, then there is no danger in any proportion or quantity of knowledge. Only the quality of knowledge, not the quantity, is venomous or malignant and could cause swelling. The "corrective spice" is charity, which the apostle added to his former clause, so that "knowledge bloweth up, but charity buildeth up,"[7] which is not unlike his comment that, if he had spoken with the tongues of men and angels but without charity, it would be "but as a tinkling cymbal."[8] If such tongues are severed from charity and "the good of men and mankind," they are hollow rather than sound.

Solomon's remarks about excess of books and writing and anxiety of the spirit and Paul's warning that we not be seduced by vain philosophy do set forth the true bounds and limitations of human knowledge, which are three: not so loving knowledge as to forget our mortality, applying knowledge for repose and contentment rather than for distaste or repining, and not presuming to attain to God's mys-

---

4. Eccles. 1.8. 5. Ibid., 3.11. 6. Prov. 20.27.
7. 1 Cor. 8.1. 8. Ibid., 13.1.

teries. Bacon refers to Solomon's remark about knowledge, darkness, and the wise man and the fool in commenting on the first limit.[9] For the second limit, Bacon says that no anxiety comes from knowledge except by accident, for all knowledge and wonder is an "impression of pleasure in itself." But when men frame conclusions out of their knowledge and minister to themselves "thereby weak fears or vast desires," then carefulness and trouble of mind arise. Then knowledge is not the *lumen siccum* that Heraclitus thought best for the soul but rather a light steeped in the moisture of the humors of the affections. Regarding the third limit, it has to be stood upon and must not be passed over lightly. If people think that by inquiry into sensible and material things they will reveal to themselves "the nature and will of God," then they have been spoiled by vain philosophy. Contemplation of God's creatures and works produces knowledge, but with regard to God, contemplation of them produces wonder, which is "broken knowledge." Therefore "one of Plato's school"[10] well said that man's sense is like the sun, which opens the terrestrial globe as it conceals the stars and the celestial globe, so that sense discloses nature but shuts up the divine. For this reason, it is true that diverse great learned men have become heretical when they fly to the secrets of God "by the waxen wings of the senses."

In questioning whether atheism and ignorance make us more dependent upon God the first cause, Bacon asks Job's question of whether one will lie to gratify God as one might for man.[11] God does nothing in nature "but by second causes," and any other view is the same as offering God an unclean lie. It is true that a little knowledge of philosophy might incline a man's mind to atheism, but further proceeding brings the mind back to religion. If the mind dwells upon second causes, this might cause oblivion of the highest cause, but when men see the "dependence of causes and the works of Providence," then according to the poet's allegory,[12] they will see easily that "the highest link of nature's chain must needs be tied to the foot of Jupiter's chair." To conclude, Bacon says that no one should be led by the conceit of sobriety or "ill applied moderation" to think that a person can go too far or can be too well studied in "the book of God's word or in the book of God's works," divinity or philosophy. Rather, men should work for "endless progress or proficience in both." But men have to apply both to charity and not to swelling, to use and not

9. Eccles. 2.13–14.　　10. Philo, *De somniis*, rec. P. Wendland (Berolini, 1898).
11. Job 13.7–9.　　12. Plato, *Theaetetus* 153c6–d5; Homer, *Iliad*, 8.18–27.

ostentation, and they must not "unwisely mingle or confound these learnings together."

In his apology to the divines, Bacon begins to show how the Christian virtue of charity causes the political problem of the present age. As we have already guessed from the dedication to the king, it is important that we learn this lesson from the classical utopian thinkers rather than from Machiavelli. In the present and most difficult part of his apology, Bacon shows what charity is in any age, whether it be Christian or not: charity is the divine cause of man's need for justice and moderation.

As we noted in the Introduction, the classical utopian thinkers considered the problem of justice to be grounded in the human openness to the divine. Specifically, the productive arts make up the city by way of the striving for divine wholeness and perfection, which is as necessary as it is politically dangerous. But the political philosopher who can criticize the arts, and so can counsel moderation, is no less a part of the city and no less open to the divine. By what other measure could the political philosopher discern that the just desire for perfect justice is a necessary but dangerous delusion? To understand the city, then, and to understand the sources and relationships between human overreaching and moderation, it is important to understand how the city is always open to two different conceptions of the divine. These two conceptions determine the difference between the striving, or eros, that moves the productive arts and calls forth philosophy and the striving that moves the political philosopher. The gods and the two kinds of eros are different, but in political life they are not and cannot be wholly separate. In his apology to the divines, Bacon shows that he understands clearly the utopian account of political life and the gods. Moreover, he shows how this account is essential for understanding the faith of the modern empire and the prince to whom he speaks.

We begin with Bacon's refutation of the divines' second charge. Bacon claims that it was not the pure, name-giving knowledge of nature and universality that led to man's fall but rather the proud knowledge of good and evil. Pride springs not just from the proud knowledge of good and evil but also from the uncharitable study of nature and God's word, which attempts to know God's will by way of his works of nature. But while Bacon says that any natural knowledge of divine will is spoiled and vain, it can be so only if God's works do not wholly exhaust his will. At the outset Bacon says nothing on this point save that God works nothing in nature except by "second causes." The scope of the "nothing" is important, however, because if it extends to nature's origin, then it would imply that God was and is not capable of working miracles and that divine work and divine will may be same. If so, then either no knowledge can be prideful or else pride is a divine sin, because though human knowledge is bound to "second causes," so too is the scope of divine will. The question would be, then, how any knowledge can be proud.

In answering the divines' third charge, Bacon says that only uncharitable knowledge is serpentine, causing man's mind to swell. But Bacon does not speak about the knowledge proffered by the pornographic snake, the knowledge of good and evil; rather, he speaks more generally about the knowledge of God and knowledge unmodified by charity. The third charge springs from Paul's complaint that *scientia inflat*,[13] which Bacon explains by distinguishing between quantity and quality. According to Bacon, any quantity of knowledge is proper for the soul as long as it is corrected or qualified by charity, which Bacon says refers knowledge to mankind's good. At the beginning of his rebuttal, Bacon clearly identifies the knowledge that swells the soul as a particular kind of knowledge, the knowledge of God. If charity, the love of God and mankind, requires that one eschew the knowledge of God, then one need only attend to charity, as Paul in fact suggests. But if charity requires knowledge of nature that includes the knowledge and contemplation of God, then Paul's complaint cannot be explained away by the claim that the soul is proportioned to any quantity of charitable *scientia*. As it turns out, everything depends on whether God began nature by way of second causes.

Now, after slandering knowledge and praising charity or love (*caritas*), Paul actually says that any claim to knowledge is not as it ought to be, and he does not say that it can be made proper by adding charity. Rather he contrasts our love of God with God's knowledge of us. Paul attacks knowledge in making a point about offerings to idols. He argues that no knowledge of the nonexistence of idols can be proper; it cannot prevent us from sinning by setting a bad example in eating the meat of sacrifices. Knowledgeably saying "it's only meat" will not help those who are weak and who do not know about idols, the universe of men other than the Jews. The only proper knowledge is God's knowledge of us, which reciprocates our love of God.[14] Therefore, lest one sin against Jesus in corrupting the gen-

13. 1 Cor. 8.1.   14. Ibid., 8.2–3.

tiles by depending on knowledge, Paul says it is better to be a vege-
tarian. If their ignorance about idols causes one's brothers to sin, then
one should abstain from knowledgeably eating meat. Knowledge
cannot prevent sin, only love can: Paul counts upon the example of
vegetarianism, not a knowledgeable discourse about the difference
between idols and God. We can see that Bacon has changed the intention and content of Paul's argument. For Paul, knowledge cannot
prevent sin, and charity, our love of God and mankind, leads us to
eschew knowledge for an example of abstinence. For Bacon, charity
requires just the use of knowledge for all mankind. But does not Bacon explain this as the difference between knowledge concerning nature and knowledge concerning divine things, like idols? The answer
would be yes were it not for two important points. First, the limitless
knowledge Bacon praises soon appears to fathom God's secret intention, that is, divine knowledge of good and evil. And second, Bacon
shows next that Paul's appeal to love presumes knowledge of nature
and divine intention even if Paul insists that it does not. Bacon makes
these points by way of his references to Solomon.

Bacon quotes Koheleth in arguing that the senses are never satisfied, which Bacon says demonstrates that the container is greater than
what it contains, that is, that the mind can never be distorted or swelled
by knowledge conveyed by the senses.[15] In referring to Koheleth,
Bacon rather remarkably changes Koheleth's intention, ignoring an
important matter that Koheleth is careful to mention. Koheleth does
not say that limitlessness of the senses is a virtue. Rather he says that
the eyes and ears are not satisfied, leading to weariness and speech-
lessness. Bacon changes the soul's defect to its virtue, as if comple-
tion and satisfaction were to be found in limitless receptivity. But in
fact, the knowing soul's activity is not limitless after all, because it can
comprehend a limited object. And what an object it is.

Referring again to Koheleth,[16] Bacon says that the mind's recep-
tion of the whole, the universal world, is delightful not just regarding difference and change but also as it is raised to the universal ordinances and decrees to be observed in difference and change. These
decrees and ordinances comprise God's works from the beginning to
the end, genesis and providence. To know the whole of genesis and
providence would reveal the divine will behind ordinance and decree that in nature would seem merely purposeless. This is the divine knowledge of good and evil, and regardless of whether such
knowledge is possible, Bacon's description of it identifies God's will

15. Eccles. 1.8.   16. Ibid., 3.11.

with his works. Moreover, nothing about the mind prevents such knowledge. Bacon leaves little doubt that in fact such knowledge is possible, or at least should not be thought impossible, for the only things preventing it have nothing to do with the limits of the mind as God created it or with the unintelligibility of divine will. Rather Bacon says that nothing would be denied to inquiry and invention, including beginnings and ends and the "inwardness of all secrets," when the impediments of short life, ill ordering of labor, poor transferring of knowledge, and many other inconveniences are overcome. And the mastery of such impediments is presented in the *New Atlantis*. Contrary to Koheleth, Bacon says that the mind can comprehend a natural and divine whole. But does knowledge of the natural and divine whole cause the soul's delight and joy, or does it distort the soul, as Paul says it must? If the latter is true, then Bacon seems to say that Paul was right about knowledge of the divine being the source of pride but was mistaken to think that charity is different from such knowledge, especially knowledge about the divine sources of nature's beginning and end. Bacon answers our question by demonstrating how knowledge of the natural and divine whole can distort the soul. He does so by referring again to Solomon and by the important omission we mentioned earlier.

Koheleth suggests why the eye and the ear cannot be satisfied: the sun can blind and the wind can deafen.[17] For him, what is contained can be greater than the container. In arguing the contrary, Bacon suggests an eye that can discern the sun and an ear that can hearken to any wind, an astonishing suggestion causing us to wonder about the kind of body that could house such organs. And it reminds us of the important matter that Koheleth mentions but that Bacon ignores. Unlike Koheleth, Bacon is silent about taste, touch, and smell, senses that are closer to the body and yet are still windows to the soul.[18] Bacon's omission suggests that the mind's ambition, its openness to the whole by eye and ear, is constrained by the animal whose body and mind taste, touch, and smell. If knowledge distorts the soul, it is because the mind as a part of the soul is sequestered in the body. Put very simply, the knowledge of beginnings and ends distorts the soul because it is not at all certain that it can tell us what to eat, but such knowledge raises this question to dangerous prominence. If this statement seems bizarre, we need only to remember the topic that concerned Paul, who warned: *scientia inflat*.

The knowledge of good and evil includes the question of what things

---

17. Cf. ibid., 1.5–8, 11.7.   18. Ibid., 2.1–11, 24, 8.15, 9.7–9, 10.1.

human beings can and may eat, a question that informs us of what Paul had to know in order to substitute the charitable example of vegetarianism for knowledge about meat and idols. Vegetarianism is possible because our abstinence from the luxury of meat is matched by our physical needs. But if so, why were we equipped from the beginning with teeth making us resemble the wolf more than the lamb, so that, as Paul makes clear, we are free for carnivorous luxury?[19] Here we see the link between knowledge and Paul's love of God. Paul of course presumes a knowledge that human beings were formed by God from the earth. But in order for God to be lovable rather than revolting, Paul must know that original beginnings were perfect. Paul's charity presumes this knowledge, and although he does not try to demonstrate it as Timaeus tried to do, his view of beginnings is not unlike Timaeus' dangerous attempt and failure to show that men were not at first created to be blood lovers, not to mention cannibals.

We need to know how Paul's view of beginnings is like Timaeus' ill-considered cosmogony. We can begin to see the similarity if we think about Bacon's words to this point regarding charity and about our knowledge of Timaeus' speech. Timaeus brought natural philosophy to the aid of the rational pretension of the productive arts, which we saw required an account of perfect beginnings. Bacon argues that charitable knowledge comprehends the natural and the divine whole—the whole including the divine knowledge of good and evil—by encompassing "second," or material and moving, causes. Now, to comprehend the whole in this way is to model it after the experience of bodily human need. And to think that knowledge of good and evil is exhausted by the knowledge of material need is exactly the pretense of the productive arts. Could it be, Bacon makes us wonder, that Paul's love of God, and his knowledge of beginnings, betrays a secret orientation according to the body's needs? And is it possible that Paul understands the human things as do the productive arts? Is this what leads Paul's love to presume knowledge of perfect and repeatable beginnings even if such repetition is projected to another world? If so, then charity is not really different from the eros of the arts and philosophy modeled on the arts. Again, if so, then charity exposes and exacerbates what the ancient utopians understood to be the necessary link between the human love for freedom and the savage harshness of acquisition. In the sequel, Bacon turns again to Koheleth, and then to the ancient utopians, to demonstrate these speculations. Throughout the whole of the treatise, he explains in detail

19. 1 Cor. 8.8.

how their truth makes it possible—and necessary—for his project for knowledge to be charitable, that is, to speak to the secret yearnings of Christian love.

The divine's fourth, fifth, and sixth objections determine the limits of human knowledge, even though such knowledge is sufficient to comprehend "all the universal nature of things." Bacon treats them all together. The first censure and the first limit (the fourth charge), concerning excess of writing and books and not forgetting our mortality, of which Koheleth speaks in Ecclesiastes 12.2, is expounded by reference to Ecclesiastes 2.13–14 regarding knowledge, fools, and mortality. But the two references to Ecclesiastes do not complement each other as Bacon would have it appear. In Ecclesiastes 12.2, Koheleth warns against pursuing knowledge beyond the sayings of the wise, that is, beyond his own sayings as the king who, besides being wise, has also taught by weighing, studying, and arranging proverbs with great care. Beyond this practical wisdom of proverbs, knowledge of all that is done under heaven is weariness of flesh, which we know springs from the ill proportion between the world and the eyes and ears. In Ecclesiastes 2.13–14, Koheleth remarks as a wise man on the superiority of wisdom to folly, but then he bemoans the vanity of such a view, because both the wise and the foolish are forgotten in their common mortality. Considered together, Koheleth's remarks do not refer to the knowledge comprehending "all the universal nature of things"; rather they condemn the practical wisdom of proverbs that precedes such universal knowledge, which is wearisome because of the limits of the eyes and ears. His complaint about mortality is not the same as the one about weariness. The limitation of mortality applies to a king's practical widom; the weariness of the eyes and the ears refers to what Bacon calls the universal nature of things. Bacon argues that the weariness can be overcome when the impediments are overcome, as they are in Bensalem. However, if Koheleth demonstrates Bacon's view, then the source of learning's unhappiness springs from practical wisdom when the active knowledge of nature attacks the limits of man's mortality. But such a goal is charitable, according to Bacon, who knows from Paul that the reward for loving God is the soul's eternal life and the concrete redemption of the body.[20]

Regarding the second censure and limit (the fifth charge), anxiety and the proper application of knowledge to repose and contentment, Bacon refers not to the Bible but to a saying of Heraclitus re-

20. Ibid., 6.19, chap. 15.

corded in Plutarch's *De esu carnium,* which treats meat eating and vegetarianism.[21] Bacon says that knowledge and wonder is an "impression of pleasure in itself" but that knowledge referred to the individual soul is infused with the "humors of the affections," unlike charitable knowledge. Bacon implies that knowledge referred to the individual soul individuates the mind according to the passionate devotion to what is one's own, which is the source of weak fear and vast desire. Contrariwise, charitable knowledge, referred to mankind, causes the individual soul to be soul as such, or soul as it would be determined by passionless, disembodied, nonindividuated mind. But Bacon defends his argument that charitable knowledge is the impression of pleasure in itself, so that pleasure could be the activity of disembodied soul or mind as such, by quoting from Plutarch's argument that pleasure, which as the handmaiden of satiety and luxury leads to meat eating, interferes with the soul's ability to know the particulars of practical activity.[22] Plutarch's reference to Heraclitus illustrates his argument that carnivorousness is not only contrary to nature but also bad for the soul's eyes and ears, which, when filled and burdened with improper food, are unable to grasp "the fine or small and hard to view ends of practice."[23]

Plutarch, of course, was a Platonist, whose treatise considers man's natural cannibalism and cosmogony, or the account of the beginning of the visible whole. He argues that there is no question that man is fitted by nature, art, and convention to be carnivorous. If man first tasted flesh because of the harsh necessity of his origins, constrained as they were by the irregularity of the elements and the absence of agriculture, technical means, and artful wisdom, it is no argument against man's natural carnivorousness to say that we kill with tools and not with our unaided bodies.[24] Like Paul, Plutarch too knows that men have canines. And like Paul, Plutarch knows that men are now free to eat meat. Plutarch argues that the perfection of art increases our ability to torture our prey for our greater pleasure as it increases the vegetable products of the fields and that there is no difference between mere killing and torture for greater delectation except as imposed by convention. He says that his opposition to meat eating springs from a mysterious and incredible source. But he will say as much as Empedocles says, which is that the soul inhabits its

21. *Moralia,* ed. Hubert and Drexler (Leipzig: Teubner, 1959), *De esu carnium* 995e. All subsequent citations of the *Moralia* refer to the several Teubner editions, identified by the name of the treatise.
22. Ibid., 993a, 995c–e, 996e–97e.   23. Ibid., 996a.
24. Ibid., 993c–94b, 994f–95b.

mortal body as punishment for animal eating and cannibalism and that, with respect to the origin of mortality, this story is older than stories about cannibalism and rebirth.[25] But this statement forces us to the conclusion implied in the argument about natural carnivorousness, art, killing, and torture: art provides our freedom to eat meat, but art does not immediately overcome necessity, the irregularity of the elements, and the absence of tools. Therefore, the original necessity, to which man was by nature suited, was cannibalism, not just mere flesh eating. Moreover, necessity and luxurious or unnecessary pleasure are *both* original grounds of cannibalism, which stands apart from mere flesh eating by a difference imposed not by nature but by convention. Even if it could be shown that man's visible, bodily origins were not in harsh necessity, so that art and freedom were simply coeval, such a cosmogony would not suffice to demonstrate the perversion of meat eating, precisely because art is not bound to the limits of mere necessity.

Plutarch next shows that the doctrine of transmigration could not help, because it proves only that meat eaters might be cannibals, not that the human body is not properly or naturally carnivorous. It is no wonder that the doctrine of transmigration turns to cosmogony, however vain such a turn may be. At this point, Plutarch shows that in grounding the saving possibility of transmigration, the wholly physical (Heraclitus' and Empedocles') and the wholly mathematical (Pythagoras') accounts of being argue that either every single thing is different, so that there are no separate kinds, or that every single thing is identical, again so that there are no separate kinds. But by these two arguments either no eating is ever cannibalistic or every kind of eating is cannibalistic, and in either case there can be no limits of delectation.[26] These doctrines also deny the natural foundations of convention. But in the absence of a sufficient cosmogony for limiting delectation, a cosmogony that could plausibly show that man was once neither needy nor artful, one must demonstrate the natural foundations of convention so that human nature can at least be habituated to a conventional, luxurious freedom from the need to eat. According to the classical utopian thinkers, such luxurious and conventional freedom is exactly the substance of the moderating moral virtues that depend, for their stability, on tales that obfuscate the truth about harsh, acquisitive beginnings. According to Plutarch, the important question of delectation concerns not the meat eating that blurs the understanding of practice but rather the proper understanding

25. Ibid., 996b–c, 997e–98f.   26. Ibid., 997e.

of convention that separates such eating from cannibalism. Only a disembodied soul need not be concerned with these matters, but Bacon says that such a soul could bear the impression of pleasure in itself. And however much the pleasures of body and soul might differ, it is hard to imagine how a disembodied, perfectly knowing soul might experience pleasure.

Such an impossible soul is really a perfected body, for whom there is no tension between need and freedom: it is really the impossible promise of the arts. Moreover, it is exactly the entity known by the God who is moved by our love. Such is the soul whose body is redeemed by Christian love, or rather, as we can see, such is the needy body whose soul is so redeemed.[27] Paul's charity is no different from the delusion of reason when it is assimilated to art; his charity is really the delusional hope of human art to produce a perfectly whole but moving human body. And just like artful hope, Paul's love is forced to account for perfect beginnings. Likewise, in failing to supply such an account, it merely reminds of their true harshness. The two elements of charity, the salvation of the soul and the resurrection of the body, are really popular forms of physical and mathematical philosophy as they are called to the service of the arts in explaining perfect beginnings. Bacon, like Paul, reminds us that Christian love serves all mankind. Charity combines the impossible mathematical hope for perfect soul as such, or perfect mind, with the physicists impossible orientation by what is material in every body. Such a view cannot abide the authority of conventions which are neither material, because they exist as authoritative principles or forms, nor simply eidetic, because they are always made. To know that conventions are mysterious, both within our command and yet beyond it, would require reflection on the similarity and difference between art and nature and between art and reason. But this is reflection that loving art, and charity, cannot abide.

From Plutarch we learn the truth beneath Paul's remarks about knowledge, love, and vegetarianism: the Christian Paul's account of beginnings is worse than Plato's noble lie and worse than Timaeus' shocking, philosophic cosmogony. In Plato's day, natural philosophy did not command public respect—Timaeus made his speech in the course of private conversation. Paul, however, speaks aloud to all mankind. The doctrine of sin expresses the two elements of charity, and as we will see, the doctrine of sin, like mathematical and materialistic natural philosophy, subverts the foundations of convention.

27. See n. 20 above.

Not all men are philosophers, but according to Paul, all should know that they are sinners. Paul's love presumes rational knowledge of perfect beginnings. But unlike Timaeus, his doctrine of sin reminds of beginnings' harshness without knowing how harsh they really are and how important it is to forget them. In loving God as he did, Paul did not imagine that God the creator might have caused harsh beginnings, even though art now makes men free to abstain from eating meat. Moreover, Paul did not think about how art and need are akin or how divine creating might be artful. Bacon thinks he should have, because, as we see in Chapter 6, art and reason come together only when men love the gods. Charity is the proximate source of pride because, like the mathematician or the physicist, it can see bodies or souls but never both. The wages of pride are great because for the prideful man, who bows to no moderating convention, there are no possibilities between the godly and the bestial. For Plato's successor Aristotle, the gods and the brutes define the limits of the conventional life of the city.[28] Charity exceeds both of the limits, and much must still be said in this and other chapters to explain why. But in the sequel, Bacon develops the point that must by now be clear: charity has its root in the very heart of political life itself.

Bacon says that he will not lightly pass over the third limitation (sixth charge) regarding Paul's warning against seduction by vain philosophy and the presumption to know divine mysteries. He does not pass it over, but he makes no further explicit mention of Paul in the discussion. Now, in Colossians 2, Paul warns not against vain philosophy but against philosophy as such, which by knowledge of tradition or the elements of the world might attempt to penetrate God's nature so as to decide matters of meat, drink, and the various divisions of times. In the context, Paul says that, just as we perish because we are, like the mortal Jesus, compounded from the elements of the world, so we should not be subject to human doctrines in matters of worship. Again he says that knowledge is futile regarding delectation and times. Again, for Paul we can count on no knowledge to regulate the particulars of practice, for our determination of these matters would depend upon knowing the unknowable source of the will governing genesis and providence, the beginning and the end of times. In discussing Paul's warning against philosophy, Bacon distinguishes between possible knowledge of God's works and creatures and the "broken" knowledge of God, but he is silent about Paul's subject, delectation and the division of times. According to Paul, precisely

---

28. *Politics* 1253a25–29.

man's inability to know God's creatures and works requires him to be ignorant about times and delectation. It is no wonder that Bacon does not refer again to Paul. Paul thinks that charitable dogma will protect against the prideful desire to know and to become as perfect as the whole that has a fixed beginning and end. But Bacon knows the truth about Paul's charity. Charity is wholly compatible with knowledge of the material whole that reveals the nature and the will of God, as if God moved nature only by material causes. In fact, charity prompts such comprehensive knowledge of beginnings; it tempts us to the knowledge of God's works and thence to knowledge of the will that moves the works and divides times between beginning and end. And Bacon has shown that such knowledge distorts the soul.

In the present context Bacon says that knowledge of God by way of His works produces wonder, or broken knowledge, and that contemplation of God's works alone produces knowledge. Bacon then says that both knowledge and wonder—"all knowledge and wonder"—is "an impression of pure pleasure in itself." This statement is puzzling, because we wonder how wonder could participate in "pure pleasure in itself." After all, wonder is a seed of knowledge, a beginning, or else it is a nonknowing end of knowledge, a mystery. And how could pure pleasure in itself have both a beginning and an end? Or to speak otherwise, how is such pleasure possible for a being that has a beginning and an end? Such pleasure is possible only on the model of art and charity, only for a perfect ensouled body that is limited by no genesis and demise. In fact, art and charity know divine will by way of divine works; they aspire to know perfectly the particulars of practice so as not to be limited by beginnings and ends. In this aspiration, charity and art confuse knowledge and wonder, which are different. In following up this point, Bacon explains how the gods move art, philosophy, and political philosophy and how these endeavors are different but never separate in any political order. We learn that charity is latent in every order of productive arts.

To illustrate his point about the difference between knowledge and wonder, Bacon quotes not Paul but Philo, again "one of Plato's school." He refers to the difficult *De somniis*, which we will have to consider briefly. The *De somniis* treats dreams that in foretelling the future cause the mind to move with the whole, appearing to be possessed and God-inspired.[29] According to Philo, God is like the sun, although His effect on the places where sense perception occurs is the opposite of the sun, for God's light darkens the objects of the senses. While the

---

29. *De somniis* 1.1–4.

sun is the only visible thing that can be likened to God, the soul is the only invisible thing likened to God.[30] Now, the sun's likeness to God is more apt than at first appears, for the sun is not itself visible; it is the source of illumination that cannot itself be seen directly by the eye. As illuminated by the sun, objects of perception are images of the sun, so that our access to the invisible ground of visibility is only through the visible things.[31] As God and the sun are alike, the sun is the source of day and night, and as God is the model of models, so the sun is the source of illumination and of the *difference* between image and what is imaged.[32] That is, the invisible sun lights up the objects of sense, but it also discloses the imagelikeness, or the darkness, of all objects of sense, which are only defective likenesses of intelligible things. According to Philo, then, the whole from beginning to end—the whole of knowing and known, creator and creature— shines through to man only by way of the not-being of images, or the togetherness of the light and the dark.

Philo argues that if the sun is akin to the soul because both are likenesses of God, then it is so only insofar as the soul is mind *(nous)*.[33] Mind knows and knows knowing, but the senses, without which the soul could not move according to its passions, are a source of darkness in comparison to mind. When the senses' rays shine, sight, hearing, taste, and smell are awake, but prudence, justice, knowledge, and wisdom are put to sleep.[34] The senses are thus like the sun's illumination of the earth, and the sun must set in order for the heavens to appear. This state of affairs is as it must be for man. If the senses are akin to the sun, then they must belong to mind as well as to soul: the eye is the mind's window to the mindlike sun, but the sunlike eye can never directly grasp the sun. The senses belong to soul and mind together. As such, the senses let the whole shine through while covering it up. Any grasping of this whole that does not also darken would lead to blindness both for the eye and the mind's eye: mind (nous) grasps the cosmos of mindlike intelligibles *(noeta)*, including itself, only by passing through the objects of the senses. For Philo, a noetic world wholly free from the distortion of images is simply unspeakable.[35] As Plato says, it is beyond being.[36] For man to "see" the whole, or to become mind, he would have to be blind to the very things that let the whole shine like the sun, the various images of sense and speech.[37] Man's openness to the virtues of mind depends on his openness to images that cover up and distort, which images always belong to the

30. Ibid., 1.72–85. 31. Ibid., 1.72–76, 1.238–42. 32. Ibid., 1.110–15.
33. Ibid., 1.77–79. 34. Ibid., 1.79–85. 35. Ibid., 1.184–91.
36. *Republic* 509b6–10. 37. *De somniis* 1.187–89.

soul that sees, hears, tastes, smells, and (as Philo modestly ignores) touches. Mind and soul are akin because they share dependence on images. The senses sense images, and only by way of sense can the soul imagine for mind. The noetic whole is always distorted by the motions of sense and imagination, which perceive the motions of sensed objects so as to reveal what is motionless, or noetic, in them.

Philo likens the noetic whole to the characteristics of the cosmic and the Olympian gods.[38] But then according to his account, these gods represent, respectively, mind and the intelligibles as they are the mysterious, motionless causes of nature's motions (the visible objects of sense), and soul and the motions of nature and of sense and imagination. Neither the perfect assimilation of soul to mind nor the mind's being perfect within such a soul is possible for man, because it would require the mind's and the soul's freedom from distorting images and the imagination. It would require freedom from the very motions through which soul is open to mind and mind reflects being. If one does not care for truth, the imagination can be free, and compared with such freedom, the human love of truth is constrained. But such perfect freedom of the imagination could never discern what it is proper to eat, so human beings are constrained to love the truth. Here we discern the importance of Bacon's reference to Philo's obscure and difficult treatise to explain the difference between wonder and knowledge. Only a cosmic god, not man, could be both a pure knower and unconstrained, because such a god is the motionless eidetic cause of motion. As truth lovers, human beings wish that their minds could become as free as their imaginations can be, but such freedom is possible only for a mysterious cosmic god that needs no imagination. The human love of truth is always itself akin to natural motion: it is always akin to the bogus Olympian gods who move, who care for men, and who yet would ever deny that they have to move. For Philo, the human desire for perfect freedom *from* the imagination is a delusion sparked by the freedom *of* the imagination, by the freedom of art. But such perfect freedom is no more possible for man than it could be for an Olympian god—an impossible god who creates men and who hearkens to and is heard by them.

Wonder and knowledge are caused by our respective stances to the cosmic and the Olympian gods. The cosmic gods inspire wonder because they represent the mystery of intelligibility, of the very appearance of problems. In particular, they represent the mystery of how the perfectly self-sufficient, eidetic form can be useful as a cause,

38. Ibid., 1.79–85.

as a final cause. The Olympian gods inspire knowledge, because in creating and caring for men, they aspire to be final causes that move: they aspire literally to *make* the mystery of intelligibility clear, so that all that can be done is grounded by all that can be known. The Olympian gods are of course impossible: when men listen to the boasts of the Olympian gods—the gods who create and care or love—they hear their own artful and truth-loving longings. The Olympian gods call forth knowledge that is always tempted to assimilate reason and art and to put an end to wonder. The Olympian gods attract the striving or the eros of the arts, engendering as well both material and mathematical natural philosophy. But the cosmic gods draw forth the love of wonder: they inspire the political philosopher's love of wisdom about the possible appearance and intelligibility of the problematic and the mysterious. And no small part of this wisdom concerns the problem of political life: the tension between theory and practice and between ends and means.

By way of Philo and Plutarch, Bacon shows that to account for man's perfect, free origins is to show man as the image of an impossible god. The desire for such an account springs from two different but related stances toward the needy human body, one that takes its bearings solely by the motions of those needs and the other that hopes to be wholly free from them. Both represent the truth-loving desire for perfect freedom and mastery. One is the physicist's dream and the other is the mathematician's dream. Both come together in the sophist's claim to perfect freedom by way of a single art, the imaginative art of appearances or images, which would elevate the imagination's truthless freedom to the noetic, cosmic god's pure knowing freedom from the imagination. According to Philo such a cosmogony must always be botched because it does not appreciate the mystery of the divine things. But according to the classical utopian teaching, human beings are always tempted to reproduce such cosmogony from their own need to produce and to know the truth, from their need to know which things to eat.

The human response to need is determined by the human orientation to the divine. As such, neediness is always tempted by the love of perfect freedom and mastery. This is the same as the necessary but dangerous and unjust love of perfect justice. We recall that Socrates fashions the best city, which requires Timaeus' botched, shocking cosmogony, in response to Thrasymachos' and Glaucon's truth-loving desire for perfect freedom and justice.[39] Political life is moved

39. *Republic* 336b1–37d2, 338c1–3, 340d1–41a4, 348c11–12, 358b1–62c8, 370a7–65.

by an Olympian god who creates, and like the Olympian god it aspires to the cosmic god's pure knowing freedom that is always forbidden to men and to gods that move. Even in its ignorance, no art is ever wholly removed from political philosophy, concerned as both are for the self-sufficient and the good. According to the classical utopian teaching, the city is always open to two different kinds of gods: the lying, bogus Olympian gods who, in promising knowledge set the productive arts in motion, and the silent cosmic gods that inspire wonder and turn the political philosopher from the practice of any art to the criticism of them all.[40]

Bacon's argument is astounding. Charity does not turn us to loving silence before Paul's God. Rather the *necessary* call of the Olympian gods causes charity, which aspires to knowing freedom and mastery by way of art and which cannot abide any merely conventional freedom from need. Paul's knowing God, who also creates and cares for men who love him, is the present face of the Olympian gods. Such a god's ridiculous knowing calls forth man's charity: the deluded eros that sets the productive arts in motion. The question we face is why, knowing better, Bacon, like the physicist, the mathematician, Paul, and the sophist, who cannot help themselves, turns his face only to the Olympian god. And in his concluding response to the divines, Bacon confirms the primacy of this question.

The seventh charge of the divines is a conclusion from the discussion of the sixth: because there is a difference between knowledge and wonder, comprehending God's secrets by Icarus' faulty tools has caused diverse great learned men to be heretical. According to Bacon's interpretation of the fable of Icarus,[41] the path to which Icarus was directed represents the moral principle of moderation, midway between excess and defect. The excessive path Icarus took, however destructive, was the better of the two worse extremes because "in excess there is something of magnanimity," which "holds kindred with the heavens" and which is found in the young. In this respect, Heraclitus' "dry light" illustrates the superiority of excess to defect, but even here "a measure must be kept," that is, the dryness must be mixed with the "moisture and humours of earth" so that the light can be subtle without catching fire. Bacon suggests that the Olympian gods are akin to youthful excess, meaning that it is not just that men mistake the gods by comprehending genesis and providence but that they become too much like excessive gods. If the gods act in excess, and

40. See Leo Strauss, *The City and Man* (Skokie, Ill.: Rand McNally, 1964), 32–34; Aristotle, *Politics* 1322b12–37, 1328b11–13.
41. *De sapientia* 754–55.

not, therefore, as one would suppose, as far beyond excess as beyond defect and moderation, then the gods themselves would never
themselves be wholly "dry" without catching fire. Charitable knowledge is open to the contentious charge of heresy because it can itself
become like a creator god, whose preference for excess over defect
is mirrored in charitable political life, where, according to the *De sapientia veterum,* moderation is "questionable and to be used with caution and judgment."[42] The rebuttal of the eighth charge would seem
to point to this speculation, for Job speaks not of second causes but
of God's works that upset human intention and justice. Job, a wise
and blameless man, is by God's intention made into a laughing-
stock.[43] No knowledge of second causes will reveal God's intention,
which is as it should be if God's causes are miraculous. But Bacon
has claimed that just such knowledge will reveal the whole of genesis
and providence, and if Job's complaint illustrates Bacon's argument,
the second-causal knowledge of beginnings and ends confounds virtue and justice to the extent that they are bounded by the moderating limits of convention. Our love of a creator god is always one or
another partisan enthusiasm.

The "poets allegory," which Bacon notes to prove that second causes
remind of God, refers at once to Homer and to Plato's discussion of
Homer's saying in the *Theaetetus.* According to Socrates in the *Theaetetus,* Homer's crowning and compelling persuasion refers to the sun
so as to demonstrate that motion preserves being, while rest destroys
being.[44] This argument expresses the implications of Theaetetus' argument that knowledge is perception, a view shared by Protagoras
the sophist and by all the philosophers except Parmenides, namely,
Heraclitus and Empedocles.[45] In the dialogue, Socrates confounds the
young mathematician Theaetetus by convincing him that *phantasia*
(appearance/imagination) is the same thing as perception. Both are
described as the same (perfectly free) motion in regard to the comprehension of large and small, light and heavy, and color.[46] The result of this argument is that all knowable things are only as they seem
and that since seeming is relative to the perceiver, nothing ever is,
but everything is always becoming. All things are different; in other
words, there can never be sameness, which means that man is no more
a measure of being than any other creature or thing. Now, in order
to refute Theaetetus and the philosophers fairly, Socrates must show

---

42. Ibid. 43. Job 12.1–4. 44. *Theaetetus* 153c6–d5.
45. Ibid., 151d7–53d5.
46. Ibid., 151d7–53d5, 182a3–e5; cf. *Republic* 523a10–24d5.

how sense differs from phantasia and from sense as it depends on phantasia; he would have to show how it is possible for him to abstract from the forms of things in order to assimilate the sense of heavy and light, large and small, and color to the faculty of the imagination. The sophist himself gives Socrates the unbridled freedom he has for his refutation, because the sophist assimilates the freedom of sense—sense as it might simply be one's own in regard to weight, size, and color—and the truth-limited freedom of image making. The sophist assimilates sense and phantasia this way so that he can be a master of appearances, making him the master not just of all the arts but of conventions that no longer limit any claims to mastery.[47]

Bacon knows that the impossible hope for such perfectly free image making forgets the mindlike form, delimitation, and constraints of the body. Perfectly free phantasia is a delusion of the desire for the perfect freedom of mastery. This desire presumes bodies that have no souls or souls that have no bodies, but neither view is sufficient to deny the most bestial forms of delectation. To refute Theaetetus' opinion, Socrates merely appears to be like the sophist, for unlike the sophist, Socrates does not fail to consider the humble constraints of need.[48] Socrates' curiosity is never witless; he knows that the imaging power of phantasia is ordered by truth and that both are constrained by the needs and limits of the body. Images are never wholly free, however much they can be fashioned at will by an image-making art. The conventional virtues have the being of just such images. They are at once willfully or freely made, and yet as they confine human longing to moderate limits, they have the authority of necessary, or needful, truth. Conventional virtue causes men to be free, but only because it balances the tension between freedom and need. And this balance is always at risk before the necessary and dangerous call of the Olympian gods. Artful men cannot but love the truth. But art can never love the truth wisely or by way of the love of wonder. Wonder reveals that truth is mysterious, that truth is always distorted by the images it constrains. Wonder reveals that we can experience and judge error and perplexity but cannot grasp the whole with the clarity required for the practical needs that call forth the love of truth in the first place. Respect for conventions stands midway between the love of truth and the love of wonder. But conventions are made by art, and the sophist's ambitions reveal the con-

47. *Sophist* 234a7–35c6.   48. *Republic* 369b4–70e3.

69

tradiction at the heart of political life: the arts are always tempted to dissolve their most precious products in the solvent of an impossible truth.

Bacon knows that knowledge answers to need. In this respect he is like Socrates. But he is unlike Socrates because he recommends charitable knowledge that would simply overcome the tension between freedom and need. Even so, Bacon knows that the attempt to be free from bodily needs, either by delusional, exclusive attention to them or by delusional abstraction from them, is the sophistical imitation of any excessive god who does not understand need. These attempts constitute charity, which is not in itself a Christian phenomenon. In discussing Juno's suitor, Bacon remarks that Jupiter assumed an ignoble and contemptible bodily shape in order to secure Juno's love.[49] This transformation teaches the "depths of moral science," that to satisfy desire may require an "outward show of abjectness and degeneracy." But Bacon also says that Jupiter's moral knowledge stands for the providential order of second causes, the knowledge of which we know to be charitable. Charity is at once abject and degenerate and the cause of dangerous pride. Charity turns our attention to an impossible cosmogony, one capable of proving that man could not originally have been omnivorous. But charity is nothing new; it is the same hope for perfect justice that prompted Timaeus to attempt his impossible cosmogony. Charity heeds the call of the Olympian gods, and as Plato and Bacon knew, it is really the divine revenge for men's hearkening to these gods. Charity is the divine revenge that is coeval with political life. Divine revenge is not necessarily just; in fact, its defective justice requires men to temper justice with moderation. We need to know why Bacon so boldly counsels charity. Bacon himself tells us why. As we will see, his answer concerns the history of Christianity, but this is because Christianity is charitable, not because charity is uniquely Christian.

49. *De sapientia* 728.

[3]

The Apology to the Politiques:
Greek Virtue and Roman Honor

[268–274]

There are four disgraces attributed to learning by the politiques. The first is that learning softens men's minds, making them unfit for honor and arms. The second is that learning weakens the disposition for government by making men curious and irresolute from too much reading, by making them too peremptory and positive from "strictness of rules and axioms," by making them immoderate and overweening from "greatness of examples," and by making them incompatible and differing from times because of preference for "dissimilitude of examples." The third disgrace is that learning diverts men from action and business to leisure and privateness, and the fourth is that learning relaxes the discipline of states, making men readier to argue than to obey and execute. Because of "this conceit" (the fourth disgrace), Cato the Censor counseled that Carneades should be expelled from Rome lest his philosophy infect the youth and alter manners and customs.[1] Likewise Virgil separated policy and government from the arts and sciences, giving in his verse one to the Romans and the other to the Greeks, to the advantage of his country and the disadvantage of his profession.[2] So too Anytus accused Socrates of diverting the youth from law and custom and of making the worse argument seem the better by force of eloquence and speech.[3]

Bacon says that the four charges against learning are unjust. Regarding the first, both in persons and in times learning and arms have

1. Plutarch, *Vitae parallelae,* ed. Lindskog and Ziegler (Leipzig: Teubner, 1960–73); *Marcus Cato* 22. All subsequent citations of the *Lives* refer to the Teubner edition.
2. *Aeneid* 6.852.   3. Plato, *Apology* 18a7–21b, 23c2–24c9.

concurred, as we see, with Alexander, Aristotle's scholar, and with Caesar, Cicero's rival in eloquence. Scholars who were great generals rather than vice versa have included Epaminondas, who abated Sparta, and Xenophon, the first to make way to overthrow the Persian monarchy. The concurrence shows better in times than in men, for in Egypt, Assyria, Persia, Graecia, and Rome, arms and learning flourished at the same time. It cannot be otherwise, because as in man strength of mind and body are coeval, with the body being strong "somewhat the more early," so too in states arms, corresponding to the body, and learning, corresponding to the soul, "have a concurrence or near sequence in times."

Regarding the second charge, Bacon says that it is improbable that learning should weaken men for policy and government. Just as it is a mistake to submit a natural body to "empiric physicians" and to rely on merely practical lawyers, so it is doubtful if states are entrusted to unlearned, "empiric statesmen." Rather "it is almost without instance contradictory, that ever any government was disastrous that was in the hands of learned governors." Ordinary political men condemn the learned by calling them *pedantes,* but many "princes in minority" have governed well because the pedantes in fact ruled, as did Seneca during Nero's minority, Misitheus during the minority of Gordianus the Younger, and women aided by teachers and preceptors during the minority of Alexander Severus. "In our times," Pius Quintus and Sextus Quintus were pedantical friars before becoming bishops of Rome, and both did greater things than those who were bred in estates and courts. Although the learned are likely to use convenience and accommodation, called by the Italians *ragioni di stato,* with which Pius Quintus had no patience (he called it immoral and irreligious), still, "to recompense that" they are perfect in the "plain grounds" of religion, justice, honor, and moral virtue. If these are well watched, there will be little use of "the other," just as a well-dieted body has little use for physic. The experience of one man's life cannot give precedents and examples for the events of another man's life, for as sometimes a grandchild or ancestor resembles a father more than a son, so present occurrences "may sort better" with ancient than with later or immediate times, and one man's wit can no more "countervail" learning than one man's means can sway the common purse.

Regarding the four particulars of the second charge, if it is granted that they may be true, it should be remembered that in each case learning gives greater remedy than cause of the indisposition or infirmity: if learning makes men perplexed and irresolute by "secret

72

operation," then it plainly teaches when and where to resolve. If learning makes men positive and regular, it teaches what is demonstrative and what is conjectural and the use of distinctions and exceptions. If learning misleads by "disproportion or dissimilitude of examples," it teaches force of circumstances, error of comparisons, and caution of application. These medicines are taught more forcefully by the "quickness and penetration of examples." Such are the errors of Clement the Seventh described by Guicciardini and the errors of Cicero described by himself in his letters to Atticus, which warn against irresolution.[4] The errors of Phocion warn against obstinacy or inflexibility, the fable of Ixion warns against being vaporous and imaginative, and the errors of Cato the Second warn against being one of the "antipodes," treading opposite to the present world.[5]

As for the third charged disgrace, learning does not lead to leisure, privateness, and sloth; it would be strange for what accustoms the mind to "perpetual motion and agitation" to lead to sloth. In fact, only the learned love business for itself rather than for profit, honor, and reputation, for fortune of pleasure or displeasure, pride, and self-conceit, or for any other end. Just as untrue valors are in other men's eyes, so such men's industries are in other men's eyes. Again, only the learned think business as natural and agreeable to the mind as exercise is to the body. If a learned man is idle in business, it springs from weakness of body or spirit, such as Seneca speaks of in describing men as so fond of shade that they worry when they are in the light, rather than from learning itself. This weakness may lead to learning, but learning does not produce it. Learning does not take up too much time or leisure, because even the busiest men have leisure, unless they be incompetent or overly ambitious for their abilities. The question is how leisure is to be spent, in pleasure or study. When Aeschines said that Demosthenes' orations "smelled of the lamp," Demosthenes replied to the pleasure lover that there is a difference between the things they did by lamp light.[6] Learning does not expel business, but it defends the mind against idleness and pleasure.

Finally, regarding the fourth charge, learning does not undermine reverence for law and government. Bacon says that blind obedience is no more superior to instructed duty and obligation than a blind man's guide is better than a seeing man's light. It is not doubted that learning makes the mind gentle and "pliant to government" in con-

---

4. Guicciardini, *Storia d'Italia*, ed. C. Panigada (Bari: G. Laterza, 1929), 16.10, 12; *Att.*, 16.7.

5. *Att.* 2.1.   6. Plutarch, *Demosthenes* 8.2.

trast to ignorance, which leads to tumult, sedition, and change. Cato's blasphemous judgement was repaid in kind when he himself turned to Greek learning in his old age,[7] showing that his censure was affected gravity rather than his true opinion. And Virgil's verse, giving to Romans the art of empire and to others "the arts of subjects," is wrong because the Romans did not ascend to empire until they were at the height of the other arts. The best poet, Virgilius Maro, the best historiographer, Titus Livius, the best antiquary, Marcus Varro, and the best or second-best orator, Marcus Cicero, lived in the time of the first two Caesars. And it is to be remembered that the accusation against Socrates was made during the bloody rule of the "thirty tyrants." As soon as the revolution was over, Socrates was made "heroical," his memory was given divine and human honor, and his discourses were acknowledged to be "sovereign medicines of the mind and manners." Bacon says that his argument should rebut the "humorous severity" or "feigned gravity" of the politiques. He says that his argument is a reproof "that is not really needed in present times" because of the two learned princes, Elizabeth, and the king, who are as Castor and Pollux, *lucida sidera*, "stars of excellent light" who have instilled in "all men of place and authority in our nation" a love and reverence toward learning.

In the previous chapter, we have seen Bacon disclose the core of his treatise in a passage that gave us the toughest sledding we will encounter. Of course much is yet to be elaborated and explained, but it will be helpful at this point to anticipate arguments to come and to fix ideas discussed so far. Bacon apologizes for his new project for mankind—the artful mastery of nature—because he knows this project to be charitable and because he knows that charity is the source of political immoderation. But charity is essential to political life; it is not simply unique to Christian faith. Rather, somehow Christian faith has liberated charity from the moderating restraints recommended by classical utopian thought. To know why we must now turn, not to these restraints, but rather to a new form of charity itself, we have to know just how this liberation came about. We need to know how charity came to be freed from the restraints of conventional moral virtue as it was understood in traditional utopian thought. We recall from the dedicatory epistle that, as Bacon is constrained by the Christian prince, he is compelled to weigh not the merits of

7. Plutarch, *Marcus Cato* 2.

philosophy and politics, as Socrates did, but the two kinds of Roman tyrant described with reference to Tacitus. It would seem, then, that to understand the compelling force of Christian charity, we must understand its origin in the empire of Rome. Also, we must understand the difference between that empire and the home of the ancient utopians. In this and the several chapters to follow, we will see how Bacon explains the difference between Rome and Greece, how he shows that this difference made Christianity possible, and how he argues that, while knowing the lessons of classical utopian thought, he must turn human effort to the charitable project he recommends.

We begin by noting that of the four disgraces mentioned by the politiques only the fourth treats the status of the laws. Bacon treats it twice, once at the beginning of the rebuttal, where he discusses Cato, Virgil, and Anytus, and again at the end, where he answers each of their specific attacks. Considered together, the three examples of the fourth charge, regarding discipline, obedience, and execution, are not consistent, because the example of Anytus contradicts the examples of Cato and Virgil. Cato and Virgil suggest that Rome is the enemy of the arts and sciences and the friend of policy and government and that Greece loves art, science, eloquence, and learning at the expense of policy, government, obedience, and execution. But the example of Anytus presents Athens as the city that condemned Socrates for his discourses, his disputation, and his profession of the sophists' and the rhetoricians' art. According to the examples, then, both Greece and Rome were the enemies of the arts and sciences when they are mixed with policy and government. According to the examples, both Greece and Rome opposed political philosophy, which was taken to be sophistical wisdom.

It is important to note what Bacon omits in his three examples. He does not mention Cato's likeness to Socrates, for according to Plutarch, Cato was not simply an enemy of philosophy nor a convert to philosophy in his old age. In his youth, Cato was reinforced in his love of restraint and simplicity by hearing the Pythagoreans speak about pleasure in the same way as did Plato.[8] Bacon acknowledges this fact later and indirectly when he notes that Cato's blasphemy was affected rather than his own opinion. Also, Bacon does not refer to the imperial arts that Virgil says will belong to Rome. In the *Aeneid*, Aeneas learns from the souls of great, politic Romans to come that Rome's imperial arts will include dictating the terms of peace, sparing the vanquished, and making war upon the proud.[9] These impe-

8. Ibid.   9. *Aeneid* 6.851–53.

rial virtues describe exactly the deeds of the prescientific Bensalemite king who defeated the old Atlantians. Finally, Bacon omits reference to Socrates' other accusers, Meletus, who represented the poets and who spoke at the trial, and Lycon, who represented the rhetoricians.[10] Anytus, who represents the artisans and the politicians and who hears Socrates' speeches in the *Meno,* did not himself speak. Bacon has the artisan-politician speak for the poet, and he distorts the exact presentation and refutation of the charges. Socrates defends himself against the charge of being a sophist only against the unnamed "first accusers," who said for years that he studied the things above and below, practiced sophistry, and must, therefore, have been an atheist and must have corrupted the youth.[11] But to the three present accusers, Socrates defends himself against the charge of corrupting the youth. He refutes the argument that he did so by teaching the young not to believe in the city's gods but in new divine things.[12] Bacon presents the trial as if it were charged that Socrates corrupted the youth with sophistry and rhetoric, but his silence reminds us that the substance of the charge against Socrates was impiety and that the rhetoricians and poets were among the accusers. At the outset, then, we are made to wonder how Greece differs from Rome, how Bensalem differs from them both, how Socrates differs from Cato, and how political philosophy differs from poetry, rhetoric, and sophistry.

As if to suggest that Greece and Rome were the same, Bacon says that Aristotle, Alexander, Cicero, and Caesar show that learning can concur with arms. He makes the same suggestion again when he discusses the human proofs of learning's dignity.[13] In the second book Bacon speaks of Aristotle and Alexander along with another Caesar, Augustus. Bacon does so when he discusses his new understanding of metaphysics.[14] There he says that, consistent with truth and the proficience of knowledge, and unlike Aristotle, he will "recede" as little as possible from ancient terms and opinions. Aristotle, "who sought only to refute ancient wisdom," reflected the "humor of his scholar" Alexander, of whom it was said by Lucan that he was a "fortunate robber" who "made a prize of nations."[15] Bacon says he will use the moderate tactic of civil government in which old terms are

---

10. Plato, *Apology* 23d–24b1.   11. Ibid., 18a7–24b1.
12. Ibid., 24b1–28a1.
13. *Advancement* 307–24; see Chap. 6 at nn. 46–52 below.
14. Ibid., 352–53; see Chap. 8 at nn. 5–8, 22–23 below.
15. Lucan, *De bello civili,* trans. Duff, Loeb Classical Library (London: Heinemann, 1928), 10.20.

used to refer to altered states of affairs, likening this practice to Tacitus' *eadem magistratuum vocabula*. The reference is to Tacitus' remarks about Augustus' time, wherein the liberty of the republic had been sacrificed to Augustus' tyranny by the giving of gratuities to the army, cheap corn to the people, and peace to the world. Under *nomine principis* Augustus gathered the world beneath his empire.[16] As Bacon has already shown, Augustus' eloquence consisted of his forthright but artful dissimulation of his ambition to emulate the tyrant Caesar, an ambition that was possible because it had been freed by empire and wealth.

Now, in the present context, Bacon does not mention that according to Plutarch, Epaminondas was one of the great Greeks whose deeds were tarnished by violence and toil, nor does he mention that Xenophon's son was said to have been the man who killed Epaminondas.[17] These omissions suggest that arms and wisdom did not coincide in Greece as they did in Rome, and if we think about the difference between Caesar and Augustus, it is not strictly true that in Caesar eloquence and arms coincided. Augustus' eloquence proved to be superior to Caesar's. And as Bacon says in discussing the architecture of private fortune, Caesar chose martial over "civil and popular greatness" because he saw that he could not be more eminent in eloquence than Cicero, Hortensius, Quintus Lutatius Catulus, and many others.[18] The coincidence of learning and arms really consists in eloquent tyranny made possible by the arms of a less eloquent tyrant whose theft of Roman liberty was requited by his premature death. We do not yet know just how Cicero and Aristotle may have served their respective thieves of liberty. But however they did, we know that Alexander's tyranny descended upon a world from without, while Tacitus tells us that Augustus' eloquent tyranny was generated, by way of Caesar's violent sacrifice, from within the free republic it enslaved.

In speaking of learning's coincidence with arms in times, Bacon refers at once to Aristotle's remark in the *Rhetoric* concerning the characters of youth, age, and the prime of life, where it is said that the body is ripe at age thirty-five and the mind at forty-nine,[19] and

16. The two phrases are translated, respectively, "the officials went by the same or the old names" and "under the name of 'foremost' or 'chief.' " In the latter case, Tacitus makes it clear that he means that Augustus appealed to Rome's origins (*a principio*) when, however, Rome was governed by kings and liberty and the consulate did not exist (*Annales* 1.1, 3).

17. Plutarch, *Timoleon* 36; Pausanias, *Graeciae descripto*, rec. Spiro (Leipzig: Teubner, 1959–64), 8.11.5–8.

18. *Advancement* 461–62.   19. *Rhetoric* 1390b.

to his own remark at the end of Essay 58, "Of Vicissitude of Things." Although Aristotle argues in the *Rhetoric* that the prime of life can be associated with age, in the *Nicomachean Ethics* he makes it clear that, with regard to teaching political science, youth and maturity are not determined simply by age. The young, whether young or old in years, who do not live according to reason, are not fit students of political science.[20] If we look to Essay 58, we find that Bacon argues that youth (arms) and middle age (learning) come "together for a time" before the "declining age of a state." The greatest vicissitude of things in men as opposed to nature is "the vicissitude of sects and religions," and wars are most likely to occur "upon the breaking and shivering of great empires" as well as upon "the great accessions and union of kingdoms." In both instances, Bacon gives the Roman Empire as an example.[21] But he says that the jealousy of the sects does not "much extinguish the memory of things" as Machiavelli says it does. To counter Machiavelli's argument about Gregory's attempt to "extinguish heathen antiquity," Bacon gives the example of "the succession of Sabinian, who did revive the former antiquities."[22] In discussing changes caused by natural catastrophe, Bacon denies that the jealousy of sects causes oblivion. Unlike Machiavelli, he does not imply that what remains of all superseded sects points to the eternity of the world or to the precedence of the world's origin to any possible memory of it. Rather Bacon remarks that a new pope simply restored all memory once again. But even so, Bacon remarks that sects and religions are subject to "great revolutions" and that their jealousy causes the greatest vicissitudes of things among men. Like Machiavelli, he lumps together sects and religions as if to ignore the universal claims of religion.[23]

We want to know how Greece differs from Rome. We learn that, while both could be hostile to philosophy, arms and learning did not coincide in Greece exactly as they did in Rome. Unlike Greece, Rome succumbed to tyranny from within her own free institutions, and unlike Greece under Aristotle's student, the robber Alexander, Rome became an armed imperial tyranny that made war upon the proud not for the sake of pride but for peace. If arms and learning come together before the decline of a state, then in Rome such a time preceded the transformation of the imperial tyranny into a new Christian empire, for sects and religions cause the greatest human vicissitudes. Rome's imperial tyranny for peace was the obvious soil for the

---

20. *Nic. Eth.* 1094b29–95a12.   21. *Essays* 515.
22. Ibid., 513; *D* 2.5.   23. *Essays* 514.

growth of the Christian empire. In the light of Bacon's discussion of youth, age, and vicissitudes, we are forced to wonder whether Rome became such an empire in part because, contrary to Aristotle's argument, she did not hesitate to teach political science to the young. The cause of the greatest changes itself has a cause that has something to do with Rome's imperial arts. And to understand the power of the turbulent Christian empire, we have to understand its power to preserve the memory of harsh beginnings. Bacon implies that his predecessor, Machiavelli, did not properly understand this fact. It would seem, then, that Bacon thought Machiavelli had missed what can be learned from the difference between Greece and Rome. It is important to keep these points in mind, for as he proceeds, Bacon develops them with great care.

Bacon's remark about government and learned governors is ambiguous. It is not clear whether learned governments are or are not for the most part disastrous, and only what is unsaid in the examples, the first of which are Roman, clarifies Bacon's meaning. Regarding the five years of Nero's minority, Bacon does not mention that, according to Tacitus, Seneca influenced Nero only in conjunction with the martial disposition of Burrus and that, when Burrus died, Seneca's position was subverted, which led to the suicide forced upon him by Nero. Neither does Bacon disclose that, according to Tacitus, the decent effect of Seneca and Burrus upon Nero's obscene dissipation depended on restraining the pride and passions of Agrippina and that as maternal power waned, Nero's debauchery increased.[24] Indeed, the ultimate waning of her power was Nero's legal matricide, aided and abetted by Seneca and Burrus, for which Seneca became the object of public disapprobation and censure. Nero refused to let Seneca retire from wealth and public life lest his retirement bring talk of Nero's avarice and cruelty to men's lips.[25] Again, although Bacon mentions Gordianus the Younger, it is impossible to compare his minority to his majority because he was assassinated by the praetorian prefect after having been proclaimed emperor by the soldiers in Rome. And yet again, Alexander Severus was ruled not directly by pedantes but only indirectly by them through the women who did rule him. Bacon does not mention that Machiavelli lists Alexander Severus among the princes who failed to escape the hatred of those upon whom they depend because they could not avoid being hated by the many. Alexander came to his sad end precisely because of his pedantic and feminine goodness: because he was considered

---

24. *Annales* 13.2, 12, 14.7–12.   25. Ibid., 14.52–56.

effeminate and dominated by his mother, to whom he adhered rather than to the people, his soldiers conspired against him and killed him.[26]

In fact, then, Bacon's Roman examples do not demonstrate the happy coherence of statesmanship and learning. Rather they suggest that the indirect rule of women and learning did nothing to moderate the broils of imperial ambition. Bacon says that the two papal examples may "sort better" with the Roman examples than with any modern examples. But if so, then they would be like indirect, feminine rule to the extent that the papal government of universal spiritual empire affects political rule. Bacon excuses the learned use of ragioni di stato by saying that it is offset by the learned's religion, justice, honor, and moral virtue. Pius Quintus condemns this reason, so by Bacon's argument he must be the best example of learned rule. But Bacon declines to say that, from his "steadfastness," Pius was the persecutor of heretics, crusader against the Turks, and the excommunicator and "deposer" of Elizabeth. He was, in fact, the Grand Inquisitor, the unnamed hero of the bloodthirsty Martius and the immoderate zealot Zebedaeus in the *Advertisement Touching an Holy War*.[27] Sextus Quintus, likewise a zealous crusader against the Turks and the English, used the power of excommunication against Henry of Navarre and the Prince of Condé. In fact, then, these papal rulers did govern indirectly, using a spiritual sword as well as steel. And their contempt for ragione di stato was not necessarily restrained by justice, honor, and moral virtue. It is important, then, to know the difference between Greece and Rome, because papal rule sorts better with Roman than with modern examples, but though the Romans conquered for peace, the popes rule peacefully for the sake of war. We must pursue this matter of the difference between the popes and the Romans, which Bacon discusses in the sequel.

Four vices are said by the politiques to give rise to the second disgrace of learning. Although Bacon claims that learned examples are remedies for the vices, the examples are not applied to the vices exactly as the vices are listed in the first presentation of the politiques' charges. The first vice is changed so that curiosity and irresolution caused by variety of reading are replaced by perplexity and irresolution caused by the secret operation of learning, and the second vice is changed so that being too peremptory or positive is replaced by being positive and regular. The examples of Clement VII and Cicero illustrate the first vice, but they refer only to irresolution, not to

---

26. *P* 19.

27. *Advertisement* 11–36. See J. Weinberger, "On Bacon's *Advertisement Touching an Holy War*," *Interpretation*, vol. 9, nos. 2–3 (September 1981):197–206.

perplexity: as the more serious perplexity replaces curiosity, the more serious vice is not cured by way of learned example. Of greater interest, the cause of this vice, the secret operation of learning, is illuminated by the examples: according to Guicciardini, Bacon's source, Clement ruled directly when he was pope and indirectly when he was closest adviser to Leo X. Clement's indirect rule accords with Bacon's account of how learning mixes with rule. When Clement was adviser to Leo, he was accounted a most malevolent but capable politician. But upon ascending to the papacy, he was revealed to have been in fact lacking in resolution and execution to match his intelligence and knowledge of affairs.[28] In this case, the indirect influence of learning upon rule absorbed the blame for Leo's malevolent but dissimulated ambition.

Even though the modern example of indirect rule sorts well with the Roman examples of Seneca and the young Nero, it differs from them in an important respect made clear by the example of Cicero. Unlike the example of Seneca and Burrus, Bacon says, the example of Cicero warns against irresolution. But according to Plutarch, Cicero's ignominious end was not so much the result of his irresolution as it was of his love of power, which induced him, hardly irresolutely, to assist Octavian to obtain the consulship.[29] After his accomplishment, Cicero realized that his friends were correct to think that he had at once ruined himself and betrayed the liberty of the people. His end was not occasioned by his cautious flight from Rome after Caesar's murder; rather it was caused by his bold alliance with the new Caesar. Now, Bacon knows this fact perfectly well. He says that Cicero condemned his own irresolution in the letter to Atticus. But Bacon surely knows that in the letter Cicero said that his actions needed no apology: he had changed his plans to leave upon hearing the latest news from Rome, and "no philosopher has ever equated a change of plan with lack of firmness."[30] The example of modern learned rule is the papacy, which despite its love of peace does not really condemn warlike ragione di stato. But it is not for this reason resolute, and in this way the ancient Roman rule differed from the modern examples that sort so well with it.

According to Guicciardini, when Clement became pope, all Italy and the world were plunged into war by a combination of the pope's irresolution and his "greed" to make Florence return to the tyranny of his family at the expense of returning Florence to good and moderate government.[31] According to Guicciardini, Clement's irresolu-

28. *Storia d'Italia* 15.6, 16.10, 12.   29. *Cicero* 45–46.   30. *Att.* 16.7.
31. *Storia d'Italia* 18, 20, 18.19, 19.15, 20.2.

tion sprang from his Christian virtues and also demonstrated the extraordinary power of the papacy. Clement plunged his country and the world into war, but he was slow to bear arms himself, and his irresolution in this matter led to France's defeat in the battle of Pavia and the later, calamitous sack of Rome at the hands of imperial troops. The whole of the misery could have been prevented had he been willing to sell ecclesiastical offices when it was timely to do so rather than later, when he finally acted against his pious scruples.[32] The pope acted according to ragione di stato only when it remedied his own incarceration, not when it might have saved Rome. But unlike Seneca and Cicero, Clement was returned to his former greatness only a few months after causing the greatest calamity suffered by the Church since it had "become great." According to Guicciardini, Clement's extraordinary recovery testifies to the superior authority and respect afforded to the papacy when compared to any Christian princes.[33]

We can now see the true likeness and difference between Roman and papal rule. In Rome, learning resolutely assisted the conquest of free peoples and princes by a resolute, steadfast, and martial imperial tyrant. In the modern world, learned popes conquer free peoples and princes and plunge the whole world into war because of pious irresolution. But unlike the case of Caesar, there is no requiting price to be paid by tyrants, who do not pay for their crimes because they profess not to be armed with the sword. In this respect, Clement seems to be more like Augustus than like Caesar. However, this is merely apparent: Rome was eventually conquered, from without by barbarians and, as we see more precisely later on, from within by Christianity, but the new Christian empire survives its greatest calamity. The problem of the modern world can be traced, then, to a fact about religion. The religious sect is the source of the greatest vicissitudes and turbulence in human affairs, and the modern empire is the empire of a sect. But while this sect causes all manner of change, so much so that it could plunge a whole world into war, it is itself immune from the changes it causes. The source of this new empire is Rome, which was the proximate cause of the popes' charitable but implacable and turbulent rule. We know that the ultimate cause of the popes' rule was the freeing of charity from the restraints of moral virtue. And now we have a clue as to how the proximate and ultimate causes fit together: we wonder if they fit because Greece differed from Rome regarding the teaching of political science to all, including the young.

32. Ibid., 18.1–2.   33. Ibid., 18.14.

In the sequel, Bacon begins to explain the difference between the Roman and the Greek application of learning to political rule.

Regarding the second vice, Bacon says the example of Phocion will warn against being obstinate or inflexible. By the time the example is presented, peremptoriness or positiveness have become obstinacy or inflexibility. The vice is changed from a mixture of a possible virtue and a vice to a combination of two vices: if the example suits the vice, then Phocion demonstrates a weakness caused by a mixture of virtue and vice that becomes a weakness of vices as such. But Phocion cannot represent the combination of two vices, because according to Bacon's source Plutarch, Phocion was, like Cato the Younger, an illustration of the godlike use of persuasion and reason to introduce what is necessary.[34] Plutarch does not support Bacon because, for Plutarch, Phocion was, like Cato, not so much an example of a course that is "too straight and opposed in all things to the popular desires" as he was an example of one whose virtue is submerged by the corruption and vice of the times. Phocion represents the convergence of learning and rule in one man: he was a student first of Plato and then of Xenocrates, and he held the office of general forty-five times. According to Plutarch, Phocion was harsh, obstinate, and inexorable only as necessary to struggle successfully against those who opposed his efforts on behalf of his country; he was called to office by the people whenever they were sobered by harsh necessity.[35] Phocion's blunder was not his obstinacy or inflexibility; rather it was his misplaced trust in Nicanor's good faith and justice, leading ultimately to his execution at the hands of the corrupt Athenians. Plutarch tells us that Phocion's fate reminded the Greeks of Socrates, because the sin and misfortune of Athens were alike in both cases. And shortly after his death, the Athenians were taught by the course of events what a caretaker and guardian of moderation and justice they had lost.[36]

Bacon's change of the second vice to a combination of two vices makes it impossible for the example of Plutarch's Phocion to cure it. Phocion, like Socrates, represents the tension, not the harmony, between political life and the life of the mind. Like Socrates, Phocion was the godlike partisan of the freedom of reason. Also like Socrates, he was only reluctantly willing to embrace the necessary constraints of political life. Therefore both Socrates and Phocion represent the combination of vice and virtue: both are models of justice and moderation and yet both display a kind of contempt for the city

---

34. *Phocion* 2.5.   35. Ibid., 4, 8, 10.   36. Ibid., 32, 38.

and its necessary pretensions to justice. Socrates admitted that his *daimonion* urged him away from politics, for the sake of his life and the private good of his fellow Athenians, although he did not wholly refrain from public activity.[37] In this regard, Phocion was like Socrates because Phocion waited for necessity to bring the Athenians to their senses and did not try to cause that necessity himself by way of public deeds. But unlike Socrates, Phocion was not a political philosopher, and although he fell victim to Athens, as did Socrates, he was not forced to apologize for a criticism of the city's various arts, even though, in his deeds, he refused to submit to the art-engendering hope to master necessity. Phocion, then, is the imitation of Socrates, a practical version of Socrates who, unlike Socrates, actually served the public needs of the city while living and acting as if he did not. He could not be philosophic because he did not seek to know himself well enough. But even so, Phocion is the model of practical virtue because of his *forgetful* freedom from need. And as we will see, there is a difference between this understanding of practical virtue, which measures the worth of glory, and Roman virtue, which consists in glory by itself.

Only the last example of Cato the Younger seems to fit the vice it is said to demonstrate, the fourth vice of being "opposite to the present world." Whereas the elder Cato resembles the philosophical Socrates, the younger Cato differs from both Socrates and Phocion in having had his hopeful desire to rule frustrated by his adherence to old-fashioned virtue. He more resembles the earlier examples of Seneca and Cicero because, like them, he ambitiously desired to rule indirectly by supporting more powerful citizens.[38] And while Bacon says that Cato the Younger represents opposition to a present time, this point is not as clear as it first appears. Although he spoke to the Romans as if they lived in Plato's Republic rather than in "Romulus' cesspool,"[39] he neither was nor pretended to be free from the desire to mix reason and rule. Socrates told tales about such a mix, never himself trying to rule,[40] and Phocion's practical virtue made him more like Socrates than like the ambitious Nicanor, who betrayed him. In fact, then, the younger Cato opposed his present world for one that was not like the world of Socrates and Phocion. In the immediate sequel, we learn that the younger Cato's past world was as uniquely Roman as his present one.

Bacon explains the difference between Greece and Rome in dis-

37. Plato, *Apology* 31c4–33b8.   38. Plutarch, *Phocion* 3.
39. *Att.* 2.1.8.   40. *Republic* 376d9–10.

cussing the politiques' third and fourth charges. Regarding the charge of leisure and privateness, Bacon appeals to Seneca to argue that learning and business or practice are perfectly compatible and that business is therefore good in itself. But Seneca does not speak of business as good in itself. Rather Seneca comments on the need to balance repose and activity, rest and motion, for delight in bustle is not industry but only the chasing about of a disquieted mind, and true repose is not judging all motion to be mere annoyance. Seneca condemns both perfect rest and perfect motion and recommends the mixture of rest and motion as night is combined with day.[41] But if rest and motion mix as night and day, then they can never be in perfect unity, because both the light and the dark cover up or distort as they disclose. By referring to Seneca, Bacon shows that hoping for the identity of theory and practice is the same as hoping for two impossible virtues, perfect freedom of mind and body. And these impossible virtues are really two possible vices: mindlessness and disproportionate attention to art. Now, to demonstrate the identity of business and mind, Bacon refers to a Greek rather than to a Roman; but he chooses a Greek, Demosthenes, who is the perfect opposite of Socrates. Socrates was a philosopher, and Demosthenes was a rhetorician; Socrates was indifferent to any kind of material gain, and Demosthenes was a corrupt lover of gold; and Socrates displayed perfect if philosophical courage (he always did his duty and retreated with grace), but Demosthenes exhorted to war and yet was always a coward in battle.[42]

The rhetorician Demosthenes represents the Athens that submerged the virtue of Phocion, which Bacon emphasizes by changing the object of Demosthenes' remark from Pytheas, as Plutarch reports, to Aeschines. In making the switch, Bacon replaces an enemy who accused Demosthenes of having a poor nature improved only by work with an enemy who said—correctly, according to Plutarch—that Demosthenes' speeches displayed a wonderful boldness.[43] Plutarch emphasizes the difference between natural boldness and courage, the artificial imitation of natural boldness, but Bacon refers to Plutarch so as to deny this distinction. As Bacon presents it, the example of Demosthenes subverts the distinction between natural virtue and natural vice, so that the artful imitation of natural virtue serves what is common to virtue and vice, as if courage were simply the artificial means to the ends common to the weak and the bold.

Belief in the identity of business and mind subverts the conven-

---

41. *Ep.* 3.   42. Plutarch, *Demosthenes* 8–9, 14, 20, 22, 25.   43. Ibid., 8, 9.

tional moral virtues as the classical utopian thinkers understood them. It subverts the moral virtues as embodied by the Socrates-like Phocion. Such a belief denies the freedom from need upon which the virtues are grounded. It looks not to the nobility or the freedom of virtuous deeds but only to the fact that virtue is produced by an art that produces convention. And it is true that those fashioned by such an art do serve the needs of the city: liberality is a virtue of acquisition, and courage is a virtue of war and victory. According to the classical utopian thinkers, practice can be virtuous, or noble, only when the needs served by the city's arts are served so as to demonstrate a freedom from the products of these arts: liberality demonstrates freedom from money, and courage demonstrates freedom from victory. Such virtues depend upon the acquisitive and victorious efforts of others, and as well, the virtuous must forget the extent to which they too merely serve the city and the extent to which they have been formed by the convention-making arts, the arts of the lawmaker and the rhetorician. Forgetful moral virtue is the sole restraint on the dangerous and necessary pretensions of the productive arts to provide freedom by way of the needs they supply. Moral virtue differs from honor, then, because honor reflects the praise of those who bestow it. When the virtuous are honored, they do not think honor valuable, especially as it is bestowed by those who practice the city's needful arts.[44]

For the classical utopians, the best regime moderates honor by virtue, depending wholly upon practiced indifference to need. And as we saw in discussing Phocion, such indifference depends upon the forgetting or ignorance of beginnings. Taking business and mind to be identical simply reflects the just love of honor. We can see now that unrestrained honor is the same as the charity at the heart of any city: a perfect identity of mind and business could exist only if men were disembodied souls (could they then have any business?) or if all human possibilities were exhausted by a perfect order of the productive arts. But these two hopes are really the twin delusions that move each and every art to serve some larger whole. In this fact we see the kinship between masterful honor and soft, loving charity. Honor is the masterful love of justice that produces the greatest injustice, but it sees all men as alike in their submission to needs that all men share. Imitating Socrates' love of private speeches, Phocion loved virtue. Consequently, he preserved the necessary tension between honor and virtue. But as we see in the sequel, the most im-

44. *Nic. Eth.* 1095b22–96a7.

portant Romans loved virtue for the sake of honor. Only Bacon could know the full importance of this difference, however, because, unlike the classical utopians and unlike Machiavelli, Bacon knew both the utopian teaching and the legacy of Rome.

When Bacon discusses the three examples of the fourth charge, he shows that the elder Cato's turn to learning in his old age was not strictly a punishment, because it could have accorded with the "inward sense of his own opinion." We are reminded of what Bacon omitted in his first mention of the elder Cato, which is Cato's debt to philosophy. But since Bacon does say that Cato was punished by his turning to Greek philosophy, we are directed to think of how Cato differed from Socrates. According to Bacon's source Plutarch, Cato's punishment was not an aged turn to philosophy but the death of his wife and son. The punishment was not complete, however, because in his old age Cato's natural powers enabled him to find a new, young wife and a new son.[45] Although Cato's love of simplicity and restraint was reinforced by hearing one of the Pythagoreans speak about pleasure as Plato did, even in his old age he was prone to indulge his sexual appetites, and although he was an inveterate enemy of luxury and wealth, in which he agreed with the philosopher Ariston, he was a moneylender who used disreputable methods and a good household manager or wealth getter. He even said that a man should be taken to be godlike if the final tally of his property showed it to be greater than his inheritance.[46] Claiming that Socrates was a talker who tried to become a tyrant by subverting his country's customs and laws, the best Cato could do was to praise Socrates for his treatment of his shrewish wife.[47]

But according to Bacon's other source, Cicero, Cato thought philosophy to be compatible with the moderate passions of old age. Moreover, Cato's wisdom helped him to enjoy old age because he followed and resigned himself to nature, which gives us old age so that we may pursue the activities of mind that do not require youthful vigor.[48] As examples of such happy old age, Cicero's Cato mentions Plato, Gorgias, the poets, and the princes of philosophy.[49] Like Socrates, Cato was not wholly deserted by his body's powers in his old age, for both had children of their old age. But unlike Socrates, who does not praise old age because of the weakening grip of the passions, Cato sounds more like Cephalos in Plato's *Republic*, who

45. *Marcus Cato* 24.   46. Ibid., 21.   47. Ibid., 23, 20.
48. *De senectute,* trans. Falconer, Loeb Classical Library (London: Heinemann, 1923), 2, 5–7.
49. Ibid., 6, 8.

claimed that the aging of the body frees the vigors of the mind.[50] As Xenophon tells us, Socrates feared old age precisely because of the connection between the waning of the body's powers and those of the mind.[51] But the most important indication of Cato's difference from Socrates is Plutarch's remark about his penuriousness: according to Plutarch, Cato treated his slaves like beasts of burden and like pots and pans to be discarded when bruised and worn out. This trait, Plutarch says, was the mark of a mean nature that knew of no tie between man and man except the tie of necessity.[52]

For Cato, mind and body can be in perfect harmony because there is no human experience other than the experience of necessity. Philosophy is simply useful and no different from any of the arts, and while the freedom of luxury is bad, there is no limit to the need to acquire and to the pain of spending. Socrates deliberated about his apology and his death so as to gild these harsh necessities with graceful freedom. But Cato forgot the physical decay of the mind because he believed the necessity of bodily desire to be the same as the necessity of truth. He could not see the inconsistency of his being at once an aged procreator and a philosopher freed by old age from passion. The elder Cato lived before the "cesspool of Romulus" in which the younger Cato lived. However, it is clear now that the two Catos are more alike than they are different: they both stand for Roman virtue, for the identity of theory and practice, for the identity of reason's freedom and the need for art, and for the identity of learning and rule. As such, they represent what was common to earlier and later Roman times: the Roman love of honor. Nevertheless, the younger Cato was softer and less harsh and though equally ambitious would have ruled indirectly, unlike the elder Cato. The younger Cato was, then, more feminine, we might even say more loving, which difference points to the link between harsh Roman honor and charitable modern times, a link Bacon will explain in much greater detail. In remarking on Virgil's verse, Bacon omits philosophy from the list of other arts that flourished along with the height of the Roman empire, and he identifies Cicero as merely the best or second-best orator. Cicero may have been second best because Caesar was his rival in eloquence. But in fact, as will be shown, Caesar sought his ultimate victory over Cicero's rhetoric by mastering the honorable arts of arms and empire.

Bacon argues that Athens was not really the enemy of philosophy

---

50. *Republic* 329a1–d6. 51. *Memorabilia* 4.8.4–9. 52. *Marcus Cato* 5.

by reminding us that Socrates was tried during the rule of the thirty tyrants. Of course Socrates was prosecuted not by the tyranny of the thirty but rather by the democracy.[53] Bacon knows that the regime that gave Socrates divine honor was in fact the one that condemned him. This final distortion fits the distortions of the first mention of the trial. In the beginning, Bacon abstracted from the charge against Socrates as it was brought by Anytus, Lycon, and Meletus, for while he presented their charge as based on the accusation of sophism, they had in fact charged Socrates with impiety. Bacon blurs the difference between two regimes and reminds us that, as Socrates could be said to resemble the sophists, so he could be accused of atheism. In Bacon's version of Socrates' apology, Anytus, the silent representative of the artists and politicians, replaces Meletus, who actually spoke on behalf of the poets. As we know, Anytus had the guilty knowledge of the laws' and the gods' dependence on the various arts that produce them. Furthermore, Bacon announces Anytus' private worry by making rhetoric part of the charge; only Bacon, not Socrates' accusers, spoke of the force of Socrates' eloquence and speech.[54] Bacon forces Anytus to speak in public, revealing what Socrates, Plato, and even Anytus saw fit to suppress: Socrates resembled the sophists because both knew the truth about virtue and honor—that honor really serves those artisans who move the regimes and the gods, that virtue can be honored, and that, as mere forms of honor, the different regimes are essentially the same.

The divine beings of which Socrates actually spoke either do not move or else they only warn against motion; they do not create, and they neither give nor receive honor.[55] Considering as he did only such gods, the only city in which Socrates might fit perfectly could not be real, because it would have no arts. An actual city has some resemblance to Socrates only to the extent that it is formed by moral virtue. But even such a city must be hostile to Socrates' knowledge of himself, the gods, and the city's arts because the virtuous cannot admit their dependence on art or on gods that themselves depend on art. For this reason Socrates practiced irony. In order to make Greece sort well with Rome and the papal examples, Bacon unmasks Socrates' irony to suggest that all gods and regimes are the same as the products of art. Socrates would not have gone so far, however, because even though any regime might kill him, there is still an impor-

---

53. Plato, *Apology* 32a4–e1.   54. Ibid., 17a1–18a6.
55. *Republic* 380c6–83b7; *Apology* 27a7–28a, 31d2.

tant difference between virtuous and nonvirtuous regimes, that is, between regimes that are modeled on Socratic freedom and regimes that are not. Socrates was no sophist, because unlike the sophists he could see the difference between virtue and honor. He knew the goodness of virtuous ignorance, and therefore he did not teach the sameness of the regimes as the knowledge necessary for the successful pursuit of honor. But he knew that in an actual city there will always be a tension between moral virtue and justice-loving honor. There will always be the likes of Alcibiades, Alexander, and Caesar in any city. But as we will see, there is a difference between the way in which Socrates and Aristotle could comport themselves toward the two former and the way in which Cicero—and ultimately Bacon— could comport themselves toward the latter. From the present context we are led to suspect that Cicero and Bacon would have to bow more to honor and that honor requires a kind of harsh candor. As Bacon will show later in detail, such candor is the link between Roman honor and Christian charity. Bacon can make Greece sort well with ancient and modern Rome only by directing this candor to Socrates' apology. But he is still enough like Socrates to have done so enigmatically.

In concluding the rebuttal of the politiques, Bacon makes a subtle change in the description of the politiques themselves. At first the disgraces charged were said to have sprung from the politiques' severity and arrogance. But in concluding he refers to their severity and feigned gravity. Feigned gravity replaces the more serious arrogance. This would seem to be inconsistent with all that has preceded were it not for the artful reference with which Bacon brings the discussion to an end. Bacon says he need not rebut the politiques in matters of present times because Elizabeth and the king are lucida sidera. This reference is to Castor and Pollux in Horace's ode to Virgil's setting out for Greece. In the ode, the Roman poet's voyage to Greece represents man's impious assault on god's purpose in separating lands by the "estranging main." Man's boldness "rushes even through forbidden wrong," while "no ascent is too steep for mortals." In our folly we seek heaven itself, and through our sin we do not let Jove "lay down his bolts of wrath."[56] The truth of political life is that honor always wars with virtue. Honor springs from the artful love of the gods. But it only provokes divine revenge; it only provokes tyranny. The truth of Christian charity is that it is the ripest form of honor. We need to know just how this statement is so. But at the

56. Horace, *Odes* 1.3.

very least we know that any charitable empire will be rent by ferocious politics. We wonder if this is likewise true for Bacon's *new* charitable empire, which is founded not on the love of Jesus but rather on the artful conquest of nature.

[4]

The Apology to the Learned Themselves:
Learning's Historical Task

Much yet remains to be learned about the difference between virtue and honor and about what Bacon learned from the ancients of the proper application of learning to political life. Already we know that Bacon was aware of Machiavelli's teaching about these matters and about the conditions that require the new project for learning. In the last chapter we saw that Bacon knew and disagreed with Machiavelli's teaching about religion and the Christian empire and that his disagreement concerned the difference between Greek virtue and Roman honor. And in Chapter 1 we saw that Bacon learned about the limits of justifying the likes of Alcibiades, Alexander, and Caesar from the ancients' critique of the sophists rather than from Machiavelli. Bacon must explain why the new project for learning must go beyond Machiavelli's political science. In his apology to the learned men themselves, Bacon continues his argument about the sources of the modern age by way of an extended critique of Machiavelli. Before we turn to that critique, it will be helpful to think briefly about Machiavelli's political science.

Bacon was not the first to think that Christianity had so changed the world as to require a new science. Before Bacon, Machiavelli argued that the rise of the otherworldly univeral Christian empire had weakened the spirit of free peoples and so undermined their political liberty. In *Discourses* 2.2, he argued that Christian humility had caused men to become prey to the tyrannical and the unjust. Now, for Machiavelli, the catastrophe of the Christian empire did not demand a return to classical political thought. On the contrary, as we noted in the Introduction, Machiavelli was the first to argue that the tradition of utopian thought had to be rejected in favor of a new, realistic political science. For Machiavelli, Christianity is the source of the present world's ills, but properly understood it reveals their causes and the principles for their cure. The Christian doctrine of sin actually reveals a natural truth: the need for acquisition underlies any pretension to virtuous freedom. The classical utopians taught that political life can be well or ill ordered by a range of good or bad regimes, judged according to the virtuously free or viciously slavish human character they claim to produce. But to Machiavelli, Christianity reveals the submerged or indirect *interest* behind every such regime claim, with each interest being ultimately the same: the harsh need to acquire and to keep what one acquires. Therefore any free people's political life has within it the impetus to imperial conquest that extinguishes freedom, and the world was therefore inevitably made ripe by Roman imperialism for the new empire based on the equal humility or neediness of all men. Such humility can not restrain the tyrannical predators, whose need to acquire is really no different from that of persons who can only humbly submit.

According to Machiavelli, to ensure human liberty it is necessary to rely not on the utopian notion of virtuous freedom but rather on the natural fact, revealed by Christian dogma, of the universal human need to acquire. Machiavelli intended to defend a common good that could consist in liberty, if not in moral virtue. The way to this end was not a particular form of rule, because every such form consists really in a pretense to a particular virtue or freedom. Rather Machiavelli prescribed a political science that would manage different forms of rule according to the common interest underlying them all: the harsh need to acquire and to keep what one acquires. In this area Machiavelli took his lesson again from the Christian empire, whose princes ruled not directly, as was the case with the several regimes understood in the classical utopian sense, but indirectly, by way of conscience, in someone else's name. The new political scientist would not himself rule, but he would manage indirectly the claims to rule of others, according to the indirect interest that really moves them. In this way, a universal political science could balance the several acquiring forces making up any political society, in particular the possessive or acquisitive interests of the many and the few, so as to preserve the common good of liberty. While moral virtue has no place in such a scheme, every kind of impetus to acquire and possess, including and especially the tyrant's, has its place in the fabric of the common good.

Now, it should be obvious from this sketch and from what we know of Bacon's account of charity that Bacon learned much from Machiavelli. But we have to understand why Bacon thought that Machiavelli's political science had at least to be complemented, if not fundamentally changed, by the new project for the conquest of nature. Our concern is not with Machiavelli as such. But we need to understand Bacon's stance toward Machiavelli. In this chapter, we will see how much Bacon learned from Machiavelli and how and why Bacon thought that Machiavelli had erred. In the course of the discussion we will pursue the themes of the previous chapters: we will learn how Roman honor gave way to Christian charity, we will learn in greater detail why this yielding has to do with teaching political science to the young, and we will learn more of why Christian charity is a political evil, requiring a new form of charity for which Bacon must apologize.

[274]

Before turning to the three parts of Bacon's apology to the learned, we need to comment on Bacon's brief general description of the discredits against learning that "groweth unto learning from learned men themselves." These spring from their fortune, their manners, or from the nature of their studies, and Bacon says that the first is not in the learned's power, the second is accidental, and only the third is "proper to be handled." However, "because we are not in hand with true measure, but with popular estimation and conceit," he says that it is not "amiss" to speak of the first two.

Now, it is not clear at the outset whether fortune, manners, and studies are the sources of ignorant charges or the grounds that support the charges. But Bacon clarifies matters immediately. The source of the charges regarding fortune and manners is the estimation and conceit (later he adds observation and traducement) of the vulgar many, while the grounds are incidental to learning, either not in the learned's power or accidental. Only the many confuse the learned's fortune and manners with their learning; by another measure it would not be proper to handle such matters. In this case, the source of the disgrace is not the ignorance of the learned but the estimation of the vulgar many, and the grounds are characteristics of the learned men that are confused by the many with their learning. But it is proper to treat the "errors and vanities" of the learned men's studies, even though Bacon is concerned with popular measure. His discussion of this proper subject does not go beyond popular measure, however, for when he discusses studies he says that he will not "make any exact animadversion of the errors and impediments in matters of learning which are more secret and remote from vulgar opinion." This passage suggests that some disgraces have their source in the learned rather than in the vulgar and have their grounds in the errors and impediments of the learned's studies. Bacon's discussion assumes that every disgrace has a ground in learning and a source in some kind of men. These sources cannot all be ignorant, however, because some kind of men must be the source of the proper but secret complaints that have grounds in learning. But if so, then a full account of the disgraces could not but include the vulgar topic of their sources in kinds of men, in this case, the learned themselves. The vulgar many's improper scrutiny of learned men rather than of their learning is essential to reveal the secret errors and impediments of learning itself, the topic reserved for the nonvulgar few. For this reason it is "not amiss" to discuss the improper and vulgar topic of learned men. Bacon defends learning by mixing the measures of the many and the few. He secretly takes his bearings by what all men have in common, and in so doing he follows the path of Machiavelli and Christian charity.

Fortune: Charitable Equality and Youthful Knowledge

[274–276]

The derogations accruing to learning from the learned's "fortune or condition" concern scarcity of means, privateness of life, and meanness of employments. Bacon begins with scarcity of means or want. It is usual for the learned to begin with little and not to grow rich as fast as others because they do not direct their labors to money. It was good "to leave the common place in commendation of poverty to some friar to handle," to whom Machiavelli attributed much when he said that the kingdom of the clergy would have ended long ago if the poverty of the friars had not "borne out" the scandal of the bishops' and prelates' "superfluities and excesses."[1] Likewise, the happiness and delicacy of princes would "long since" have turned to barbarism had not the poverty of learning "kept up the civility and honor of life." Even without these advantages, it is good to note how revered and honored poverty was in Rome, which was "nevertheless

1. *D* 3.1.

95

a state without paradoxes." The political importance of poverty is shown by Livy, who introduced his history by arguing that if he is not deceived by his affection for his subject, no state has better examples than Rome and no state saw avarice and luxury enter so late or poverty honored for so long.[2] After Rome had degenerated, the person who sought to advise Caesar told him to restore the state by the single act of removing the estimation of wealth, which would happen when magistracies and vulgar desires were no longer to be had by money.[3] To conclude, Bacon quotes Diogenes by saying that as a blush is virtue's color, even though it comes from vice, so poverty is virtue's fortune, even though it comes sometimes from misgovernment and accident.[4] Surely Solomon says in censure and in precept that one who hastens to wealth will not be innocent and that, as one ought not to buy and sell the truth, so one ought not to buy and sell wisdom and knowledge.[5] This advice shows that means should be spent on learning and not vice versa.

Bacon says nothing about poverty and learning as such; rather he speaks about poverty and the kingdom of the clergy, or modern Rome, poverty and princes and great persons, poverty and ancient Rome, and poverty and virtue and innocence. Only in the last references to Solomon does Bacon speak about means, truth, wisdom, and knowledge. But Solomon speaks about the proper disposition of means, not about poverty. Bacon actually considers the effect of poverty and wealth on various kinds of political bodies, and in considering private fortune, a subject according only with vulgar measure, he treats fortune and the felicity and delicacy of princes, a subject that is secret and retired and presumably, then, according only with learned measure.

Bacon defends the learned's poverty by referring to Machiavelli's praise of poverty's effect on the kingdom of the clergy. It is important for us to pay close attention to the context of Machiavelli's remark. According to Machiavelli, every political body is a "mixed body," a whole made up of separate parts. And from the necessary conflict of its parts—the many and the few—every political body is com-

2. Livy 1, *Praefatio*.
3. Sallust, *Epistulae ad Caesarem senem de re publica*, ed. Kurfess (Leipzig: Teubner, 1962), 1.8.
4. Diogenes Laertius, trans. Hicks, Loeb Classical Library (London: Heinemann, 1965–66), 6.54.
5. Prov. 28.20, 23.23.

pelled to imperial conquest. But just such conquest leads ultimately to the exhaustion of the freedom-loving spirit that gives it its force. Therefore the task of the new political science is to renew degenerated political bodies by reminding their parts of the harsh need to acquire. This task was to be undertaken so as to revive the tension of their possible relations, or "orders," and also so as to confine the resulting energy to a scope narrower than enervating empire. Now, the specific context to which Bacon refers is Machiavelli's remark, in *Discourses* 3.1, that St. Francis' and St. Dominic's return of the Church to its origins is an example of the renewal of corrupted "mixed bodies," which include republics, sects, and monarchies.[6] All such bodies have their courses, which are limited by demise, but demise can be postponed by renewals that cure the corruptions that develop with time. Two kinds of renewals are possible, with both serving to remind of harsh beginnings: renewals from external accident and renewals from internal prudence. The former is represented by the Gauls' conquest of Rome, which, when the city was retaken, caused the city's "orders"—especially religion and the Senate—again to be observed and which showed not only the importance of religion and justice but also the importance of honoring good citizens and valuing their virtue more than their comforts.

Machiavelli says that internal accidents (he no longer speaks of internal prudence) are the result of law or some man. Actually, they require both, or rather the deeds of a virtuous man who will animate the laws by executing the laws' transgressors. Machiavelli lists these important executions, which happened before and after the Gauls' capture of Rome. In fact, then, internal accident has the same harshness of external accident, but the former can be managed by prudence. These executions must occur every ten years so as to remind men effectively of the harshness of origins: the necessities of punishment, fear, and sacrifice. Machiavelli says that such harsh reminders can be produced by the example of a single man's simple virtue, apart from any law or execution. However, of the examples Machiavelli gives of such men, all but the elder Cato denied themselves gain, sacrificed or risked the sacrifice of all or a part of themselves, or met a cruel and unjust end. Again, all returns to origins remind of the sacrifice and cruel injustice that accompanies harsh necessity. Machiavelli says that if such signal executions or noble examples had occurred in Rome every ten years, Rome would never have become

---

6. See Harvey C. Mansfield, Jr., *Machiavelli's New Modes and Orders: A Study of the "Discourses on Livy"* (Ithaca: Cornell University Press, 1979), 299–305.

corrupted. The two Catos would have been noble examples, but they were too far removed from the last such example and too far removed from each other.

For Machiavelli, that sects also need such renewals is shown by "our own" religion. The poverty of Saint Francis and Saint Dominic and their example of Christ's life established new orders saving the religion that, from the dishonesty of its prelates and heads (*prelati e capi*), had become almost extinct in men's minds. But these revivers taught the people that it is evil to speak evil of evil and that the people should obey and should leave punishment to God. Therefore the noble examples of the friars revived the religion as it suborned the corruption of its prelates and heads. The mixed body of the Christian sect is not like other mixed bodies: unlike the Roman Senate, the Church is not chastened in its nobility, and there is no need for examples that must occur every ten years. On the contrary, these two examples remain until the present day. The Christian examples instill fear only in the minds of the people, because the corrupt prelates and heads do not fear punishments they cannot see. God's punishments are, of course, invisible. However weak the example of Jesus' poverty may be with the prelates (what about the example of his execution?), it is much more lasting with believing people than any Roman examples were with the Romans. Consequently the Church can survive despite the corruption of its leaders who do not believe. Presumably to reform the leaders harsher examples would have to be used, but for the Church as a mixed body such a harsher example is not necessary. Machiavelli says that monarchies too need renewals, as shown by the kingdom of France, in which the parliament executes the laws against the nobles and even against the king himself. It is necessary, then, for sects, republics, and monarchies to be returned to their beginnings, which is better done by internal than by external means because, however successful they can be, external means are so dangerous as to be undesirable. Only the proper use of internal examples can manage the natural course of corruption. Therefore the conquest of fortune has its limits.

Now, as if to emphasize what is less explicit in Machiavelli, Bacon refers to the Church as the kingdom of the clergy, which Machiavelli does not do in *Discourses* 3.1. In so doing Bacon forces us to the conclusion that if the Church is in fact a kingdom, then as in the kingdom of France, the renewing executions should be the most extreme, for they are directed not only against the nobles but even against the king himself. Bacon has presented an example of *the* Christian prince who survived the most extraordinary external accident, which demonstrated the extraordinary power of the papacy and its influence over secular Christian princes. The forces that sacked Rome were of course themselves Christian, so in this case the most extraordinary calamity was both internal and external. Bacon forces us to conclude that the Christian kingdom does not need to be protected from its own corruption. According to Bacon the Church is not wholly corrupt: its strength is located in popular belief shared by the pope as well as all other Christian princes, for Clement's catastrophe sprang not only from his obvious tyrannical ambition but also from his scrupulous reluctance to bear his own arms and to sell the offices of the Church. Bacon of course knows that the Church can be reformed and that its power can rest firmly on belief. The Church survives, robust and martial despite its corruption, by the spontaneous management of internal calamities.

The Church founded on belief in fact survives the most extraordinary external calamities. Moreover, its powerful belief is compatible with tyrannical ambition. Bacon rather boldly suggests, then, that the modern Roman prince has wisely but scrupulously benefited from the original execution of Jesus, the first Christian prince, which could be said neither to be internal nor external. In contrast to Machiavelli, Bacon suggests that all the inhabitants of the Christian kingdom, its peoples and its princes, are reminded of renewing necessity by belief in a harsh execution. Bacon mentions the friars' voluntary poverty, but he does not mention the life of Jesus as does Machiavelli. Machiavelli demonstrates the weakness of the example of Jesus for the heads of the Church, but Bacon, in his silence, emphasizes its extraordinary strength. Bacon's impiety is enormous, compared with that of Machiavelli. For Bacon, the example of Jesus surpasses the examples of Saints Francis and Dominic because it produces a self-managing mixed body that can accommodate the most extraordinary ambition and can conquer the dangers of external accidents. With the example of the tyrannical and scrupulous Clement, Bacon subverts Machiavelli's distinction between believing peoples and unbelieving princes, whether they be secular or papal. We must not forget that in the whole of the present section, Bacon mixes the measures of the vulgar many and the learned few.

Bacon's first disagreement with Machiavelli concerns the power of Christian belief. It would seem that Machiavelli underestimated its power because he underestimated its support by the egalitarianism it frees. Like Machiavelli, Bacon looks to what is common to the many and the few, but unlike Machiavelli, this common ground extends beyond hidden interest to the possibility of belief. For Bacon, the

Christian empire had already accomplished by itself what Machiavelli had hoped to do for entirely different ends. For Bacon, the lesson of charitable belief—of Christian sacrifice—is really a harsh, renewing agent for all the orders of the Christian kingdom, peoples and princes alike. And so renewed, the Christian kingdom is immune from the changes that might spring from internal or external accidents. In particular, it may be that no political science modeled on Christianity can combine the force of internal and external accidents and so conquer the kingdom of the clergy from within and without, because charitable belief always mixes the many and the few, peoples and princes, into an indissoluble "mixed body" by reminding of the harshest necessity: Christian sacrifice reminds of the universal need to acquire. This recollection saves the Christian kingdom. But it also fuels the robust jealousy of the sects so that the source of the greatest vicissitudes is itself immune to vicissitudes, even those hatched by conspiratorial, or indirect, political science. There is a difference, then, between Bacon's and Machiavelli's reference to the elder Cato, even though Bacon has already explained how Cato can fit with Machiavelli's examples of extraordinary, noble men while not having sacrificed himself or his gain or having met a violent end. Cato suits Machiavelli's purpose because, as Bacon has shown, Cato thought that the only tie between men was harsh necessity. The elder Cato was an accuser—one who called for executions—but for Bacon he represents much more than an internal accident that failed to reform a corrupted order. Rather both Catos taken together show the Roman origins of the preternaturally enduring mixed body, the kingdom of the clergy. Because extraordinary renewals are now constant, the Christian kingdom is not so much corrupt as rent by jealous warfare. The learned must attend to the fact of Christian equality, and the key to understanding its power is to be found in its Roman origin.

The example that suits Bacon's consideration of poverty is Sallust's advice to Caesar after Rome had degenerated and was no longer herself. Bacon refers to Sallust's advice to Caesar. But Sallust did not simply recommend that Caesar renew respect for poverty. He advised that Caesar renew such respect so as to renovate the Roman plebs by granting citizenship to the great multitude conquered by Caesar. Fostering poverty among old and new citizens would protect Caesar against the expected objections of the nobles. If, as Sallust said, Rome was divided into two bodies, the patricians and the plebs, the proper combination of universal citizenship and poverty would help

Caesar lead the ascension of the plebs.[7] This advice sorts well with Christian poverty, which reminds all men, whether nobles or plebs, of their membership in a kingdom that grants citizenship to a whole world.

Bacon concludes the discussion of poverty by referring to Diogenes' remark about virtue's color. Diogenes Laertius reports Diogenes' remarks right after recording two of Plato's remarks about Diogenes. When Diogenes said that he could see cups and tables but neither cup nor table as such, Plato said the reason was that, although Diogenes had eyes to see, he lacked mind to discern. When he was asked what kind of man he thought Diogenes, Plato called him a mad Socrates.[8] Diogenes lacked Socrates' divine madness because he believed only in what can be seen, but this characteristic produced another madness that induced Diogenes to oppose courage to fortune, nature to convention, and reason to passion, as if courage were not somehow conventional, convention were not somehow grounded in nature, and reason were simply free from the soul's motions.[9] According to Plato, the extreme poverty from which Diogenes affected these oppositions is just a different kind of pride.[10] By Diogenes Laertius' description of Diogenes, those who believe only in what they see, like Machiavelli's corrupt and unbelieving prelates and heads, pridefully believe that all men ought to cultivate a perfect freedom of mind that can never soften the harsh necessity to which every human body is always and equally subject. According to this prideful humility, there is no practical gracefulness that is not to be revealed as pretense. Bacon has suggested that such a view of mind and body and freedom and necessity is at the heart of charitable belief. According to Bacon's reference, Christian belief encompasses present unbelief because both share the same egalitarian view of necessity and freedom. The Christian empire will never be conquered by managing belief and unbelief because they are already harmonious within the bounds of Christian empire. Bacon shows that, by Machiavelli's measure, the Christian kingdom is a perfect mixed body. It remains to be seen whether this unintended consequence of Machiavelli's teaching is true. To whatever degree it is not, the cure for the contentious jealousy of the sects may depend not on realistic political science but on the believing hope for another world. And Bacon learned the reason for this not from Machiavelli but from Plato.

---

7. Sallust, *Epistulae ad Caesarem senem de re publica* 1.8, 2.2, 5–9.
8. Diogenes Laertius 6.53–54.   9. Ibid., 6.38, 71.   10. Ibid., 6.26.

[276–277]

Regarding "privateness or obscureness," as the vulgar esteem it, Bacon says that abstemious private life has so often been extolled over civil life for safety, liberty, pleasure, and dignity, or at least freedom from indignity, that all who treat it do so well. To this statement he will add only that learned men who are forgotten in states are "like the images of Cassius and Brutus in the funeral of Junia, of which not being represented, as many others were, Tacitus saith, *Eo ipso praefulgebant, quod non visebantur.*"[11] And as for "meanness of employment," the employment "most traduced to contempt" is the government of youth. Because the young have little authority, this lack is projected on the employments in which youth is "conversant" and on those that are "conversant about youth." The attribution can be seen to be unjust if one "will reduce things from popularity of opinion to measure of reason," which will show that men are "more curious" about what is put into a new vessel than they are about what is put into a "vessel seasoned"—and more curious about the "mould" put around a new plant than about what is put around a "plant corroborate," so that "the weakest terms and times of all things use to have the best applications and helps." Bacon asks if we will not listen to the "Hebrew Rabbins" who say that "your young men shall see visions, and your old men shall dream dreams,"[12] maintaining that youth is superior to old age because the young are closer to "apparitions of God" than are dreams. Bacon asks us to note that, however much the conditions of life of pedantes have been mocked in theatres as the "ape of tyranny," and however much in modern times little care is used in the choice of "school-masters and tutors," the "ancient wisdom of the best times did always make a just complaint that states were too busy with their laws and too negligent in point of education." The ancients' lesson has been revived in the "colleges of the Jesuits," although because of their superstition, Bacon remarks, as did Diogenes, "the better the worse,"[13] and then, as Agesilaus did to his enemy Pharnabazus, "they are so good that I wish they were on our side."[14]

---

11. *Annales* 3.76. *Sed praefulgebant Cassius atque Brutus eu ipso quod effigies eorum non visebantur* ("They [Cassius and Brutus] had the preeminence over all—in being left out" [Spedding's trans.], *Advancement* 276).

12. Joel 2.28.

13. Diogenes Laertius 6.46, Spedding's trans. This remark is omitted from the *De augmentis*. See *Advancement* 277, n. 1.

14. Plutarch, *Agesilaus* 12.5, Spedding's trans.

Whatever Bacon knows about the Christian kingdom includes what Machiavelli knew and also what is to be learned from the likes of Plato. In the present section, we see how Bacon thought it necessary to look beyond the scope of Machiavelli's political science by looking backward to traditional utopian thought. We begin by noting Bacon's reference to Tacitus to defend the privateness of learning. In the *Annales,* Tacitus says that the effigies of Brutus and Cassius were omitted from Junia's funeral procession as Tiberius' subtle revenge for having been insultingly ignored in Junia's will. But learning's invisible presence cannot affect public affairs as the invisible effigies did if Bacon aims to say, as indeed he does, that learning can have an indirect effect on political rule. Rather, if the example from Tacitus shows the powerful effect of learning on rule, then learning must be like the imperial tyranny that triumphed over the remnants of senatorial resistance (Caius Cassius and Marcus Brutus) and the scruples of the younger Cato, who was Junia's illustrious uncle. But if so, then the indirect effect of learning cannot be to confine the impetus to empire, as Machiavelli believed. For some reason, the fortune of charitable learning is no freer from universal, imperial ambition than are charitable peoples and princes. This fact, which Bacon learned from sources other than Machiavelli, determines the scope and the limits of Bacon's new project for learning. But what did Bacon know that Machiavelli did not? We receive our clue from the discussion of learning's mean employment with the young.

Bacon argues that the learned must attend to the young because the young are weak and because such weakness deserves the best help. To prove his point, he asks that we listen to the Hebrew rabbis who demonstrate the superiority of the youth whose weakness the learned will serve. But the rabbis do not simply assume that the young are weak. What they say, as Bacon notes, is that youth is superior to old age because the young's visions are closer to apparitions of God than are the dreams of old men. In mentioning the rabbinical interpretation of Joel 2.28, Bacon refers to Abrabanel's commentary on Joel, where Abrabanel argues that the young are closer to God because their visions are clearer than dreams.[15] In Essay 42, "Of Youth and Age," Bacon refers again to this commentary, but there he is careful to note that the argument was made by "a certain rabbin,"[16] while in the present context he refers to the "Hebrew Rabbins." Bacon forces us to think for a moment about the difference between the words of

15. Abrabanel, *Perush ʿal neviim acharonim* (Amsterdam, 1641), Joel, Gimel, vehayah.
16. *Essays* 477–78.

one rabbi and two rabbis about Joel, dreams, and visions. Abrabanel comments on Joel, but he also comments on another rabbi's comment on Joel. That is, Abrabanel comments on Maimonides' remark about Joel in his (Abrabanel's) commentary on Maimonides' *Guide of the Perplexed.*[17] Bacon casually treats these two rabbis as if they agree about dreams and visions. But at least at first glance, they do not.

In the *Guide,* Maimonides argues that the text of Joel does not show that all of Israel's sons and daughters will prophesy but that various kinds of prognostication by means of soothsaying, divinations, and veridical dreams will occur. We should understand this point so that we are not led to think that the young might prophesy. According to Maimonides, because prophecy is a natural perfection, requiring proper preparation, it is impossible for the young.[18] Abrabanel denies that prophecy is a natural perfection; rather it is a miracle and, therefore, dreams and visions, the two kinds of prophecy, do not require preparation and can be experienced by the young.[19] And if the young have visions while the old merely dream, then by way of miracles the prophesying young can be closer to God than the prophesying old.

According to Maimonides, while divine speeches are heard only in dreams, in visions we have actual knowledge of what is open to science and speculation. Abrabanel agrees that in dreams the imagination is richer and distorts more than in visions. But he denies that visions could be any kind of science or speculation, because even though they are clearer than dreams, they are still miraculous motions of the imagination.[20] If Maimonides and Abrabanel are simply taken to be the same, as Bacon would have it, the implication is that the young are superior to the old, as science and speculation are to the hearkening to God's speech. Bacon boldly suggests that Abrabanel may have been more philosophical than pious, and there is a remark in the commentary on Joel that could be taken to imply the same thing.[21] But whatever the actual case may be, Bacon's artful reference asserts the superiority of youthful knowing of God's nature to the older hearkening that must precede obedience. Charitable knowledge is youthful, appearing not to require obedience to di-

---

17. Abrabanel, *Perush ʿal sepher moreh nevukhim* 2.32, 69r.b.41, trans. in Alvin J. Reines, *Maimonides and Abrabanel on Prophecy* (Cincinnati: Hebrew Union College Press, 1970).

18. Maimonides, *Guide of the Perplexed,* trans. Shlomo Pines (Chicago: University of Chicago Press, 1969), 2.32.

19. See n. 15 above; *Perush ʿal sepher moreh nevukhim* 2.32, 66v.a.43.

20. Maimonides, *Guide of the Perplexed* 2.45; Abrabanel, *Perush ʿal sepher moreh nevukhim* 2.32,69r.b.41.

21. See n. 15 above.

vine command, although it remains to be seen whether the new knowledge is every wholly free from divine will. Charitable knowledge is aimed at all men who are young rather than old. And at least to the extent that the Jesuits are a model of the rabbis' preference for the young, it would seem that charitable learning differs from the ancients regarding the subjects to be taught to the young. Reminding us that the Jesuits teach "human learning" and "moral matters" to the young, Bacon forces us to wonder if there is a dangerous connection between charitable education and youthful lawlessness.

As we noted, Bacon's reference to the rabbis directs us to Essay 42 on the nature of youth and old age. Bacon begins the essay by arguing that a "youth may be old in hours if he have lost no time." But such a youth is rare; generally it is like first thoughts, "not so wise as the second." However, the young have more lively invention than the old, and their imaginations are more divine. Regarding action, men whose natures "have much heat and great and violent desires" are not ripe until they are "past the meridian of their years." As examples of such men, Bacon mentions Caesar and also Septimius Severus, of whom Spartianus said *Juventutem egit erroribus, imo furoribus, plenam,*[22] but who was the "ablest emperor almost" of all the list. Bacon refers to no list that he has presented, but he does remind us of two lists of emperors given by Machiavelli in *Discourses* 1.10, where he identifies Caesar as an emperor, and in *The Prince* 19, where he praises Severus as a new prince who successfully combined the qualities of the fox and the lion. Machiavelli knows that, however much Caesar founded an empire, he was not himself emperor, because in *Discourses* 1.10, he speaks of the twenty-six emperors from Caesar to Maximinius, and if Caesar were included, the list would total twenty-seven. Machiavelli says that Caesar's tyranny did not result in posthumous execration, because the power of the empire he founded compelled writers not to speak their minds about him. As a result, to learn what Caesar was really like one must read the praises of his enemy Brutus.[23] A founder will be praised as long as his tyranny is indirect, as with Caesar—that is, if he does not simply become an imperial master but if the empire he founds commands praise and blame.

But according to *The Prince* 19 and *Discourses* 1.10, Caesar's founding, indirect command over praise and blame came at the inequitable expense of those who praise and blame, however much Caesar himself in his life paid an equitable price in not becoming emperor

22. "He passed a youth full of errors, yea, of madness" (Spedding's trans.), *Essays* 477; *Historiae*, Spartianus, *Severus* 2.1–9.
23. Mansfield, 66–69.

and not dying normally. The reason was the problem of succession coupled with the unique problem of the Roman Empire, which was that Roman emperors, however noble and strong of character, had to contend with the political force of the army. For the Roman emperors, it was not enough to be loved by the people to be free from the danger of internal conspiracy, because they had always to satisfy the cruelty and avarice of the soldiers, whose desires are inconsistent with those of the people. Nature does not always cooperate with the principle of hereditary succession: even a good emperor like Marcus, who as a hereditary emperor was free from the soldiers and the people, so failed to educate his son Commodus that Commodus' cruel and bestial disposition was allowed to flourish. Commodus so corrupted the army that the good Pertinax came to a bad end at their hands. Considered in the light of his posterity, then, Marcus suffered the same inconveniences as any new man in need of extraordinary favors. The bad analogue of the good Marcus was Severus, who was bad but whose great reputation defended him against the people who otherwise would have been induced by his rapacity, beloved by the soldiers, to conspire against him. But Antoninus lacked his father's extraordinary virtue, so he came to ruin although he was not himself a new man. In the Roman Empire, all princes shared the defect of the new man, and therefore Caesar's immortal fame proved ultimately to be unjust with respect to peoples, however satisfactory it was with respect to the occasional good and bad emperor or with respect to the likes of Marcus and Severus.

Machiavelli suggests a solution to the problem of the empire when he remarks in *Discourses* 1.10 that some good emperors were killed by the corrupt effects of their predecessors, that of the emperors who died normally some were bad but virtuous, like Severus, and that all the bad emperors but one were hereditary and all the adopted emperors were good. By this argument, even the good Marcus was bad precisely because of his corrupt natural offspring, and the way to solve the problem of succession is to kill bad rulers (whether they be individually bad or good) so that adopted rulers will be good or to kill the natural offspring of bad rulers for the same end.[24] Where the soldiers are politically important, justice can be had for the people by imitating natural succession by the practice of adoption, but doing so requires injustice toward good princes or possibly good sons. In *The Prince* 19, Machiavelli says the likes of this practice can be seen

---

24. Ibid.; Leo Strauss, *Thoughts on Machiavelli* (Glencoe: Free Press, 1958), 52, 227.

among the Turks, in the empire of the sultan, and in the Christian pontificate, which are neither hereditary nor new but in which the sons of dead princes are not heirs and the adopted have the power of an ancient authority. Although the princes of our time do not have to attend to the soldiers as did the Roman emperors, the army is still important for "the Turk and the sultan," whose regimes resemble in their "order" the Christian pontificate.

The difference of the Christian pontificate from other mixed bodies is that it can be returned to its beginning without reforming the prelates and heads, that is, without reforming their weakening effect upon peoples and armies who become the prey of evil men. Machiavelli's intention is to reinvigorate the power of armies and to manage the tension between armies and peoples by the artful imitation of nature, which can be learned from the Christian pontificate itself. Doing so requires the management of extraordinary executions so as to remind the mixed body of the harsh necessity it experienced when new. The artful renewal of mixed bodies demands harsh, unjust executions of princes. Machiavelli's intention is to replace Caesar as a founder and to be, unlike Caesar, the godlike source, rather than the commander, of praise and blame. And unlike Caesar, he will turn the founding harshness of Severus against the preserving goodness of Marcus, so that peoples are salved by Severus' rapacity rather than by uncertain princely goodness and so that more direct conspiracy can be avoided as modern peoples are reinvigorated.

According to Bacon, Severus is the best "almost" of a list of emperors who had youthful natures that were hot, great, and filled with violent desire and perturbation and who were made ripe by older age. Augustus, the truly first emperor and the only other emperor mentioned, is said to represent a reposed, and by implication older, nature that "does well in youth." Along with Augustus, as examples of such a nature, Bacon mentions Cosmus, duke of Florence, and Gaston de Fois. However, we know that Augustus' repose was in fact the artful dissembling of his tyrannical ambition, and according to Machiavelli, Cosimo, whose triumph in Florence began the tyrannical suppression of republican liberty, ruled not when he was young but only after the "meridian of his years," when he was no younger than forty years old.[25] Finally, however reposed he may have been, Gaston died in his extreme youth at the battle of Ravenna where,

25. *Essays* 477; Machiavelli, *Istorie Fiorentine*, ed. F. Flora and C. Cordié (Rome: Mondadori, 1967), 8.1.

according to Guicciardini, he had urged on his troops by promising them booty from the sack of Rome,[26] an event that was the greatest calamity of the Christian kingdom and yet was survived effortlessly by the pope and the papacy. Bacon's examples of young, reposed natures mix heat and repose and confuse the old and the young. They suggest, then, that Severus, Caesar, and Augustus each combined the natures of youth and age or the new and the old. Furthermore, there is really no difference between any Roman example and the modern examples of Cosimo and Gaston. Bacon's examples agree with Machiavelli's argument that the wise management of succession must ignore the difference between legitimate antiquity and illegitimate youth. And Bacon links the Roman examples to the modern examples from the Christian kingdom.

Although tyranny over Florence was a papal ambition, the pope's ambition was immune from the harsh effects of rapacious, booty-loving armies. In the new, Christian kingdom, armies are not important in the same way that Machiavelli says they were in Rome: the violence of armies has been combined with self-regulating, efficient succession, to which all Christian princes will bow. As Machiavelli misjudged the egalitarianism of Christian belief, so he did not see that all peoples are potential soldiers in the violent armies of the sects. Machiavelli's indirect conspiracy is otiose because the problem of the Christian kingdom is not the weakness of its peoples but their charitable armies that combine youthful ferocity with respect for the most legitimate and oldest of princes. Whatever Bacon's solution to the modern world may be, Bacon agrees with Machiavelli that it requires accommodating peoples to youthful or new rule, to a tyranny armed with a perfect tool. Bacon calls Caesar an emperor more openly than does Machiavelli—the question will be whether Bacon thinks it possible to justify the tyrant by denying him the full measure of his tyranny. Like Machiavelli and unlike Aristotle, Bacon suggests that the tyrant is always necessary. But unlike Machiavelli, Bacon suggests that the tyrant may not be justified by the realistic management of his recurrent founding by the true, learned founder. We know that he may not, because Bacon has warned about the danger of youthful lawlessness. Charitable knowledge teaches political science to the young, and as we see in the sequel and beyond, Bacon attends to older arguments about the danger of such teaching.

Bacon's discussion of youth and age differs from Aristotle's discussion in the *Rhetoric* in a decisive respect. In the *Rhetoric*, Aristotle

26. *Storia d'Italia* 10.

describes the young as lovers of the noble and the old as lovers of the useful, so that the young are unjust from insolence, while the old are unjust from vice.[27] Since the prime of life is a mixture of both kinds of love, it is not overly fond of the noble or the useful.[28] In the *Rhetoric,* the possible injustice of the very desire for the noble— for moral virtue—comes to light. For this reason, the *Rhetoric* treats the moral phenomena of youth and age realistically. The *Rhetoric* treats the way the young can be made to act nobly; it shows the way that moral virtue is produced by an art. But to this extent, the *Rhetoric* reveals that the self-sufficient or the noble depends on a productive art, and such knowledge, if open to the young who love the noble, merely inflames the youthful desire for mastery. Moral virtue is forgetful; therefore, the *Ethics* tells us that if the *Ethics* itself is not a fit subject for the young, still less is the *Rhetoric.*[29] The *Rhetoric* demonstrates the tension between the self-sufficient worth of the noble and the desire that is open to the noble. Although the young are unfit auditors of political science, the effect of proper education is to bend youthful wills and affections to the noble so that the noble can be consistent with the just.

In Essay 42, Bacon apes Aristotle's argument so as to emphasize his difference from it. He argues that "for the moral part, perhaps youth will have the pre-eminence, as age hath for the politic." And after praising the young by referring to a "certain rabbin," he says that "the more a man drinketh of the world, the more it intoxicateth: and age doth profit rather in the powers of understanding, than in virtues of the will and affections." By this argument, the young, or rather their wills and affections, grow hotter and intoxicated with age just as the understanding grows, which means that reason can rule the will and affections only by force, not by way of art and experience that might render them moderate and graceful. Moreover, the knowledge that gains as it ages and rules is, as the reference to Abrabanel discloses, a youthful knowledge that conquers and does not itself hearken to law. But in differing from Aristotle about the young, Bacon implicitly grants Aristotle's argument that the young are not themselves governed by reason. The problem is not that Aristotle was wrong about the young and the noble but rather that Christian charity causes the young to be intractable to reason because it reveals too readily the dangerous truth of the *Rhetoric:* the moral virtues and the noble are products of a necessary art. For Ba-

27. *Rhetoric* 1388b31–90a23.   28. Ibid., 1390a28–b13.
29. *Nic. Eth.* 1094b27–95a11.

con, such charity inclines the young to the greed and rapacity of the conscientious soldier. Bacon does not flinch from disclosing that his new charitable knowledge will pursue the means for satisfying such greedy desire; he lives in a world in which Jesus has corrupted the young.

It remains to be shown how Christian charity itself arose from the teaching of political science to the young. But it is already clear that Bacon thinks that, in the kingdom of the clergy, all men know the harsh trade secrets of Machiavelli's political science. Likewise, it remains to be seen whether Machiavelli's political science can be wholly replaced by a new campaign against nature's penury or whether there will always be a tension between them. But whatever the case, Bacon knows that both are at risk from their necessary, but dangerous, dependence upon the young. Bacon relies for this knowledge upon the ancients, who understood the tension between virtue and the common, or political, good. He did not learn it from Machiavelli, for whom a realistic political science was to satisfy the youthful hope that moral virtue and the common good, the *need* for art, might be the same. Bacon seems to know that for Machiavelli to justify the tyrant by reference to a perfected common good, he must presume a natural, as opposed to political or conventional, commensurability between the many and the few. It is hard to see how such a presumed commensurability could be compatible with mankind's division into separate, self-sufficient political bodies. Bacon suggests that Machiavelli's political science is unjust because it secretly looks beyond mankind's political nature without being able to overcome it. We see this more clearly in the rest of the present chapter, but already we suspect that Machiavelli did not understand the hold that charity had on even his thought.

Manners: Honor, Avarice, and Empire

At this point we know the rudiments of Bacon's agreement and disagreement with Machiavelli. But in order to understand them more deeply, we have to grasp Bacon's account of the limits of Machiavelli's political science. We have to see why Bacon thought that, while these limits required his own new project for learning, they reveal the limits and dangers of the new project itself. Bacon knows that his teaching, like Machiavelli's, depends dangerously upon youthful hope, and to understand why, Bacon probes more deeply into the particular causes of Christian charity. Ultimately, Machiavelli missed the true causes because he was too quick to turn from the lessons of classical utopian thought. Bacon works through these questions in his apology for learned manners and learned fortune. In this part of Chapter 4, we examine the discussion of learned manners. In the next part, we examine the discussion of learned studies.

[277–278]

The first fault of manners, supposed of Demosthenes, Cicero, Cato the Second, Seneca, and many others, is that, "because the times they read of are commonly better than the times they live in, and the duties taught better than the duties practised, they contend sometimes too far to bring things to perfection, and to reduce the corruption of manners to honesty of precepts or examples of too great height." This fault of learning has "caveats enough in their own walks," by which Bacon means that the very learned examples causing the learned to set their sights too high provide lessons that cure the fault springing from the examples. Corresponding to the examples of Demosthenes, Cicero, Cato the Second, and Seneca are the examples of Solon, Plato, Caesar's "counselor," and Cicero himself. When asked if he had given his citizens the best laws, Solon said they were the best they could receive.[30] When Plato refused office in corrupt Athens, he did so believing that one's country should be used as one's parents, with persuasion rather than contestations.[31] Caesar's counselor warned him against trying to return Rome to her origins now that she had become corrupt,[32] and Cicero made the same point in writing to Atticus that Cato erred in considering Rome Plato's Republic rather than the dregs of Romulus.[33] The "same Cicero" excused the philosophers by saying that they aim high so as to have men hit the mark[34] and "yet himself might have said, *Monitis sum minor ipse meis;* for it was his own fault, though not in so extreme a degree."[35]

Bacon begins by returning to his distinction between Greece and Rome. The reference to Demosthenes, Cicero, Cato the Second, and Seneca reminds us of the politiques' complaint that learning perverts

30. Plutarch, *Solon* 15.   31. Plato, *Epistle* 733 1a5–d5.
32. Sallust, *Epistulae ad Caesarem senem de re publica* 1.5.
33. *Att.* 2.1.8.   34. Cicero, *Pro Murena* 31.
35. Ovid, *Ars amatoria*, trans. Mozley, Loeb Classical Library (London: Heinemann, 1979), 2.548.

men's minds for policy and government. We recall that, in the earlier context, the younger Cato's opposition to a present world showed his likeness to the elder Cato and that the two Catos together showed the difference between the Catos', Seneca's, and Cicero's Rome and the Greece of Socrates and Phocion. This difference is elaborated by the examples Bacon says provide corrective "caveats enough." The examples of Solon and Plato, taken from Plutarch and the famous Seventh Letter, warn against too great expectations less than they illustrate the likeness and difference between Solon, Plato, and Socrates. Solon, a practical lawgiver not given to speculation about the whole, was willing to combine persuasion and compulsion or to "combine violence and justice together." Plato, one who would advise but not rule and who wrote about Socrates as having questioned the things above and below as well as human practice, would compel a slave but never a father, a mother, or one's own city.[36]

It is not difficult to see why Solon would have to combine force with persuasion. We can learn the reason by thinking about Plato's, and Plato's Socrates', relation to the laws. Plato speaks about the laws by animating Socrates' telling of a tale within a tale. In such a tale it is easy to guarantee that stories about fine beginnings will be believed. But Solon's much more difficult task is revealed no more clearly than in the *Crito,* where in defending themselves against one who would have Socrates violate them, the Athenian laws press the absurd claim to have engendered Socrates.[37] In a real city, such a fiction could be sustained only by persuasion *and* force. But the nature of such force is incompatible with the city's claim to be like one's natural family. For the laws to be revered, they must use force, and yet such force must be forgotten. Both Plato and Solon knew that harsh necessity could not be overcome in human affairs. As Plutarch reminds us, they both gave accounts of Atlantis—the lost city that tried and failed to conquer natural and divine necessity. And according to Plutarch, necessity, not lack of leisure, prevented both from finishing their accounts.[38]

The philosophic Plato could playfully imitate force. But the poetic lawgiver Solon had to use force and to see that it was forgotten. Solon's legislation required a mindful production of others' self-forgetting, which must have been produced by a rhetoric like the one at the heart of Aristotle's political science.[39] Solon bowed to necessity,

---

36. *Solon* 3,15.   37. *Crito* 50d1–3.   38. *Solon* 31–32.
39. *Politics* 1282b14–84a3.

unlike Socrates, but he was not for this reason an unknowing imitation of Socrates, because he had to know how to produce the likes of Phocion, who was. Solon, the statesman, is the best practical human being, with his difference from the political philosopher being his willingness to use persuasion and force to fashion in the virtuous an unphilosophic oblivion of the link between freedom and harsh necessity. The difference between the philosopher and the statesman points to the imperfection of the just statesman, but the statesman's knowledge of this imperfection guarantees that his artful justice will be bent to virtue rather than to art as such. The statesman must know that art and reason are not the same in order for his art to be reasonable. Such an art must direct the useful to the self-sufficient, yet without reducing one to the other.

Bacon does not identify explicitly the source from which he derives his reference to Plato's remark about persuasion and force. One would of course think it to be the Seventh Letter. But another source is possible: Cicero's letter to Lentulus, in which Cicero defends his support for the Triumvirate and especially, in the context, for Caesar. After referring to Plato's remark about persuasion and force, Cicero contrasts himself to Plato, saying that, while Plato thought that the Athenians had reached senility, the Romans of his day had not done so. Unlike Plato, he is not at liberty to decide whether or not to engage in politics; rather he was constrained.[40] Bacon shows that in fact Cicero did not think that Cato had erred in regarding Rome as Plato's Republic rather than as the dregs of Romulus. This point is just what we should expect, for as Bacon has shown, Cato did not think reason free enough to be the limiting measure of practical needs. On the contrary and unlike Plato, Cato would have thought the Republic possible because he considered reason the mere handmaid of practical necessity. Solon and Plato represent the way philosophy might partially order politics without itself being political, by way of the self-knowing statesman who, with proper shame, will stoop to persuade and coerce the likes of Phocion. Cicero, however, proclaims the total subjection of philosophy or mind to the necessities of politics. Socrates' apology is constrained, while his deliberation is perfectly free, but Cicero can do no more than bend to the necessity that ultimately determined his unfree and ignominious end. In fact, as Cicero explains later in the letter, were he at liberty, he would have been "not otherwise [involved] in the public interest" than he is. Moreover, he

---

40. *Fam.* 1.9.18.

says that his constrained participation in public business is as useful to himself as it is defended as good by any righteous man one pleases.[41]

According to Cicero, the Roman, there is no possibility of philosophy that is free from public business requiring both persuasion and force. But if so, then there is no human possibility before which the statesman should feel shame in coercing others; there is no model in the light of which the statesman can discern the difference between art and reason. Without such a model there is no difference between the statesman and any claim to artful mastery. As we know, in the absence of the political philosopher, one cannot turn to a creator god to be the humbling model for the statesman. Cicero could only have been drawn to the tyrant Caesar, rather than vice versa, as we will see was the case with Socrates and Aristotle and the likes of Alcibiades and Alexander. Cicero and Rome put an end to the possibility of the statesman, and in the absence of the statesman, there is no practical way to blunt the claims of the productive arts.

The counselor of whom Bacon speaks was Sallust. In the context to which Bacon refers, Sallust says that Caesar has not just Rome to put in order but rather the "whole world." This ordering must concern not virtue or the nobles and the Senate but the harmonizing of the young's and the commons' money making and spending.[42] The earlier quotation from Sallust suggested the egalitarian foundations of universal empire. Bacon here suggests that the principle of the new equality of princes and peoples is not poverty but the regulation of money making, as if avarice were the passion liberated by the belief in human equality. And he suggests that this avarice has its roots in Rome, not in the dregs of Romulus, but in the likes of the two Catos who stood for Roman honor. The dregs of Romulus sprang from Roman honor as it aspired to empire. Bacon's argument suggests that Cicero's Rome and Cicero's view of necessity and freedom take human desire to be wholly satisfied by money, a principle of the equality of the many and the few. Cicero was correct to say that necessity, the public good, and his private good were simply compatible, because desire so understood subverts the distinction between public and private as it assimilates the noble and the base. If Roman honor makes way for Roman empire, and if Roman empire makes way for Christian charity, then the implication is that according to Roman honor and Christian charity there is nothing that money can-

41. Ibid., 1.9.21, 18.
42. Sallust, *Ad Caesarem senem de re publica* 1.5, 2.10; *Att.* 2.1.8.

not buy. Certainly by this corollary what is one's own and private is always commensurable with what is common and public, a principle dear to the tyrant.

According to Bacon, Cicero understood that philosophers measure duty too strictly, so that, in missing the measure, men will hit the proper mark. To illustrate, Bacon quotes the *Pro Murena*, where Cicero berates Cato for a graceless interpretation of Stoic doctrines. Cato would push duty beyond nature, so that there is no room for influence or mercy and no distinction between kinds of crimes and so kinds of punishments.[43] Cicero condemns Cato's assumption that all necessities are the same and that Rome cannot have time for both luxury and toil, and he condemns his failure to understand that one pushes duty beyond nature so as to end at nature's intended place.[44] But according to Bacon, it must be said that Cicero himself fell short of his own precepts, "for it was his own fault, though not in so extreme a degree." Bacon's remark is curious because, by Cicero's argument, which Bacon has approved, Cicero's missing the mark of his precepts should not be reprehensible; rather it would be his hitting the mark intended by nature. In describing Cicero's faultily missing his own measure, Bacon quotes from Ovid's *Ars amatori,* where the poet counsels submission to base and humiliating things for the sake of the object of one's erotic desire.[45] This example explains Bacon's criticism: Cicero's fault was precisely his failure to hit the moderate mean; his fault was his equating of the high and the low for the sake of necessary desire.

We now have the decisive clue to Bacon's critique of Machiavelli. To see the limits of Machiavelli's political science, we must see the difference between moral virtue and Roman honor and the concomitant difference between political science that accommodates the statesman, who artfully orders and mixes incommensurable goods, and political science that in accommodating the tyrant presumes that all goods are commensurable by nature. The interpretation of modern times depends upon understanding these differences, for Christian charity is the legacy of Roman honor. As we see in the sequel, Machiavelli correctly discerned that, despite the Christian respect for poverty, the real link between Roman honor and Christian charity is avarice. Bacon has suggested that in the world of Christian charity, common avarice is the real tie between the rational pretense of art and the ambition of the acquiring tyrant. Therefore, to understand

---

43. *Pro Murena* 30–31.   44. Ibid., 31.65, 35.74.
45. Ovid, *Ars amatoria* 2.510–600.

Machiavelli's mistakes and to see the fault of charitable learning, we must understand the nature and the limits of avarice.

[278–280]

Another fault thought to be incident to the learned is that "they have esteemed the preservation, good, and honor of their countries or masters before their own fortunes or safeties." This trait is shown by Demosthenes' remark in the *Chersonese* that his counsels do not increase his standing among the Athenians as they diminish the Athenians' standing among the Greeks but rather are sometimes not good for him to give, although they are always good for the Athenians to follow.[46] Bacon then mentions Seneca's "honest and loyal course of good and free counsel" during Nero's five-year minority and after Nero had become "extremely corrupt in his government." This point cannot be otherwise, because learning teaches men of "the frailty of their persons, the casualty of their fortunes, and the dignity of their soul and vocation." For this reason, it is impossible for the learned to consider the greatness of their own fortune as the end of their "being and ordainment," so that they desire "to give their account" to God and to kings and states in the words of Jesus' parable: *Ecce tibi lucrefeci* and not *Ecce mihi lucrefeci*.[47] In contrast to the learned, the "corrupter sort of mere politiques," who are not "established by learning" in love and knowledge of duty and who never "look abroad into universality," "refer all things to themselves." These corrupter politiques "thrust themselves into the centre of the world" as if "all lines" coverge in their fortunes, not caring about their states so "they may save themselves in the cockboat of their own fortune." Men who feel the "weight of duty" and who "know the limits of self-love" live up to their "place and duties" even at their own risk. If these lovers of duty involved themselves in "seditious and violent alterations," they do so because of the respect the "adverse parts do give to honesty" rather than because of any consideration of their own advantage. However fortune may tax it and the corrupt many may despise it, this "point of tender sense and fast obligation of duty" endowed by learning in the mind receives an "open allowance," and for this reason it "needs the less disproof or excusation."

It is more probably defended than denied that learned men sometimes fail to apply themselves "to particular persons." This want of "exact applications" springs from two causes. One cause is that the

46. Demosthenes, *Chersonese* 106–7.
47. Matt. 25.20. "Lo, I have gained for thee," not "Lo, I have gained for myself" (Spedding's trans.).

learned's "largeness of mind" can "hardly confine itself to dwell in the exquisite observation or examination of the nature and customs of one person." It is the speech of a lover and not a wise man, *Satis magnum alter alteri theatrum sumus*,[48] but even so, Bacon will "yield" that one who cannot "contract the sight of his mind as well as disperse and dilate it, wanteth a great faculty." The second cause is not an inability but a chosen rejection. The "honest and just bounds of observation" of one person by another should extend no further than what is necessary "to understand him sufficiently," not to give offense, to give faithful counsel, and "to stand upon reasonable guard and caution in respect of a man's self." To know another so as to be able to work, wind, or govern him springs from a false heart, which in friendship is "want of integrity" and toward princes or superiors is "want of duty." The Levantine custom of not fixing the eyes on princes is barbarous, but the moral lesson is good, because men ought not to penetrate the hearts of kings, which "the Scripture hath declared to be inscrutable."[49]

We want to know how understanding avarice illuminates the limits of Machiavelli's political science and Bacon's new project for learning. To start, we observe that Bacon's argument in this section rests on the distinction between the private fortune of the learned and the wider considerations of God, prince, and state. But in the discussion of the learned's fortune, private fortune concerned the broadest political matters as well as the relationship of the new knowledge to obedience to God's law: the fortune of the learned is youthful conquest that does not hearken to divine law, subverts the distinction between peoples and princes, and subverts the possibility of taking the measure of the tyrant. In the context of the *Chersonese*, from which Bacon quotes, Demosthenes refers to and defends his own courage. Although he is not bold, brutal, or impudent, he has more courage than the politicians, because his advice is often good for the Athenians and bad for himself. With this example, Bacon recurs to the objection of the politiques that learning disposes men to leisure and privateness, where he referred to Demosthenes' remark about pleasure as reported by Plutarch. In that context, Bacon artfully emphasized the difference between Demosthenes' natural boldness and his physical courage so as to subvert the distinction between natural

48. "Each is to other a theatre large enough" (*Ep.* 1.7.11–12, Spedding's trans.).
49. Prov. 25.3.

virtue, natural vice, and the artificial imitation of natural virtue. In this vein, we can see that the present reference to Seneca's counsel recurs to the politiques' objection that learning perverts men's disposition to government. In the present context Bacon remarks that Seneca held a course of honest and free counsel after Nero had grown corrupt. But in the earlier context Seneca's later counsel to Nero was not free but was wholly constrained because of his part in the vile murder of Agrippina. In the present context, the first two allusions remind that for the learned to prefer public or universal obligation is not different from their preferring their own private fortunes.

The learned serve universal ends as did the servants of the parable of *Matthew* 25.20, but if so, then according to the biblical reference, they serve universal fortune as master money makers.[50] The learned serve human interests insofar as they can be measured by lucre. Since God's command is not heard, the command of the money-loving tyrant is. The tendency of the learned to prefer divine or common to private fortune is "likewise much of the kind" of their tendency to set their sights too high by reason of high examples. And yet the vice of preferring common to private fortune is taken to be a vice only by the many "in the depth of their corrupt principles." But in fact, this supposed vice "needs the less disproof or excusation" precisely because it is counted a virtue by common opinion or, as Bacon says, by an "open allowance." The new common fortune is morally problematic because the gift it promises is at once private, common, and apparently limitless. And the danger of this gift becomes clearer if we consider Epicurus' saying that each man is a sufficient theater to another.

As Seneca reports this message, its advice is directed not to a lover but to the wise as opposed to those whose souls are debased by the extreme passions of the many. Seneca's advice to Lucilius is to avoid the boisterous mob, which is best revealed by its voluptuous bestiality at the games. If this sporting mob is not bad enough, Lucilius should think about what happens when the world at large confronts us. Rather than imitate or loathe the world, Lucilius should learn for himself from the few who can make him better, just as Epicurus said to one of the partners of his study (*uni ex consortibus studiorum*) that his writing was not for the many, but for him, each of them being a sufficient audience for the other.[51] Bacon says that the wise man must rise to the world at large, but he is silent about the context of Epicurus' saying that identifies the passions of the world, or the univer-

50. Matt. 25.14–30.   51. *Ep.* 1.7.

sal many, as the source of the most bestial vices. If the example illustrates the point, the learned are not lovers of the few wise and
good; rather they are lovers of the many and their boisterous, bestial
passions. Bacon repeats his reference to Epicurus in Essay 10, "Of
Love," where he explains how the passions of the many can be depraved precisely as they are served by the attentions of the wise.[52]
By way of the reference, Bacon turns our attention to the essay.

In the essay, Bacon says that, on the stage, love is at home, but in
life, love "doth much mischief; sometimes like a fury." Men of great
spirit and business are not "transported to the mad degree of love."
But even with them love can find its way even into a well-fortified
heart as well as an open one, for two exceptions to the rule are Marcus Antonius, "the half partner of the empire of Rome," who was a
"voluptuous man, and inordinate," and Appius Claudius, the "de-
cemvir and lawgiver," who was "an austere and wise man." According to Bacon's source Plutarch, Marcus Antonius was a tyrant who
enslaved Rome and who was so great that men thought him worthy
of greater deeds than he desired. But Marcus Antonius was so
impeded by his submission to Cleopatra that his insolence and love
of luxury reduced him to a condition paralleling Hercules' enslave-
ment and debasement at the hands of Omphale.[53] Livy, Bacon's second source, informs us that the other great lover, Appius Claudius,
was a prideful tyrant who first presented himself as the protector of
the plebs and who then let the burden of his tyranny fall upon them.
His end was occasioned by the outrageous effects of his extreme desire for the maiden daughter of Lucius Verginius. However guilty
Appius was of threatening the wives and daughters of the plebs, the
plebs had misgivings at jailing him because they saw that their punishment of so great a man made their own liberty seem to have become excessive.[54]

Bacon speaks as if Appius' fault were his lust rather than his op-
pressive tyranny. However, in accordance with the argument at hand,
Appius' fault sprang from his love of an individual rather than from
his love for the many, for the wise are lovers of the many, not the
few or one. But then the problem of the many's passions is that, as
they are served by the wise, every man who has a wife and daughter
can be the lover of all wives and daughters. These desires are boisterous and bestial because, however liberated they are by the gifts of
the wise, they do not transcend the tension between what is common

---

52. *Essays* 397–98.   53. Plutarch, *Antonius* 88–91.
54. Livy 3.32–36, 44, 57.

and what is one's own. It is not just that the ability to "contract the sight of the mind" is a "great faculty," as Bacon says in the present context, but rather that the tension between "contraction" and "dilation" makes man "the subject of the mouth (as beasts are)," because all men do "kneel before a little idol." The eye, which is not our master but the tool "for higher purposes,"[55] has a well-known relation to the stomach.

In the essay, Bacon borrows from Plutarch to argue that the lover is a greater flatterer even than the arch flatterer, a man's self, so that indeed it is impossible to love and to be wise. But if the lover is a greater flatterer than the self, then according to Plutarch, the lover even more than a man's self is likely to call the king's vices virtues. While the fiercest of the wild animals is the tyrant, the fiercest of the domesticated animals is the flatterer.[56] Bacon, speaking on behalf of the wise, has just praised the tyrant Appius Claudius, so that not only are the wise lovers, but Bacon emphasizes this truth by his own imitation of the wise lover. The lover, Bacon says, is the object of secret contempt, but also, and he quotes Ovid's sixteenth *Heroide*, "he that preferred Helena, quitted the gifts of Juno and Pallas." Bacon explains this statement to mean that "whosoever esteemeth too much of amorous affection quitteth both riches and wisdom." But Ovid describes the gifts of Juno and Pallas not as riches and wisdom but rather as thrones and might in war, respectively.[57] Bacon replaces empire and war with riches and wisdom, as if money were universal power and wisdom the means to that power. He suggests that the wise are like warriors and, by implication, that both are like lovers. We know that the wise are lovers of the many as their interests are measured by money, and in the essay, Bacon gives an important indication as to why this must be so. It is precisely because the warrior is indeed a lover, or rather, because of what warlike lovers become.

In the essay Bacon says that the warrior is a lover, referring obviously to Aristotle's discussion of war lovers in the *Politics*.[58] There Aristotle speaks not of Solon's Athenians but of the warlike Spartans. The Spartans' love of war led to their love of men and women, which led to a defective love of money that corrupted the constitution toward democracy. The love of honor led to egalitarian cupidity, which lesson, taught by Aristotle, must be understood as Aris-

55. *Essays* 397.
56. Plutarch, *Quomodo adulator ab amico internoscatur* 56f, 61b–d.
57. Ovid, *Heroides*, trans. Showerman, Loeb Classical Library (London: Heinemann, 1914) 16.81.
58. *Politics* 1269b29–1270b6.

totle did if the problem of charitable learning is to be grasped fully. The link between honor and Christian charity is avarice, and Machiavelli erred because he did not see this point clearly. Honor is the first seed of empire. But imperial honor softens to avarice, which is the secret truth of Christian charity. Learning has no choice but to serve the boisterous passions freed by avarice. Unlike Machiavelli, Bacon knows that such service is dangerous, and for this reason he must apologize on learning's behalf. It remains to be seen just how honor, empire, and avarice are related and just how honor could give birth to Christian charity. But when we do so in the next section, we see how Machiavelli mistook these relations and so mistook the scope and manageability of empire.

[280]

The fault with which Bacon concludes his discussion of manners is that the learned often "fail to observe decency and discretion in their behavior and carriage, and commit errors in small and ordinary points of action." This fault would presumably be of no importance were it not that the vulgar judge the learned's capacities in "greater matters" according to "that which they find wanting in them in smaller." This kind of judgment often deceives, and to cure the deception of the vulgar Bacon refers not to a learned man but to the saying of Themistocles, who when asked to play the lute responded that "he could not fiddle, but he could make a small town a great state."[59] Although this remark was arrogant and uncivil, as he himself said about himself, it is pertinent and just when applied to "the general state of this question." Bacon does not doubt that there are many who are poor in "little and punctual occasions" but are yet "well seen in the passages of government and policy." Bacon then refers these vulgar capacities to "that which Plato saith of his master Socrates," whom he described as like a "gallypot," which was ugly on the outside but "contained within sovereign and precious liquors and confections."[60] By this comment Plato acknowledged that by "external report" Socrates was not "without superficial levities and deformities" but that inwardly he was "replenished with excellent virtues and powers." With this comment, Bacon finishes treating the "point of manners of learned men."

---

59. Plutarch, *Themistocles* 2.4; *Cimon* 9.1.
60. Plato, *Symposium* 215a4–16c3; Xenophon, *Symposium* 4.19.

Our subject now is the relationship between honor, avarice, and Christian charity, and the scope and manageability of empire. According to the argument developed so far, the vulgar understanding of the difference between great and small matters can be corrected by the examples of Themistocles and Socrates, as if the vulgar can appreciate the invisible, inward "virtues and powers" of the soul as easily as they can be awed by the visible signs of empire. Bacon does not rest with this farfetched claim, however. He knows perfectly well that no judgment of the vulgar prompted Themistocles to his arrogant, uncivil, yet pertinent and just boast. Rather it was, according to Bacon's source Plutarch, precisely those reputed to be educated who taunted Themistocles, who had been poor at gracious learning.[61] Likewise, Bacon speaks as if it were simply Plato who described Socrates, but again he knows that, according to Plato, it was the tyrannical Alcibiades who praised Socrates. For Alcibiades, Socrates is a more marvelous piper than the satyr because he can excite to ravishment without instruments, and yet inside he is full of sobriety that shuns physical beauty, wealth, or any sort of honor that is deemed to be enviable by the many.[62] Of course Socrates' sober interior was not enough to moderate Alcibiades, whose frenzied excesses played no small part in the fatal accusation brought against Socrates by the city of Athens.

By way of his silence, Bacon suggests the limited power of learning over the likes of Themistocles and Alcibiades at the same time that he asserts the vulgar to be more impressed with learning than with empire. Insofar as learning serves and is appreciated by the many, it apes the likes of Themistocles and Alcibiades, both of whom aimed not to be learned or virtuous but to make small states great. Bacon knows perfectly well that, despite his failure to tame Alcibiades, Socrates humbled him. Likewise he knows that Alcibiades was drawn to Socrates more than Socrates was drawn to him. But in the charitable age bequeathed by Rome, learning loves the many and embraces the tyrant. We now know something of how such a connection has to do with avarice. At this point, Bacon develops these matters by calling our attention to the essay that treats Themistocles' ambition, the essay on the true greatness of kingdoms and estates. We must turn our attention to this essay, in which Bacon proceeds to the heart of his criticism of Machiavelli.

According to the essay, the speech of Themistocles, haughty and arrogant for himself but "grave and wise applied to others," dem-

---

61. *Themistocles* 4.   62. Plato, *Symposium* 215a4–16c3.

onstrates the two kinds of men who "deal in business of estate."[63] There are those who can make a small state great but cannot fiddle, and there are those who fiddle very well but "bring a great and flourishing state to ruin and decay." By fiddling, Bacon means the "degenerate arts and shifts" that counselors use to curry favor with masters and estimation with the vulgar. There are also "middling counselors" who are "sufficient" (*negotiis pares*) to manage affairs but who are unable to "raise and amplify an estate in power, means, and fortune."[64] Bacon speaks here as if the Themistocles story represents simply the several abilities to enlarge or ruin an estate. But again, Bacon is silent about the identity of the individuals who desired Themistocles to touch the lute.

According to Plutarch, it was not just people learned in graceful studies who goaded the rude Themistocles; rather Themistocles made his famous remark after his aristocratic rival Cimon had sung so well that the guests at the occasion found him to be more clever than Themistocles.[65] The difference between Themistocles and Cimon was not between one who enlarged and one who ruined an estate. Both were enlargers, although both ultimately failed to subdue the Great King and were ostracized by the Athenians. Rather the difference between these two enlargers was between democracy and aristocracy, and their common fate, as well as the fact of Athens' having become a popular democracy after Cimon's death,[66] suggests that the alternative to monarchical empire is imperial democracy that is subject to internal broils. This popular empire is not, strictly speaking, a democracy, because it contains regimes other than democracy. While Bacon begins his treatise on the proper measure of empire by abstracting from the difference between democracy and aristocracy, this course of action does not point to a moderate, mixed alternative, unless there is an alternative to empire itself. At the outset this latter possibility seems remote, for the two examples of middling counselors and governors (artfully referred to by Bacon's parenthetical remark *negotiis pares*) are a counselor who survived the executions following Tiberius' execution of Sejanus and the base wastrel Petronius who, as a member of Nero's corrupt entourage, was forced to kill himself, which deed he performed only weakly.[67] Before Bacon turns from the identity of the "workmen" to the "work" itself, it already appears that the workmen of the new empire must fashion a popular empire which, however much its formation depends on the proper

---

63. *Essays* 444–52.   64. Ibid., 444.   65. *Cimon* 9.1.
66. *Themistocles* 4, 22; *Cimon* 10, 15, 17, 19.   67. Tacitus, *Annales* 6.39, 16.18.

proportion of commons to nobles, is directed to an end that can be gauged only by a quantitative measure.

Bacon addresses the "work" of kingdoms and estates by considering what must be counted in order to pursue empire. A proper count is important so that great and mighty princes will not "by overmeasuring their forces . . . leese themselves in vain enterprises" or "by undervaluing them . . . descend to fearful and pusillanimous counsels." Bacon does not consider the possibility of eschewing empire except as the result of an opinion of weakness. Therefore he speaks as if empire were a political necessity as inevitable as the kingdom of heaven. He says that the capacity for empire is not to be measured simply by quantity, for by such a measure the power and forces of an estate can be misjudged. Just as heaven is compared to a mustard seed, so large states may not expand and small states may be the "foundations of great monarchies." Rather the capacity for empire is moral; in particular, it depends on the stout and warlike disposition of a people. Numbers are nothing if a people is cowardly, for as Virgil says, "it never troubles a wolf how many the sheep be." Bacon comments that this point is made by Alexander's remark that "he would not pilfer the victory" by attacking the numerically superior Persians at night and by the remark of Tigranes the Armenian, who mocked the numerically inferior Romans but who was defeated by them.

Bacon's references are instructive. The quotation of Virgil comes from the seventh *Eclogue*, where it is a line of the poem composed by the poet who loses a battle of poets.[68] Alexander's remark comes from two sources, Plutarch and Arrian. In both case, it was particularly Parmenio, Alexander's lieutenant, who offered the advice to attack at night. And according to Arrian, Alexander chose not to do so because he thought a night attack too risky.[69] Tigranes' remark comes from Plutarch's life of Lucullus, the Roman general who defeated the hapless Armenian. According to Plutarch, Lucullus caused Rome more harm than good because his trophies in Armenia incited Crassus to attack Asia, and his ignominious turn from public life to an unseemly combination of philosophy and pleasure, in the face of Pompey's democratic tyranny, demonstrated (according to some) that a political cycle has its dissolution, a lesson that might well have been heeded by Cicero and Scipio. This fact may apply to empires as well as to persons, for Plutarch notes that some also say that Lucullus

68. *Eclogues* 7.52.
69. Plutarch, *Alexander* 31; Arrian, *Anabasis,* trans. Robson, Loeb Classical Library (London: Heinemann, 1929–33), 3.10.

quitted politics because public affairs had become diseased and beyond proper control.[70] Of course Plutarch compares Lucullus to Cimon, whom Bacon has compared to Themistocles so as to dissolve
the political distinction between them. Bacon now compares Lucullus and Alexander, who were alike as generals and who both ended
their days in a perverse relationship to philosophy: Lucullus combined prodigious dissolution with reverence for the Old Academy,
and Alexander came to suspect philosophy so much that he put many
philosophers to death and was said by some to have been poisoned
at the behest of his former teacher, Aristotle.[71] Socrates humbled Alcibiades, and Aristotle was no more in the grip of a tyrannical thief
of liberty. How different was the case for Cicero. Bacon's comparisons tell us that empire is ultimately popular, that the moral foundation of empire is more the wise calculation of risk than courage,
and that the learned are the captains of the new empire only as long
as learning is unlike the political philosophy of Plato and Aristotle.
The questions that arise are whether in present times a learned empire, severed from political philosophy, is necessary, whether and how
such an empire is subject to cyclical dissolution or decay, and whether
learning can manage the risks of imperial ambition. With these questions, we come to Bacon's critique of Machiavelli's hope to limit and
manage the necessary tendency to empire.

If it is correct that courage is the essential foundation of empire,
then money cannot be the sinew of war. But Bacon seems to differ
from Machiavelli's apparent view in *Discourses* 11.10 when he (Bacon) says equivocally that money is not the sinew of war "where the
sinews of men's arms, in base and effeminate people, are failing."[72]
That is, according to Bacon, money is the sinew of war when arms
are strong enough. Bacon points explicitly to Machiavelli when he
recalls Solon's famous words to Croesus about the relative worths of
gold and iron,[73] when he warns princes to know their own strengths,
and when he warns against the use of mercenary forces. Machiavelli
does not himself believe his categorical rejection of the false opinion
about money and arms. He attributes the saying to Quintus Curtius,
who concluded the opinion from the case of the Spartan king's misfortune in his war with Antipater. According to Machiavelli, while
the Spartan king was forced to come to battle because he lacked
money, two days after the battle, which the Spartans lost, Alexander

70. *Lucullus* 27, 36, 38–43.   71. Ibid., 41–42; *Alexander* 55, 59, 77.
72. *Essays* 446.
73. Lucian, *Charon*, trans. Harmon, Loeb Classical Library (London: Heinemann,
1968), 12; cf. Diogenes Laertius 4.48.

died. If the battle could have been postponed for these two days, the Spartan king would have been the victor without fighting and would not have had to try his fortune in battle. But in fact Quintus Curtius does not tell this story, which is not about a Spartan war with the Macedonians before Alexander's death. The story actually occurs in Plutarch's *Agis and Cleomenes*, where Plutarch tells of Cleomenes' battle with Antigonus. It was not Alexander's death that would have saved the day (it could not have been); rather it was an Illyrian invasion of Macedonia that summoned Antigonus away.[74]

To prove the false opinion, drawn from the false story, to be false, Machiavelli cites the examples of Darius and Alexander, the Greeks and the Romans, Charles the Bold and the Swiss, and recently, the pope, the Florentines, and Francesco Maria in the war of Urbino. He refers to the story of Solon and Croesus, to the example of the Macedonians who by demonstrating their gold to the Gaulic ambassadors incited them to break a treaty and despoil the king, and to the Venetians, who recently lost their whole state despite their wealth. But the examples do not equally support Machiavelli's categorical rejection of the opinion. In the first place, the pope and the Florentines won with some difficulty, unlike the rich losers with whom they are ordered in Machiavelli's list. The pope, then, must have been a rich winner whose soldiers fought a difficult fight against rich but not weak or degenerate losers. Second, while the Gauls were incited to conquest and plunder by the sight of Macedonian gold, Solon, seeing the gold, gave Croesus good advice that, according to Plutarch, saved Croesus when he fell into the hands of Cyrus, so that Croesus in the end had his money and his life and his honor.[75] Finally, although Machiavelli cavalierly says that the Venetians met defeat despite their full treasury, he argued in *Discourses* 1.53 that the Venetians failed because they failed to use their treasury. At the beginning of the chapter,[76] Machiavelli remarks that in deciding whether or not to engage in any enterprise, a prince must not measure his resources by money, geographical conditions, or the goodwill of his people, unless the prince has his own arms. But "one's own arms" really means the faithful arms of others. After repeating the three conditions twice, Machiavelli proceeds in fact to discuss the subject of money and arms, dropping the subject of a people's goodwill as well as the grounds of soldiers' good faith. So far it is clear that money is often useful in war, although it is not by itself sufficient. But this

74. *Agis and Cleomenes* 27; Mansfield, 215–19.
75. *Solon* 27–28; Herodotus 1.30–33, 86–88.   76. *D* 2.10.

assertion is not surprising and does not by itself warrant Machiavelli's caution, because he explicitly modifies his intransigent rejection of the opinion when he says that money *is* necessary but that it is secondary and that its lack can be overcome by good soldiers. The question is how one has good soldiers if one's own arms are in fact the faithful arms of others.

After the last example of the Venetians, Machiavelli says that gold will not procure good soldiers but good soldiers will procure gold. The Romans could not have used even the whole world's treasure to accomplish what they did because of the scope and difficulty of the enterprises they attempted. The Romans made war with iron and did not want for gold because it was brought to them—even into their camp—by people who feared them. To conquer the world the Romans could not have benefited from having the world's treasury because their aim was command over the entire world's treasury. It turns out, then, that money is not the sinew of war, but the lust for money is the guarantee of the faithfulness of others' arms, not faith or goodwill itself. A general will always choose to fight and so will try his fortune rather than flee and lose for sure, and lack of money can be one of many necessities that forces such a choice. But the essential requirement for the successful willingness to joust with fortune, which may be necessary because of poverty and may be aided by the proper use of money, is the promise of money or riches. In this light, Machiavelli's example of the pope is important. According to his list, the pope and his Florentine arms represent a rich winner whose soldiers fought well against rich, good, but losing soldiers and who was, for all that, still a loser. In the papal example, then, there is no difference between winner and loser, a feature that harmonizes with the unstated fact that both the pope's forces and Francesco Maria's forces were Christian. For Machiavelli, the new Roman Empire loses as it wins. It is as weakened and corrupted by its own success as was the old Roman Empire, which, after all, replaced hope for the world's wealth with its possession by universal conquests.

As Machiavelli argues elsewhere, there is no successful alternative to empire.[77] But the problem of empire is that it is corrupted by its own success. And when the universal prize is not gold but an otherworldly salvation promised to all by a gracious gift, there is not even the temporary benefit from the acquisition of corrupting wealth. Machiavelli's intention was to overcome the inevitable, enfeebling, corrupting, and nonpolitical Christian empire by imitating its prin-

---

77. Ibid., 1.5–6, 2.3.

ciples of indirect rule. His intention was to resurrect the possibility of the good soldier in place of the Christian soldier who hopes to be resurrected. By imitating the Christian corruption of Rome, he intended to solve the problem of empire itself. But Machiavelli's solution, while modeled on the two most important empires, is not itself imperial because his soldier must always have enemies who do not corrupt by virtue of being conquered. Machiavelli's intention, then, was to overcome the cycle of empire by the proper application of the principles of the universal Christian empire to man as the political, not the imperial, animal. The need for empire is always necessary, but empire and the consequences of empire are not: the mastery of the imperial cycle is never a universal freedom from war.

We know that Bacon thinks Machiavelli cannot conquer the Christian empire, because he believes that Machiavelli underestimated the power of Christian belief and Christian egalitarianism. The consequence of this failure is that the Christian soldier has become not enfeebled but rather a much more dangerous and unmanageable exaggeration of Machiavelli's soldier. The solution to the problem of the Christian empire may not be political, as Machiavelli had hoped; it may rather depend on a more universal promise of the new learning. It is not surprising, then, that in referring to *Discourses* 2.10 Bacon starts where Machiavelli finished, with a conditional rejection of the opinion about money and arms. Likewise it is important that he mentions only the example of Solon and Croesus. For Croesus, Solon's advice about virtue proved to be the perfect complement to his wealth. Until he was defeated by Cyrus he was rich, and after the defeat, by virtue of advice he received free of charge, he retained his life and his honor. Plutarch remarks that Solon saved one king and instructed another. But of course Cyrus never quitted his crown or his wealth.[78] According to Bacon's example, then, money is a perfect substitute for arms. As we will see, the solution to the problem of universal empire is not to bend the desire for wealth to the proper health of arms; rather it is to bend the use of arms to the proper pursuit of wealth, which only the new learning can make possible.

When Bacon warns in the essay that the prince must know his own strength even when he has "subjects of a martial disposition," he knows that arms—or rather wealth—can in one sense be even less the prince's own than Machiavelli thinks it, first because there is ultimately no difference between peoples and princes and second because, as measured by money, what is one's own is commensurable with every-

78. *Solon* 28.

thing that is not one's own. But of course it does not follow that men will be free from the love of their own. We pursue this issue by noting that, after warning the prince to know his own strength, Bacon follows Machiavelli in warning against the use of mercenary forces. With such forces, a prince may "spread his feathers for a time, but he will mew them soon after."[79] Now, Bacon's warning taken from Machiavelli seems to be inadequate for its purpose; Bacon mentions the use of mercenary forces but fails to mention auxiliary forces, which Machiavelli says in *Prince* 13 are more dangerous than mercenaries because auxiliary troops are united, owe their obedience to others, and so are courageous, unlike mercenaries, who are cowardly. Machiavelli's argument fits well with the argument about money and arms, for while mercenaries must be paid in advance, explaining their cowardice and reluctance to fight, auxiliary troops fight, as do any well-managed troops, in the hope for future gain. But the argument does not fit well with Machiavelli's claim in *Prince* 12 that Italy had been ruined by the use of mercenary troops, which was increased as the pope's reputation in temporal matters increased. One would expect the ruin of a whole country—not just that of a prince or republic—to be caused by the more dangerous military evil. Machiavelli's acknowledgment that Italy had been divided into many states, along with the pope's political ascendency, might satisfy this objection, but Machiavelli suggests the way in which mercenaries might in fact be used safely.

It is true that the problem of mercenary captains is not different from that of *any* captain: if they are not good, they lead to ruin, and if they are good they lead to ruin because they aspire to their own greatness by turning against their master or by attacking others against the master's will. In order to avoid this disaster, a prince must be the captain himself and a republic must send its own citizens as captains but must restrain them by means of the law. Because all captains must fight with arms that are, however faithful, not simply their own, a captain must be able to promise future gain. If the law depends ultimately on extraordinary executions, delegated captains must be killed before their success incites them to their own greatness. All good troops are, then, well-managed mercenaries, that is, mercenaries who are not paid in advance and whose captains are killed in order that they may be well managed. Machiavelli's example of Cesare Borgia bears out this point: his use of both mercenaries and auxiliaries did not lead to his demise; rather it was the prelude or the means to his

79. *Essays* 446.

ever-increasing fortune. After first using auxiliaries to take Imola and Forli, he had recourse to mercenaries, hiring the Orsini and the Vitelli. Then, when they became unmanageable, unfaithful, and dangerous, he "dispensed" with them *(le spense)* and relied on "his own," which means in fact the faithful arms of others. The important difference between troops, then, is not so much the difference between mercenaries and auxiliaries as it is between ill-managed and well-managed mercenaries. Italy could be ruined by the lesser of two dangers, not because she was sundered into many states, but because there is no real difference between mercenaries and auxiliaries.

In this light, the example of the pope in *Discourses* 2.10 and the example of the Capuans in *Discourses* 2.20 come together. If there is no difference between mercenaries and auxiliaries, then auxiliaries must be as ubiquitous as the mercenaries who everywhere spoiled Italy in proportion to the growth of papal political power. In the example of the pope and Francesco Maria,[80] both were identified as losers. Both forces were Christian, of course, and so in an important way *all* forces are auxiliaries of the pope's Christian empire, meaning for Machiavelli that they are all ill-managed mercenaries. The problem with Christian arms is not simply that they debase the use of arms. Rather for Machiavelli the problem of Christian arms is that Christianity spreads its empire without knowing how to manage mercenary troops, or rather troops as such. The Capuans did not request auxiliary troops; they accepted them after having been conquered by the Romans, to whom they were then forced to give tribute.[81] This tribute saved them after their military defeats, a case in which money served better than arms. The auxiliary troops were preparing to attack Rome, as auxiliaries are wont to do, since they forget their own countries as soon as they forget the country they protect. Machiavelli says that the Romans suppressed and punished the "auxiliaries'" conspiracy against the Capuans, so the Romans themselves guaranteed the success of the Capuans' wise use of money in place of arms. But in the previous chapter, Machiavelli cites the Roman acquisition of Capua as an example of an acquisition that can be harmful even to a well-managed republic. The Capuans' moral corruptions would have ruined the Roman republic had Capua been at a greater distance from Rome and had the soldiers' excesses not been quickly corrected, or had Rome herself been in some part corrupt. According to Livy, cities like Capua corrupt and so conquer their

---

80. *D* 2.10; Mansfield, 215–19.   81. *D* 2.20; Mansfield, 251–53.

conquerers,[82] and Rome was eventually conquered by the very tribute she commanded.

We begin to see the core of the teaching that Bacon learned from Machiavelli. As Livy said, Rome was conquered by the tribute she commanded. But this tribute was necessary for the successful Roman control over Roman arms. The vigor of the Roman Empire was softened by this tribute, so that Rome could be conquered by the soft Christian empire, which itself must be conquered by the proper management of tribute and arms.

However, unlike Machiavelli, Bacon mentions the danger of mercenaries and is silent about auxiliaries and about their similarity and difference. On the one hand, by suggesting that a single reference suffices for both, Bacon emphasizes the important fact about mercenaries insofar as they are auxiliaries in the Christian empire. This kind of mercenary is simply a Christian soldier, the model of the auxiliary, and as such always the soldier of another country that might lose as it wins. On the other hand, by condemning mercenaries and being silent about auxiliaries, Bacon refuses to condemn auxiliaries as Machiavelli did; in the Christian empire, this silence approves masterless auxiliaries and Christian soldiers.

Bacon does not intend to revive the fortunes of any *patria,* even patria as Machiavelli understood it, an exclusive, free patria or common good governed by universal principles of management. For Bacon, the solution to the problem of Christian empire extends beyond the limits of exclusive political bodies. At least apparently, Bacon imitates the principles of Christian charity much more than does Machiavelli. Insofar as the *New Atlantis* is a model, the solution to the problem of tribute is to fashion the means for its limitless production and distribution, transforming rapacious soldiers into sheep who could not be more softened by even more booty. But this apparent freedom from patria does not free men from the limits of what might properly be one's own. The new knowledge assimilates peoples and princes in accordance with Christian charity, but Bacon's silence about auxiliaries follows his warning to an *individual* prince. In the sequel Bacon argues that the blessing of Judah and Issachar[83] will never meet, because "no people over-charged with tribute is fit for empire." The example is perfect for the context, for the lowly ass's strength is matched by its stubbornness, grounding the dependence of the one who is borne on the one who bears. If such dependence

82. *D* 2.19; Livy 7.38.   83. Gen. 49.8–14.

ultimately brought the victory of Christianity over Rome, the new learning can conquer Christianity only by promising, falsely, to accommodate the lion and the lamb. The bearer of this promise is the king whose empire pursues the blessing of Zebulun. Zebulun, standing between Judah and Issachar, was destined to become the haven of the sea.

However much a perfected new empire might produce the tribute it receives, the way to this empire depends upon the political fortune of the present English king. In the essay Bacon offers advice to assist him on his way. As long as the new empire is not yet self-sustaining, tribute and tax must be levied on its soldiers. This constraint for the sake of freedom must be consistent with the martial valor leading to sheeplike peace. Bacon speaks "now of the heart and not of the purse." All tax or tribute is a weight upon the purse, but "although the same tribute and tax, laid by consent or by imposing, be all one to the purse, yet it works diversely upon the courage." In this regard, the policies of the Low Countries and England are exemplary. Bacon says that a similar observation can be made regarding the orders of society. Imperial policy must attend to the proper proportion of nobility and gentlemen to commons. If the gentlemen are too many, the commons will be too base, weakening the infantry, which is "the nerve of an army." The cases of England and France exemplify this rule. The superiority of the English soldiers to the French can be traced to the wise policy of Henry VII, who "made farms and houses of husbandry of a standard." By ensuring that subjects had enough land, he ensured that his subjects would not be of "servile condition." Bacon then speaks directly to the king: "and thus indeed you shall attain to Virgil's character which he gives to ancient Italy: *terra potens armis atque ubere glebae.*" These are the words spoken by Dido as she let the Trojans proceed to Italy.[84] Therefore according to the reference to Virgil, the policies Bacon recommends for his own king and country suit a voyage to a new land, that is, to a new universal empire.

In the *New Atlantis* the sailors who voyage to Bensalem represent the seafaring, world-exploring peoples of Europe. Moreover, the narrator of the story speaks in English. It is no surprise, then, for Bacon to say that only in England and "perhaps in Poland" is the wise policy to be found that encourages "the state of free servants and attendants upon noblemen and gentlemen; which are no ways

---

84. "A land powerful in arms and in productiveness of soil" (Spedding's trans., *Essays* 447); Virgil, *Aeneid* 1.531.

inferior unto the yeomanry for arms." By this device the splendor, magnificence, great retinues, and hospitality of noblemen and gentlemen "doth much conduce unto martial greatness," whereas, contrariwise, the "close and reserved living of noblemen and gentlemen causeth a penury of military forces." The proper policy for managing the proportion between nobles and commons tends to equality of station and fortune, and England is the vehicle that will make such equality universal. But if the new learning must promise such a universal empire, Bacon is under no illusion that it can outstrip the horizon of the exclusive patria. In the sequel, then, he discloses the natural limits of empire.

Bacon warns that any state aspiring to empire must have a proper number of "natural subjects of the crown or state" in proportion to "the stranger subjects that they govern." Bacon refers to the proportions the trunk of a tree must bear to its branches and boughs, reminding us of Machiavelli's use of the same example for the same subject in *Discourses* 1.6 and 2.3. But while Machiavelli speaks simply of the tree, Bacon refers to "Nebuchadnezzar's tree of monarchy," so that he recalls Daniel 4.10 at the same time that he recalls Machiavelli. Bacon says that because states can imitate nature by creating natural subjects, by naturalization, "all states that are liberal of naturalization towards strangers are fit for empire." The foundations of empire are not exclusively moral, then, as the example of the Spartans demonstrates. The Spartans were "a nice people in point of naturalization." But for this reason their acquisition of empire caused them to "become a windfall upon the sudden." The Romans, on the contrary, were unlike any state in being "open to receive strangers," helping them to grow "to the greatest monarchy." The Romans granted all the rights of citizenship to persons, families, cities, and even sometimes to nations. If we compare the two constitutions, it is evident that "it was not the Romans that spread upon the world, but it was the world that spread upon the Romans." The Spaniards seem to be a marvelous exception to the rule, "containing so large dominions with so few natural Spaniards," but there are reasons for the Spaniards' success: Spain's "whole compass" is a "very great body of a tree," and they have done what is "next" to allowing liberal naturalization; they have "employed almost indifferently all nations in their militia of ordinary soldiers; yea, and sometimes in their highest commands." Even with these policies, however, the Spaniards are aware of their "want of natives," as the Pragmatic Sanction demonstrates.

According to Machiavelli in *Discourses* 2.3, if the pursuit of empire requires the conventional or political imitation of the tree's natural proportion, it does not follow that this imitation can imitate the natural harmony of the tree within its natural limits. The tree dies according to nature, but it is not for that reason always diseased. But empire is a political disease because it produces disunity in the imperial power, and as we see, there is no alternative to such proportion, according to the argument in *Discourses* 1.6. Here Machiavelli says that unlike imperial Rome, Sparta and Venice were both unified cities.[85] The Venetian way to unity was a combination of nature and convention, for Venice was founded on a site that permitted the rapid development of numbers sufficient to require laws. At the moment of its political origin, all the inhabitants were participants in government. Henceforth it was forbidden for newcomers to participate. Those denied power were only the new; the old came to be called gentlemen by the new when the new became numerous enough. The nobles could govern the plebs, so that any old inhabitant could govern any new one, because the new had been deprived of nothing, so that they had no reason to cause disturbances, lacking the means as well, because they were not used in the army. By this account, Machiavelli describes beginnings where nature and convention are in perfect harmony: at the outset, Venice could be founded *without* having to deprive anyone of anything (the primordial Venetians were no cannibals), and the noble is the noble simply by the application of a name to an all that differs from another all merely by the accident of time.

The Spartan source of unity was purely conventional. At first Machiavelli presents a distorted picture of the Spartan regime, in which he ignores the second king and the ephors. Then he comments that Sparta had few inhabitants, permitted no newcomers, and obeyed Lycurgus' laws establishing equality in fortunes and inequality in rank. Therefore, equal poverty prevailed in Sparta, where the many were not ambitious because officers were few and good treatment by the nobles did not inflame them. This last factor depended on the Spartan kings, who lived between the two orders and yet among the nobility. These kings had no better way to maintain their own dignity than by protecting the many from injury. But after speaking correctly about two kings, Machiavelli says that there were two, not three, causes of Spartan unity: small population and the prohibition against admitting strangers. He omits Lycurgus' laws prescribing equality of fortune or equal poverty. Machiavelli concludes that the Romans

85. *D* 1.6; cf. Mansfield, 48–53.

would have had to do one of two things to enjoy the unity known in Sparta and Venice: not to use the plebs in the army, as the Venetians did, or not to allow the admittance of strangers, as the Spartans did. Machiavelli drops another of the causes for unity in Sparta (small population) and looks past the natural accident that allowed the Venetian nobility to deny military service to the commons and to obtain possession of most of Italy not by war but by money and cleverness.

In the sequel Machiavelli shows why he proceeds as he does: first, the account of a naturally harmonious beginning is no better than a myth that is dangerous, because nature's necessity is harsher than any supposed benevolence of nature giving perfect salutary freedom to human convention. Second, the key to a proper understanding of Sparta is not a grasp of her regime's animus to the admission of strangers but comprehension of just how it is impossible to pursue such a policy, because natural increase will never suffice to accommodate the necessity for growth in size. The Romans followed neither the Spartan nor the Venetian course, so the Roman republic was wracked by disturbances. However, had Rome been quiet, she would have been feeble, so in human affairs one can never simply avoid disturbances. To think of founding a republic, one must choose between narrow limits and empire, between Sparta-Venice and Rome, but Rome cannot be quiet and Sparta-Venice cannot expand successfully. A middle course between these two possibilities would be the true political life (*il vero vivere politico*), which would be like Sparta or Venice, sufficiently powerful to be difficult to conquer and yet not so powerful as to frighten neighbors. These two characteristics would remove the two causes of war. But this middle, true way is impossible because the human things are in constant motion and because necessity compels so that reason cannot resist. The way of this necessity is not war but peace, which if heaven-sent simply causes a city to become weak or subject to disturbances, both of which lead to ruin. Rome is superior to any other model, for there is no middle course between all possibilities and Rome. However, Machiavelli says that there is an honorable way, which is to devise a way of keeping what one cannot help acquiring. Obviously, this honorable way is not to be found in the examples of Sparta and Venice.

But in these examples, a powerful clue about the honorable way is to be found. Since the Venetians grew by means of their money, they were easily deprived of their acquisitions because of their improper reliance upon their wealth. As Machiavelli says, they failed to use their treasury at the right time, so the Venetians failed because they did not manage money well.[86] The same lesson is to be learned from the Spartans. The Spartans were defeated by more than just Pelopidas, and were not entirely but only "almost" ruined, unlike the Venetians, who lost everything in a single battle. Of the causes of Spartan unity, Machiavelli drops references to population and Lycurgus' laws, focusing only on the prohibition of immigration, because, as Aristotle argued, the Spartans, being a courageous, warlike few, were eventually ruled by women, had a consequent love of money and disproportion of fortune, and showed a consequent tendency to extreme democracy.[87]

The Spartans' use of leather money discouraged the influx of strangers. But such austerity served their courage only to corrupt them by producing an ill-managed love of money. The account of the Spartan constitution is as much a dangerous myth as the account of the Venetians' experience of natural bounty. The lesson of Venice and Sparta is that nature affords no bounty, so that man must by necessity be an acquirer. For Machiavelli, man is unlike the tree because man is diseased by the effects of nature's limits. By nature there is no alternative to empire; it is the tendency of every free patria, but empire in itself is, unlike a natural tree, always diseased and dangerous to freedom. The way to imitate the tree is not by means of the conventional moral virtues but rather by means of the wise management of man's inescapable, necessary, but potentially corrupting avarice. The "true political life" is a middle way only by virtue of its having escaped the need to expand that characterizes the two extremes between which it stands. If it is impossible, then in fact the only alternatives are the realistic possibilities of Sparta-Venice and Rome. But this conclusion is on second thought too hasty. As an impossible freedom from necessity, the "true political life" is essentially different from the realistic, or nonmythical extremes. Therefore it is not even commensurable with the extremes by which its being a middle is to be measured. The implication is that Machiavelli's argument about the lack of a middle way is less than serious and that there might be an alternative to Venice and Sparta and Rome. This middle way is Machiavelli's honorable course.

The honorable course is to follow Rome, not simply, but so as to keep what one acquires. This course cannot be honorable, however, because it requires the substitution of well-managed avarice, money love, for the love of honor. Neither the Spartans nor the Venetians could increase their populations, because honor, whether it is taken

86. *D* 1.53.   87. *Politics* 1269a24–71b19.

to be natural or conventional, depends on the proper proportion of the few to the many. Sparta and Venice suffered from a common disease: their dependence on *honor* made them incapable of managing the expansion they could not avoid. Even the Venetians, who expanded by virtue of money and cleverness rather than war, foundered on the shoal of honor, which distinguished between old nobles and new plebs. The orientation by honor is the real source of fatal fiscal incompetence. Rome too was diseased, and Rome differs from Machiavelli's way because she did not keep what she acquired but rather was conquered by her conquests. The honorable Romans did not understand Machiavelli's way, which limits empire by the proper management of avarice rather than by the exercise of honor. The Roman inability to manage the internal disturbance was simply the symptom of the real disease, the Roman failure to manage her imperial conquests. As long as the tension between princes and peoples, the Senate and the plebs, is understood as the tension between honorable and base rather than between those who desire to take and those who more modestly desire to keep, it will not be possible to manage the enervating consequences of successful conquest. The real difference between Rome and Venice and Sparta was not between disease and unity but between two instances of the same disease, honorable fiscal incompetence, where one case is more visible by symptoms than the other. Machiavelli follows Rome less than he learns from her modern counterpart.

According to the present argument, the Romans as honor-loving conquerors should be more like the Spartans than different from them. But Bacon's demonstration thus far indicates that the Romans as conquerors should also have been, like the Spartans, lovers, who succumbed to the love of money as they submitted to the crack of the femine whip. And indeed, in *Discourses* 2.3, Machiavelli argues that Rome maintained the treelike proportions of her empire by combining force and love. In destroying cities and compelling populations to come to her, Rome softly welcomed foreigners. While the unadulterated, forceful courage of the exclusive Spartans and the Athenians did not suffice to support their empires, the Roman use of the two methods did. But Rome, too, ultimately lost her empire, for the same reason as Sparta. Just as Spartan courage collapsed into love and enervating avarice, Rome's forceful policy of empire enervated Rome as it lovingly embraced the entire world. Rome was made feminine by the tribute she honorably commanded, which led to the replacement of the empire of force and love by the empire that only loves its enemies. Machiavelli mentions honor in the title of *Discourses*

2.3, but he mentions it nowhere in the chapter itself. The Roman orientation by honor made it impossible for them to see the truth about force and love, which has to do with the proper management of tribute. Machiavelli's way cures the disease of imperial expansion by managing avarice so as to sever its kinship to love. To make avarice hard rather than soft, the founder must produce the good effects of excess acquisition without also producing the ill effects of actually having too much. The wise founder, then, must know how to remind of the harshest necessity—the harsh penury of beginnings—which means that he (or she) must regulate successful getting by guaranteeing a periodical, indirect taking from those who get. Politics is unlike nature in being able to expand beyond its natural limits, and the more it does, the more it is certain that an unnatural disease will lead to decay. But the disease of man's political character can be cured by wise pruning.[88]

Machiavelli knows that as freedom-loving honor is martial, it depends upon avarice—this was true for Sparta, Venice, and Rome. And when honor is properly understood, the dishonorable management of avarice can be used to encourage the vigor of imperial conquest and yet bind it within the limits of the free patria. In this way, the dishonorable understanding of honor makes it possible for honor to flourish. The Christian empire is weak because it does not know how to manage the money it loves and professes to hate. This empire's contempt for honor betrays the link between honor and avarice: its money-loving contempt for honor is the decayed remnant of honor. By way of avarice, honor is the ultimate source of the charitable, enfeebling Christian empire. The rise of Christianity is a model of the problem of politics, or honor. And a person must learn the Christian contempt for honor in order to preserve the ultimate source of that contempt, which is honor itself.

Bacon knows with Machiavelli that honor is the root of Christian charity. But in the account of the ways of empire, Bacon makes a long detour through Machiavelli in order to demonstrate his trans-Machiavellian Machiavellianism. Machiavelli's political science promises not just the imitation of nature but the political conquest of the natural cycle of growth and decay. But as we know, Bacon thinks Machiavelli underestimated the extraordinary power of Christian love and egalitarianism to coexist with the fiercest and most unmanageable, the most honorable, martial passions. Machiavelli's political science is political because it works by frustrating the possibility of per-

88. Cf. Mansfield, 199.

fect empire. For Machiavelli, the universal phenomenon of avarice must never be allowed to subvert lovingly the political distinction between citizen (friend) and foreigner (enemy), just as it cannot be allowed to subvert the political difference between princes and peoples.

Bacon's example of Spain is crucial in the present context, because it points to his difference from Machiavelli. Spain made up for its reluctance to admit foreigners by employing foreigners in its armies, by using a kind of auxiliary, and by encouraging the natural increase of its own citizens. The reader cannot help seeing the similarity of Bensalem to Spain in this regard. The Bensalemites enlist the visitors in their efforts to conquer the world for science, and their society praises fecundity above all other virtues.[89] The difference between Spain and Bensalem is that while Spain must rely on the cooperation of spouses and heaven's gifts, that is, sufficient food, lack of pestilence, and so forth, Bensalem relies on its spouses and on the scientific conquest of nature. Bacon's reference to Nebuchadnezzar's tree is bolder than Machiavelli's more subtle teaching that one cannot rely even on the heavenly gift of peace. Bacon's boldness about our freedom from God's gift of empire is matched by his intention to master and control the force of nature's increase. In Bensalem, the scientific promise of infinite bounty appears to obviate the need to manage the relations between force, the love of money or acquisition, and the softness of love. Human force applied to nature, rather than to man, serves a universal empire of enervated, money-loving sheep without succumbing to the cycle of natural decay. Bacon's new empire conquers by receiving rather than by dominating, and its use of auxiliary forces demonstrates that the world will spread over it for the sake of a world, not for any patria.

But at this point we can see that, despite their apparent difference, Bacon's universal empire was modeled on Machiavelli's political science. Moreover, Bacon thought that Machiavelli did not fully understand how his political science could be transformed into Bacon's universal project, and because he did not, Machiavelli could not have understood why Bacon's new project was necessary and why it would have the same problems as his own political science. In both cases, Machiavelli suffered from having turned too quickly from the political teaching of the ancient utopians.

Machiavelli could not have seen the necessity for Bacon's project because he misjudged and overestimated the difference between

89. *New Atlantis* 147, 166.

honor, Christian charity, and avarice, even though his political science depended on his knowledge of their likeness. Machiavelli erred in this regard because he did not know the lesson to be learned from the difference between virtue and honor. He did not appreciate the difference between Socrates and Phocion and the likes of Themistocles and Alcibiades. Machiavelli knew that in devoting themselves to honor, the Romans took men to be tied together by need rather than by need and moral virtue, and from his knowledge of Christianity he understood how such a view leads from empire to enervating love. But he did not understand how such a view rebounds spontaneously to ferocious honor and how it fuels the youthful desire for mastery. Not having learned from the ancient utopians, Machiavelli did not know that honor severed from virtue is the delusion of the productive arts that aspire to perfectly free mastery. Such aspiration is delusional because it looks past the limits of conventional political life, as if the order of the arts were not political at all, being simply spontaneous rather than being fashioned by a chastened art of statesmanship. Machiavelli did not know that the universalism of Christian charity is simply the present form of the charity that always looks beyond political life as it animates every political order. He did not know that such charity always appears in practice as the love of honor. Had Machiavelli not rejected classical utopian teaching, he would have known that Christian charity would be both avaricious *and* honorably ferocious. It would be avaricious because, reflecting Roman honor, it takes all men to be bound by nothing but need. And it would be honorably ferocious because it inflames the artful desire that in serving need dreams of freedom from every restraint, from every convention that might be the handiwork of another's art.

Had he not rejected the ancient utopians, Machiavelli would have known that his honorless service of honor would only inflame the unmanageable ferocity of those he took to be too weak. But how could Bacon's more promising *universal* project be fashioned upon Machiavelli's so apparently *political* teaching? Again, Machiavelli would have known this deepest truth about his own teaching if he had not rejected the ancient utopians. Machiavelli's intention was to reverse the corrupting effects of empire—which had spawned Christian weakness—by the honorless and unjust deflection of avarice toward political honor. To do so, he had to justify the tyrant. That is, Machiavelli taught that the common political good is not to be fashioned according to conventional virtue but rather can be discovered—and so managed—in the natural economy of acquiring and possessing. By denying the tyrant the full measure of his acquiring, the possessing of others can be served and guaranteed so that each in his own way, and the political whole, can exercise honorable self-assertion and freedom. But to justify the tyrant, it is necessary that all human desire be taken to be commensurable, so that one simply adjusts an economy of desires in which the parts, individual desires, are nothing but the parts of a limitless whole. By such an account, there is nothing that is by nature either self-sufficient or irreducibly private or both. Bacon knows with Plato that at the very least the human body is irreducibly private. Likewise he knows that no body-serving art moves without being drawn to the self-sufficient. These two facts cause human beings to be political animals, to be always potentially divided by different and exclusive claims to rule for different and exclusive conceptions of a common good. These claims call for the statesman's art because each such claim by itself looks beyond its political delimitation by other claims and, of course, its need for the statesman's art. Human society is an order of productive arts, with every art tending to produce an extreme, or tyrannical, political claim because every art can presume its separate perfection to cohere with a natural commensurability of every human desire, so that all separate needs can be perfected by the perfection of one: by way of a single art's mastery of a single need, all neediness can be served.

We can now see the astounding face of Bacon's teaching about Machiavelli: the new project for learning is modeled on Machiavelli's political teaching because, like that teaching, it presumes the transpolitical perspective inherent in every productive art. Bacon thinks that Machiavelli simply did not understand how and why man is the political animal, and the lurking secret of Machiavelli's science for political health is the transpolitical hopes of every political claim that arise when these claims are not managed by the statesman. The tyrant always *seems* to be justifiable, even when the statesman's art succeeds, because his art uses rhetoric and well-formed convention to obscure the incommensurability of goods as well as making it appear: without such artful mixing to produce virtue, the necessary but incommensurable energies of the city—the love of the useful and the love of the self-sufficient—cannot at once cohere and be moderately ordered. In taking the tyrant to be justifiable, Machiavelli acted as if, knowing with the political philosopher and the sophist that the virtues are less self-sufficient than the virtuous take them to be, he could replace virtue with knowledge. Not knowing the ancient utopian teaching about the tension between practice and theory, Machiavelli was unaware of how far his intention was in the grip of the sophists' artful hope and dangerous Christian charity.

Apart from the statesman's art, every good citizen can always honor one of two extreme possibilities, each of which is one side of the same coin: one who is beyond all needs because he has mastered them all or one who is only all of his needs because he does nothing but strive. Avarice is nothing but these two possibilities ascribed to all men: all must always make useful money, and money making is self-sufficient. Avarice is always possibly ferocious, because while it tries to, it can never rise above the limits of political life. Bacon knows that without the statesman it will do no good to bend avarice to honor, and knowing that in the present world Christian charity silences the statesman, Bacon teaches that avarice must be bent to the transpolitical conquest of nature's bounty. But precisely because he understood classical utopian thought, Bacon knew that the very reasons for the need to replace Machiavelli's project disclosed the limits and dangers of his own. In the charitable age, Bacon's project could not but be modeled on the secret transpolitical hope of Machiavelli's political science. Unlike Machiavelli, Bacon argues that the kingdom of the clergy must be replaced by an empire of like proportions, one whose proportions are quantitative rather than qualitative. Machiavelli erred because, by respecting honor, he *had* to succumb to the impossible transpolitical hopes of honor. Without the statesman's art, such hopes could not be managed and certainly not by a science like Machiavelli's that did not know how much it was infected by them.

But to the extent that Bacon's project is modeled on Machiavelli's, Bacon knew that he too bowed to the love of honor. According to the *Novum organum,* "the benefits of discoveries may extend to the whole race of man, civil benefits only to particular places; the latter last not beyond a few ages, the former through all time."[90] For this reason, the authors of inventions are awarded divine honors, while those who "did good service in the state (such as founders of cities and empires, legislators, saviors of their country from long endured evils, quellers of tyrannies, and the like) receive no higher honors than heroic." Bacon himself admits to being one who deserves the highest honor, which is inevitable because, for all its curative power, the new learning is still charitable. Charity is still the impossible attempt of honor-loving political men to outstrip the limits of political life. Bacon's gift deserves divine honor; and no wonder: in serving

90. *Novum organum* 1.129.

charity he serves what he knows to be a heavenly gift. Bacon does not deny Machiavelli's bold claim that divine gifts are dangerous; rather, unlike Machiavelli, he knows that the most dangerous form of such a gift is the belief that practical life can be ordered without it. In the age of ferocious Christian charity, the best that Bacon can do is to turn its avarice against nature rather than men, so that men will for a time pursue as universal what is less narrowly their own. But nature's increase requires spouses as well as material bounty. And unlike the other things that money can buy, a spouse can be taken to be *only* one's own.

Bacon knows that as the learned serve to free the desires as if they were both common and limitless, they serve the likes of Themistocles and Alcibiades, another gifted turncoat, just as they open these possibilities to the many. Themistocles' understanding of great and small, the proper measure of empire, is "pertinent and just" when applied to "the general state of the question" but is arrogant and uncivil "being applied to himself out of his own mouth." Precisely because the new, learned empire strives to transcend the limits of political life, it embraces the likes of Themistocles and Alcibiades, who can be faulted as tyrannical in comparison to the moderate, contemplative Socrates. For this fact Bacon must apologize. Bacon's criticism of Themistocles, and so by implication his criticism of Alcibiades, is all too serious. However necessary Bacon thinks it is for the learned to serve the world by satisfying and justifying the tyrant, Bacon himself, unlike Machiavelli, never abandons the horizon of the moderate but not necessarily honorable Socrates.

We conclude our discussion of the essay by noting Bacon's comment that it is important for a nation to "profess arms as their principal honor, study, and occupation" if it is to "aspire to empire and greatness." This profession, says Bacon, is shown by the saying of Romulus who, as is either feigned or reported, "sent a present to the Romans that above all they should intend arms; and then they should prove the greatest empire of the world." Now, according to Plutarch, Romulus recommended not just arms but the combination of courage and self-restraint.[91] Plutarch remarks that such myths as the one about Romulus are important because, while it is impious and base to reject the divine source of virtue, it is foolish to mix heaven and earth. He says that the soul must be understood to be a divine image of life (*zoon*) that returns from the body to the gods, so that according to Heraclitus, a dry soul is best. Such myths are important be-

91. *Romulus* 28.

cause it is important not to violate nature by sending the bodies of the good with them to the heavens, but we must believe that their virtues and their souls ascend, according to nature and divine justice, from men to heroes, then to demigods, and finally after they are wholly free from death and sense, to gods, not according to the city's laws, but according to truth and right reason.[92]

For Plutarch, political life depends on myths that combine the fact of the soul's being the principle of a body that moves, so that the soul is what it is only as some possibility of sense, with the contradictory requirement of the *nomoi* for the perfect incorporeal divinity of the soul. Only imagelike myths or conventions can do so, and of course they must have the certainty of divine law. The laws, even as they are divine, ground the possibility of practical freedom. Such freedom requires a rank of freedom and submission, but this rank is never more than a difference between the good and the bad who are only more or less free from the body's harsh demands. The city's conventions require that neither the body nor the soul be wholly forgotten, for the sake of a rank that is always partly mythic or conventional. If charity subverts all conventional claims to rank, then the tension between body and soul is broken. And if only well-formed convention stands between the noble and the most bestial possibilities, then charity exacerbates the need for the self-restraint it subverts. This is ultimately no less true for the new charitable learning, because however much it can produce bounty, it can never reproduce Jesus' universal body. While it strives to, it can never overcome the tension between what is one's own, and private, and what is common, and public. Unlike Machiavelli, Bacon knows that his beginning is not simply new and not without risk. But however dark its true limits may be, there appears to be no alternative to pursuing a new universal empire. The essay ends with dark confirmation of the teaching about the darkest limits of empire. Bacon says that while Jesus warns that the body cannot be made to grow, the power of princes and estates can be made great with the counsels he has given.[93] But Jesus actually warns against concern for possessions. We cannot be fully hopeful about an empire that embodies Jesus' contradictory teaching, that at once condemns and then promises material possessions as if, like loaves and fishes, they can be shared without limit. The new empire cannot be likened to a tree, because it aspires rather to "treeness" that knows no natural limit. In such an empire individuals are likely to forget that human bodies are more like nature's trees.

92. Ibid., 28.6–8.
93. Matt. 6.27; Luke 12.25; *Essays* 452; cf. *New Atlantis* 147–56.

[280–282]

After detouring through the essay on empire, Bacon considers the question of the learned's manners to be completed, but not quite, for he does not intend "to give allowance to some conditions and courses base and unworthy, wherein diverse professors of learning have wronged themselves and gone too far." As examples of such "trencher philosophers," Bacon mentions the humiliation of Thesmopolis at the hands of the transvestite in Lucian's *De mercede conductis,* and, above all the rest, "the gross and palpable flattery whereunto many (not unlearned) have abased and abused their wits and pens, turning (as Du Bartas saith) Hecuba into Helena and Faustina into Lucretia." In addition to this, the customary dedication of books and writings to patrons is to be commended because the only patron should be "truth and reason." The old custom was to dedicate books to "private and equal friends," or "if to kings and great persons, it was to some such as the argument of the book was fit and proper for." These faults of manners cannot be defended because they "deserve rather reprehension."

In his brief conclusion, Bacon recapitulates the whole of his argument. In describing the learned, Bacon speaks first of "diverse professors of learning," then of "trencher philosophers," and then of "many not unlearned" as if there were no difference between philosophy and other kinds of learning. He is silent about the difference between philosophy and other kinds of learning, as Lucian is careful to note. According to Lucian, the philosopher suffers greater indignity when he chooses to become a hired companion, because he is thought to be no worthier than his fellows by his vulgar employer.[94] Bacon commits the very deed condemned by Lucian, but that he does so is not surprising, because the new learning does serve the great in their likeness to the vulgar, the vulgar in their likeness to the great, and both in their likeness to the learned themselves. Du Bartas warns against poets who use their gifts to write lascivious verse to seduce the passions of the young.[95] But the commensurability of the great, the many and the learned—the fact that learning can, as Bacon says here, have both a "price and an estimation"—collapses the noble and the base, so that it is not the poet-lovers but the learned

94. Lucian, *De mercede conductis,* trans. Harmon, Loeb Classical Library (London: Heinemann, 1960), 4.

95. Du Bartas, *Divine Weeks,* trans. J. Sylvester (Oxford: Clarendon Press, 1979), "The Second Day of the First Week" 1–28.

who are lovers, who in fact equate the beauty with the bitch and licentiousness with outraged innocence. Bacon's own dedication of his work to the king is proper because the argument is "fit and proper" for the king, but if so, then it is proper for a mind that is a universal tool that fits all things and equates the whole of knowledge with the whole of the productive arts.

Bacon's final remark about manners is to refuse to condemn "the morigeration or application of learned men to men in fortune"; he approves the tart obeisance of philosophy to the tyrant Dionysius[96] and the yielding of a learned man "to him that commanded thirty legions."[97] The examples of the learned men are Diogenes, Aristippus, and the rhetorician Favorinus, and Bacon attributes a remark to Diogenes that in fact belongs to Aristippus. The one who asked about philosophers and the rich was Dionysius, while the source of the tart answer was Aristippus, the pleasure-loving sophist whose teaching led to the views that the pleasures of the body are superior to those of mind and that theft, adultery, and sacrilege are allowable upon occasion because they are not by nature base.[98] Bacon is silent about the fact that Favorinus did not simply defer to Hadrian on a matter of usage but that he accounted Hadrian, who bestowed honor on musicians, tragedians, comedians, grammarians, and rhetoricians, the most learned of all men.[99] According to the new light of the new learning, the likeness of the great, the vulgar, and the learned unites the sophistical, licentious lover of pleasure and the ridiculous slave of necessity, both of whom are citizens of a world but never of a patria, and both of whom are in common bondage to the tyrant. Such application is outwardly base, but it is "submission to the occasion and not to the person." But the necessity of such occasion is universal and inescapable; in deference to it, the new learning wears the several but identical garbs of Aristippus, Diogenes, and Favorinus. Bacon knows that the new learning apes the soft forms of honor, for which reason it is blind to the forms of moderation that could restrain the harshest forms of honor.

Studies: Belief and the True Sources of Christian Faith

By this point, the core of Bacon's argument about the task set for learning is coming to light. But much more remains to be explained

96. Diogenes Laertius 2.69, 79. 97. *Historiae*, Spartianus, *De vita Hadriani* 15.
98. Diogenes Laertius 2.73–75, 99. 99. See n. 97 above.

in greater detail. We do not yet know precisely how Christian belief is charitable in the sense that Bacon understood from the ancients, we do not yet know the details of Bacon's understanding of the ancients' understanding of virtue and the science of politics grounded on it, and we do not know precisely how Christian charity leads to dangerous politics. The present section deals with this last matter, and in working it out, Bacon presents his penultimate critique of Machiavelli. Machiavelli missed the power of Christian egalitarianism because he misjudged the power of Christian belief. He did not understand just how people who believe only what they see are really like those who believe in invisible punishments and rewards. And again, we will see that he missed this point because he turned too quickly from an older teaching about the call of the gods.

[282–285]

In the present section, Bacon apologizes for the "errors and vanities which have intervened amongst the studies themselves of the learned." Bacon says that it is not his intention to "make any exact animadversion of the errors and impediments in the manners of learning which are more secret and remote from vulgar opinion." But we learned earlier that discussing learning itself from the standpoint of vulgar and untrue measure revealed the secret truth about learning's faults. These faults concern the kinds of belief to which learning is subject. Therefore, to separate what is sound from what deserves vulgar censure, we must remember that men "do scandalize and deprave that which retaineth state and virtue, by taking advantage upon that which is corrupt and degenerate," and we should take our bearings in this regard from the example of the "Heathens in the primitive church" who "used to blemish and taint the Christians with the faults and corruptions of heretics."

There are three vanities in studies "which by learning hath been most traduced." Things are vain that are either false or frivolous, on the one hand, or truthless or useless, on the other, and persons are thought vain who are either credulous or curious, with curiosity pertaining either to matter or to words. Corresponding to these three vanities, in "reason as well as experience," are the three "distempers" of learning: fantastical learning, continuous learning, and delicate learning, which spring, respectively, from vain imaginations, vain altercations, and vain affectations.

Bacon begins with the third distemper, which, according to his description in the sequel, must be the least malignant of the three. Martin Luther, governed by providence but yet knowing rationally "what a

147

province" he had undertaken against the pope and the Church and finding that he was alone and not aided by the "opinions of his own time," was forced to enlist the aid of ancient authors. For this reason, the ancient authors, who had for "a long time slept in libraries," came to be read and "revolved" and men came to make a "more exquisite travail" in the ancient languages. From such practice grew delight in the manner of style and phrase, which was intensified by the new religious partisans' enmity toward the schoolmen, who were generally "of the contrary part." While the new partisans' opinions were "primitive but seeming new," the schoolmen for the most part spoke for the Catholic Church of Rome. "The great labour then was with the people," and in order to win and persuade them, power of eloquence and variety of discourse came to be important for access to vulgar wits. In speaking of the people, Bacon refers to John 7.49, where the Pharisees reported that those sent to arrest Jesus, and who had never before been addressed as Jesus addressed them, were *execrabilis ista turba, quae non novit legem.*[100] Admiration of the ancients, hatred of the schoolmen, the exact study of languages, and the efficacy of preaching, "did bring in an affectionate study of eloquence and copie of speech," soon growing to an excess, so that style took precedence over "weight of matter, worth of subject, soundness of argument, life of invention and depth of judgement." From this excess sprang Sturmius, who was bound to Cicero and Hermogenes, Car of Cambridge, and Ascham, who almost deified Cicero and Demosthenes and who lured the young to study them. All this provoked the comic remark of the echo to the young man who admitted to having spent ten years in the study of Cicero. When told by the youth that he had spent ten years studying Cicero's Latin, the echo replied not in Latin but in Greek: *"one, Asine"*[101] However much the learning of the schools came to be despised as barbarous, the "whole inclination and bent of those times was rather towards copie than weight."

Bacon next remarks that although he has given an example of this distemper "of late times," it has been and will be *"secundum majus et minus*[102] in all times," a comment he chooses to make in exquisite Latin

100. "The wretched crowd that has not known the law" (Spedding's trans.), *Advancement* 283. In the *De augmentis,* this discussion is truncated, with reference to providence, the Church's traditions, the study of ancient authors, and new opinions being omitted. This statement too is omitted.

101. Bacon transliterates the Greek for "ass" (ὄνε) and then translates into Latin. Erasmus, *Opera omnia,* ed. Halkin, Bierlair, and Hoven (Amsterdam: North Holland), *Echo* 99–100.

102. Either more or less.

rather than in his own vernacular. This defect cannot but discredit learning "even with vulgar capacities." Also, it seems to Bacon that "Pygmalion's frenzy is a good emblem or portraiture of this vanity" because words are "but the images of matter," and unless they have "life of reason and invention, to fall in love with them is all one as to fall in love with a picture." It is not simply proper to condemn "sensible and plausible elocution" that can adorn the obscurities of philosophy, good examples of which are to be seen in Xenophon, Cicero, Seneca, Plutarch, and, to a degree, in Plato. Although it can hinder the inquisition into truth and progress into philosophy, such adornment as is to be seen in these authors is useful for the application of truth and philosophy to "civil occasions" of conference, counsel, persuasion, discourse, or the like. But the excess of this art is so contemptible that those who are "more severe and laborous" inquirers into truth will despise it as incapable of divinity, just as Hercules said of Adonis, *nil sacri est.*

The learned preference for delicacy can be traced to delight in ancient letters caused by the perturbational jealousy of the Christian sects. Bacon says in particular that this jealousy was caused by Luther's revolt against the Church of Rome, because the use of ancient literature against the schoolmen was a part of the conflict between Catholic and Protestant dogmas. But if we look carefully, it is clear that the vice is not simply Protestant, for the "Portugal bishop" is listed along with the dissenting scholars and Erasmus. Furthermore, Bacon says that this vice is akin to Pygmalion's frenzy. But according to Bacon's source Ovid, Pygmalion's love was hardly a frenzy, if it is measured by its happy outcome, for with the proper supplications to the gods, Pygmalion's beloved three-dimensional image was transformed into a living body.[103] Pygmalion's frenzy lasted as long as he desired an image, but when he received the gift of the gods, his desire was perfectly normal. Bacon does not tell us whether he thinks Pygmalion was frenzied before and after the gift from the gods. We must not forget that we are now discovering learning's secret vice; somehow it has to do with frenzied, or normal, hopes for impossible or real bodies—and somehow such hopes are related to schisms within Christian belief.

We have to pursue this beginning by noting peculiarities in Bacon's references to classical sources: Erasmus' echo criticizes the ten-

103. Ovid, *Metamorphoses*, trans. Miller, Loeb Classical Library (London: Heinemann, 1964), 10.243–97.

year study of Cicero by speaking in Greek, and in the course of discussing the distempers Bacon criticizes Seneca, a Latin author, on his own account and repeats the echo's censure of the practice of reading Cicero, but he makes no such remarks about the Greeks, Xenophon, Plutarch, and Plato, or about Aristotle, who suffers not from what he did himself but from the labors of others, namely, the Catholic schoolmen, who have made him their dictator. In censuring the Romans but not the Greeks, with the exception of Demosthenes who, along with Cicero, represents the "delicate and polished kind of learning" to which those corrupted by the distemper would allure the young, Bacon reminds us of the difference between Greece and Rome. As we know, this difference has to do with the difference between the classical utopian teaching about virtue and Christian universal charity. But far from obliterating the moral and political wisdom of ancient utopian thought, Christian charity caused the revival of that wisdom. The secret vice of the new learning concerns belief about possible and impossible bodies, and this vice requires a revival of the utopian as opposed to the Roman wisdom. Christian belief, which certainly has opinions about such bodies, has imprisoned Aristotle, the utopian wisdom's most practical spokesman. But it was not simply Christian belief that called for and imprisoned the old wisdom. Rather it was a schism within this belief. Moreover, Bacon suggests that such a revival may have happened more than once, for the ancient Church suffered from the same perturbations as the Church rent by Luther's province "against the Bishop of Rome." The secret vice of learning concerns the schisms within Christian belief about possible and impossible bodies and the relationship of such schismatic belief to the memory of prior times.

Now, Bacon discusses this subject in detail in Essay 58, "Of Vicissitude of Things," where the subject is how Christian belief is like or unlike religious belief in general with regard to preserving the memory of prior times. In the essay, Bacon remarks on Machiavelli's discussion of Christianity in *Discourses* 2.5, where Machiavelli compares Christianity and the ancients' religions and says that the Christians' need for the Latin tongue preserved the record of ancient, pagan deeds. Essay 58 is the last essay; it is also the last of the four essays in which Bacon refers explicitly to Machiavelli. As it turns out, the four essays in which Bacon refers to Machiavelli by name, Essays 13, 15, 39, and 58, form a whole that treats the nature of Christian belief; in order to understand the secret vice of learning, we have to examine Machiavelli's account of the unique power of Christian belief.

In Essay 58, Bacon disagrees with Machiavelli's claim that the jealousy of the sects "doth much extinguish the memory of things." Rather Bacon says that floods and earthquakes cause oblivion of past times.[104] Against Machiavelli's argument about Gregory the Great,[105] Bacon remarks that Gregory's zeal to "extinguish all heathen antiquities" did not have great effects and did not last long, "as it appeared in the succession of Sabinian, who did revive the former antiquities." Here Bacon confirms our earlier conjecture: Luther's revival of antiquity was not the first. Now, in disagreeing with Machiavelli, Bacon in fact confirms Machiavelli's subtle Averroistic argument in *Discourses* 2.5.[106] There Machiavelli shows that to explain the oblivion of former times is to defend the argument that the world is eternal and not created. Those who say that the world was created note that if it were eternal there ought to be memory dating back farther than the mere five thousand years men can remember. Machiavelli explains how such memory can be interrupted and so supports Averroes and Aristotle, who thought the world eternal, when he says that the Christians and the Romans were both unlike sects in general that wish to extirpate the religions they replace. Whereas the Romans might have destroyed the Tuscan religion but did not and in fact brought its goddess to Rome,[107] the Christians would have extirpated the religions they replaced but were unable to do so. Having to rely on unarmed, loving persuasion, they were forced to use the Latin language, which in fact preserved the memory of gentile deeds. Bacon's example of Sabinian shows that he agrees with Machiavelli about the possible eternity of the world, and therefore Bacon agrees with Machiavelli that Christian belief aims to comprehend an impossible beginning. Furthermore, Bacon agrees with Machiavelli about the Christian sect's need to remember preceding times. It remains to be seen whether he really disagrees with Machiavelli about sects and whether he really thinks the Christian sect differs from sects in general.

Machiavelli's intention was to disrupt the changes of the sects, to effect a true beginning, as it were, which required breaking the cycle of vigor and decline that afflicts all "mixed" or political bodies, of which sects are one kind. Now, according to Machiavelli in *Discourses* 2.5, natural catastrophes can be seen not only as a possible reason why the world is eternal while memory is short but also as gifts of nature that purge conventional or mixed bodies in the same way that nature purges her own simple bodies. Machiavelli says that the three heavenly causes of oblivion are pestilence, famine, or inundation, of

104. *Essays* 513.   105. *D* 2.5.   106. See Strauss, 175, 202–3.
107. *D* 1.12.

which the latter is more important because it is more universal and leaves only ignorant remnants of men behind. But if nature purges by extinguishing the old and making way for new beginnings, then new sects can imitate nature. Thus there are three ways of purging mixed bodies: by men in the changing of sects, which like floods exterminate the old and yet pave the way for extraordinary new beginnings; by heavenly causes or divine grace, as the Bible describes the greatest flood; or by natural disaster such as floods, which may be great but are never as great as the one that freed Noah for a new beginning. But Machiavelli equates changes from heaven with natural accident *(l' accidente)*, as if divine revenge is not to be counted on for the purgation of diseased mixed bodies, and the wise management of sects (and languages) must be imitations of nature's purges that are superior to nature precisely because they are *not* accidental. Machiavelli never denies the difference between natural or simple and conventional or "mixed" bodies. He says there is a difference between a simple body such as a natural living being and mixed bodies such as a "world or a human generation." The latter are larger wholes of which simple bodies are parts and which "human craft and malice" *(la astuzia e la malignità umana)* order. Nature's purges of natural, simple bodies require no art or malice, for which reason they can never be wholly appropriate models for mixed bodies, because nature's purges can never be as certain as human practice requires. Nature's simple bodies, human individuals, can be parts of conventional wholes, but conventional wholes are never satisfactorily ordered by nature. Furthermore, there is no individual perfection apart from mixed or political bodies, and the health of any political body depends upon the wise management of art and malice or contentious differences.

Now, because he insists upon the difference between simple and mixed bodies, between conventional political wholes that are different from their natural parts, it would seem that for Machiavelli the art of politics should not be reducible to a science of homogeneous matter in motion. But from what we have learned about his hope to justify the tyrant, we must doubt this appearance, even if Machiavelli does not. In supposing an economy of commensurable human desires, Machiavelli differs from Aristotle, who thinks that the natural perfection of an individual might order the artificial political whole. To be more precise, as a lover of honor, Machiavelli does not think that the political whole is ordered by an individual transpolitical whole or good. In Essay 58 Bacon agrees in part with Machiavelli by arguing that the "great winding sheets that bring all things in oblivion" are deluges and earthquakes but that, in the case of the West Indies, it is more likely that flood was the cause of the youth of their people.[108] In this case, the West Indians did not suffer the fate of the island of Atlantis as it was reported by the Egyptian priest to Solon. In the *New Atlantis*, Bacon follows Machiavelli in equating divine revenge with main accidents of time, contrasting the pious antiquity of the Egyptians with the scientifically guaranteed antiquity of the new, as opposed to the old, Atlantians.[109] In the *New Atlantis*, the description of the "feast of the Tirsan" shows the goodness of Bensalem to consist in the mere generation of human bodies: the elaborate account of rank and merit is subordinated to the virtue of mere fecundity as such, so that the difference between one who is and one who is not honored as a Tirsan is simply the difference between twenty-nine and thirty sheeplike descendants. Like Machiavelli's unintended teaching, Bacon presents the political whole as a collection of homogeneous, commensurable parts. Ultimately, then, like Machiavelli he likens "mixed bodies" to the simple bodies that are its parts.

The universal scientific conquest of natural accident or divine revenge requires that mixed bodies—the conventional or political things—be understood to be like any natural body. Bacon knows that, on Machiavelli's own grounds, it is impossible to preserve political distinctions by reference to honor alone or, in justifying the tyrant, by imitating nature's purges of simple bodies. Despite his intention to the contrary, Machiavelli's concern for an honorable common good forced his thought *beyond* the horizon of the differences that constitute political life. Machiavelli blurs the difference between vicissitudes of nature and vicissitudes of human things in his vain attempt to preserve the distinction between natural and conventional bodies. Machiavelli unwittingly looked beyond the boundaries of any patria no less than did the disciples of Christian love, and Bacon can now disclose the reason.

In imagining that the political whole can be ordered apart from the standard of an individual, transpolitical good, Machiavelli actually blurred the difference between the political whole and its standard. He therefore unknowingly imagined a political body as free from the limits of political life as Socrates, who, though he belonged to the city, could never be understood solely in terms of its possibilities. As measured by moral virtue, itself modeled on the political

108. *Essays* 513.
109. *New Atlantis* 141–47; see Introduction at nn. 23–25 above.

153

philosopher's freedom, the several conventional regimes could be ordered, grounding the differences between their corresponding kinds of citizens. Machiavelli imagined a political body in which the good ordering the common good does not preserve the difference between the free and the needy. But such a view takes all citizens to be the same and subverts the exclusive claims of every particular polity. For all his emphasis on craft and malice, Machiavelli confuses individual bodies, the larger artificial bodies that attract their motions, and the kind of being that orders the latter. For all his tough realism, Machiavelli is a believer in impossible bodies. His belief is ultimately not different from Paul's, and whatever the problems of Christian belief might be, we cannot trust Machiavelli to lay them bare skeptically or realistically.

Now, in the essay, Bacon first clearly distinguishes between the vicissitudes of nature and the vicissitudes of the human things, but then just as clearly he follows Machiavelli in blurring their difference.[110] Again, Bacon's new project follows Machiavelli's charitable lead in ministering to extravagant transpolitical hopes. But unlike Machiavelli, Bacon knows that with the tools bequeathed by Christian love, it will not be possible to overcome the danger of extravagant beliefs. Bacon argues more boldly than Machiavelli that the Christian sect preserves the memory of ancient deeds and virtue. But it does so not just because its love forced it to use Latin and Greek, as Machiavelli says. Rather we are forced to this memory because Christian belief strains the difference between natural and conventional bodies. In rejecting the ancient utopian teaching, Machiavelli simply did the work of the believing Christians, who, if they could not extirpate the ancient wisdom, did their best to imprison it in the web of Christian dogmatics. As we see in the rest of this section, Machiavelli misjudged the danger of Christian politics because, believing as he did, he could not have judged the true power of charitable belief.

In Essay 58 Bacon says that "the true religion is built upon the rock; the rest are tossed upon the waves of time." But in fact, the true religion was at its beginning rent by schism, and thus it was at its beginning rent by the "discord" that is one cause of the rise of a new sect. The other two causes of new sects are the decay and scandal of the "professors of religion" and the stupidity, ignorance, and barbarity of the times. When these causes are accompanied by the rise of an "extravagant and strange spirit" such as Mohammad, then the rise of a new sect is to be feared, but only if two conditions obtain:

110. *Essays* 512–14.

"the supplanting or opposing of authority established, which is always most popular, and the giving of license to pleasure and a voluptuous life." The best way to stop the rising of new sects is to reform abuses, to compound the smaller differences, to proceed mildly rather than with sanguinary persecutions, and "rather to take off the principal authors by winning and advancing them, than to enrage them by violence and bitterness."

The true religion does not come into being as a rock free from the possibility of vicissitudes, because it is marked by schism; but Bacon has already suggested that the Christian sect is more resilient than any sect and that while it is itself the source of vicissitudes, it is not itself subject to them, because its corruption does not cause its decay. The reform of the Christian sect merely strengthens the sect of changeable sects. And what is more, Bacon has suggested that the Christian soldier is in fact a lover of all the things that money can buy. The Christian contempt for the body and for honor is just the kind of honor that can revere "pleasure and a voluptuous life." The question is how this soft reverence can be the source of political perturbations. Nothing is more popular than the "supplanting of authority established," and as the rest of the Machiavelli essays show, the Christian sect more than any other serves this popular passion.

In Essay 13 Bacon refers to Machiavelli's bold criticism of Christianity in *Discourses* 2.2. Machiavelli complains that Christianity subverts the honor of the world, by which he means that it subverts the possibility of ferocious self-defense. Bacon refers to this argument as he praises philanthropy, or humanity, which, he says, "answers to the theological virtue of charity and admits no excess, but error."[111] We know already that Bacon regards just the opposite of Machiavelli's argument as true. The reason for its truth is the relationship between Christian charity and avarice, as Bacon indicates by a clever reference to Busbechius. We know, Bacon says, that when charity is not directed toward men, it is so strong that it will be directed toward other living creatures, as can be demonstrated by Busbechius' story about the "Christian boy," the cruel Turks, and the maltreatment of a duck. This reference proves to be a merry, if serious, joke, however, because the one Bacon calls the "Christian boy" is in fact a Christian goldsmith.[112] The problem of Christian belief is the ferocity of its self-defense, which can be traced to the close link between charity and avarice.

111. Ibid., 403–5.
112. Busbechius, *Turkish Letters*, trans. E. Forster (Oxford: Clarendon Press, 1968), 115.

In Essay 15, "Of Seditions and Troubles,"[113] Bacon warns that tempests in the state are greatest "when things grow to equality." To demonstrate the signs of such tempests he refers first to Virgil, who says that one sign is the "fame," and to Tacitus, from whom we learn that sure signs are offense taken at both bad and good actions and obedience that is ready to serve but is more inclined to interpret commands than to obey them. In actual fact, Virgil speaks of fames when, in bemoaning the state of Rome without the murdered Caesar, he says that they spurred the dallying Aeneas on his way. The fame was not simply bad, then, but it was when it was the sign of the murder of one who aspired to be emperor.[114] A closer examination of Tacitus' comment shows that the Romans were truculent because everything in the state was for sale. Moreover, Tacitus' remark about bad obedience follows his argument that the greed for power grew as the empire waxed great.[115] The real sign of tempests in the state is the ubiquity of avarice, which is the true legacy of empire, for avarice liberates the passion that Bacon says is always "most popular"— the "supplanting of authority established." To the extent that Christian charity is avaricious, then, it feeds the fires of political ambition. And as Bacon shows in the sequel, Christian conscience is the perfect vehicle for such ambition.

In the essay, Bacon refers next to Machiavelli's remarks about princes and parties in *Discourses* 3.27. Bacon notes Machiavelli's argument that all cities are by nature divided and that since modern weakness makes us judge ancient practice as inhuman and impossible, modern (Christian) weakness makes the ferocious governing of naturally divided cities impossible. Machiavelli's point is that Christian charity so enervates as to prevent the ferocity required to keep cities acquired by imperial conquest. Both Bacon and Machiavelli refer to French kings, but while Machiavelli speaks of Louis XII, who spoke (indirectly) of the harsh punishment of partisans, Bacon refers to Henry III as a king whose siding with the Catholic League led to his being assassinated by Friar Clement, the Catholic monk.[116] Bacon mentions Henry III indirectly in Essay 39, "Of Custom and Education,"[117] where he comments on Machiavelli's famous discussion of conspiracies in *Discourses* 3.6. Bacon argues that Machiavelli speaks well when he says that, to have a conspiracy executed, a person must not trust in the "fierceness of any man's nature, or his res-

---

113. *Essays* 406–12.   114. *Georgics* 1.465; *Aeneid* 4.179.
115. Tacitus, *Historiae* 1.7, 2.38–39.   116. *Essays* 408.   117. Ibid., 470–72.

olute undertakings, but take such an one as hath had his hands formerly in blood." But he says that Machiavelli is not correct when judged by Bacon's times, for he could not have known "a Friar Clement, nor a Ravillac, nor a Jaureguy, nor a Baltazar Gerard." Machiavelli's rule that nature and words are not as strong as custom "holdeth still," but not really, because now "superstition is so well advanced that men of the first blood are as firm as butchers by occupation."

Machiavelli's point is not just that one must be accustomed to blood and the sword to be a good executor of conspiracy; rather his point is that the successful assassin must overcome the confusion caused by conscience, which prompts assassins to cry out oaths of justification that give them away prematurely.[118] Furthermore, Machiavelli argues that conspiracies of one are not really conspiracies because they lack the complicated problem of managing other men's arms and consciences. Bacon tells us that the new assassins are either enthusiastic clerics or businessmen's clerks: Machiavelli did not know the likes of Friar Clement or Ravillac, who killed the tolerant Henry IV, for whom Paris was worth a mass, nor did he know the zealot or the merchant who both, for different reasons, executed the conscience of the Catholic Philip,[119] who warred against the tolerant but Protestant William of Orange. For Bacon, the problem of the sects is not that they enervate so as to weaken the management of natural parties. Rather, as we see here and in greater depth in the sequel, the problem of the sects is the unmanageable ferocity of Christian conspiracies, which are many, because Christian avarice persuades to equality. The war of sect against sect is always open, but the Christian assassin is always alone because he is not confused by conscience that reveres what is great or deserving in superior men. On the contrary, only *his* conscience is his guide, and conscience is always perfectly invisible.

In the rest of Essay 15, Bacon argues that when "the great ones in their own particular motion move violently and are as Tacitus expresseth it well *liberius quam ut imperantium meminissent*, it is a sign that orbs are out of frame."[120] The materials of sedition are extreme poverty and much discontentment, as Bacon shows by referring to

118. Mansfield, 334–43.

119. See John L. Motley, *The Rise of the Dutch Republic* (New York: Harper, 1864), III, pp. 538, 608.

120. *Essays* 408, "unrestrained by reverence for the government" (Spedding's trans.), *Annales* 3.4.

Lucan's description of Rome before the Civil War, where because estates were eaten up by usury, *"multis utile bellum."*[121] If this poverty is joined to want and necessity in the many, then "the danger is imminent and great." The motives of sedition are several, but of cures one must speak about some general preservatives: reducing want and poverty, which requires proper limits to population, a proper proportion of nobility to commons, and a proper number of clergy; wise management of foreign trade because "the increase of any estate must be upon the foreigner"; and proper distribution of wealth. For the removing of discontentments, Bacon says that it is important that the "nobless and the commonality" not both be discontented at the same time, for the danger is "when the greater sort do but wait for the troubling of the waters amongst the meaner." The people are as the poets describe Briareus, whose hundred hands saved Jupiter from the wrath of the rest of the gods.[122] It is better to give "moderate liberty" to griefs and discontentments "so that they may evaporate rather than turning them in upon themselves," and in the case of discontentments, "the part of Epimetheus ought well become Prometheus," for "Epimetheus, when griefs and evils flew abroad, at last shut the lid, and kept hope in the bottom of the vessel." Bacon argues that the "politic and artificial nourishing and entertaining of hopes, and carrying men from hopes to hopes is one of the best antidotes against the poison of discontents," which is "less hard" because "both particular persons and factions are apt enough to flatter themselves, or at least to brave that they believe not." It is also wise to see that there is no man of "greatness and reputation" to whom the discontented might turn, either by reconciling them to the state or by turning a part of the faction against them. It is important for princes to watch their tongues so as not to "give fire to sedition," as Caesar, Galba, and Prebus learned, and finally, princes should have near to them against all events "some great man or men of military valor who can repress seditions in their beginnings." Without such men, the state is endangered as Tacitus said, so that a few would attempt mischief, more desired such mischief, and all would allow it.[123] But these great men must be "assured, and well reputed of, rather than factious and popular; holding also good correspondence with the other great men in the state; or else the remedy is worse than the disease."

Here Bacon argues that faction can be managed if poverty and

121. "War a gain to many" (Spedding's trans.), Lucan, *De bello civili*, 1.181.
122. Homer, *Iliad* 1.398, cf. *Advancement* 345.
123. *Essays* 412; Tacitus, *Historiae* 1.28.

wealth, and number and proportion of the nobles and the vulgar many can be managed by princes who do not speak loosely, who can manage belief and hope, and who can rely on a great captain to nip discontents and factions in the bud. The prince must be able to manage and also to conspire against conspiracies, but as a Christian prince he must rely on one whose conscience can never be known by reference to prior experience of great and bloody deeds. Furthermore, Tacitus spoke not of motions of the great but rather of the people as they are ordered by tribes, and Lucan spoke not of much poverty but of the dislocations caused by imperial wealth.[124] When discussing the nature of discontentments, Bacon says that their danger is not to be measured by their justice and injustice, because this would be to judge people to be too reasonable, for they "do often spurn at their own good." They are not to be measured according to the size of the "griefs whereupon they rise," because fears are ever greater than actual feelings, an expression that Bacon takes from an epistle in which Pliny worries about his endless fear of a flood that will destroy all in its path, whether great or humble.[125]

By attending to Bacon's examples, we see that the causes of the vicissitudes of the sects and the causes of seditions are the same: the dislocations caused by the love of money, subversion of the difference between the noble and the vulgar, and the combination of fear and credulity. Furthermore, the wise silence of a prince, or his trust in a single or a few martial men, is confounded by the perfect silence of the conscience. And the population of lone Christian conspirators knows no bounds. According to Bacon, Christian belief, even reformed Christian belief, is no proof against sect and sedition. On the contrary, it makes sedition more ferocious and unmanageable because it serves the passions of men as if they are commensurable, so that every ambitious tyrant can see the justice of his hopes. In the light of Christian charity, men's passions are commensurable by reference to the gold that silent conscience shows to be the measure of every practical claim to virtue or grace. Conscience confounds the management of sedition because it frees men from awe as it causes them to become invisible.

To follow Bacon's argument about Christian belief further, we note his comment that in the case of discontentments, Epimetheus might become Prometheus, by which he means that hopes can be used to manage discontentments. Bacon makes this comment directly after

124. *Annales* 3.4; *De bello civili* 1.158–82.
125. Pliny (the Younger), *Epistulae* 8.17.

confusing Pallas and Thetis, as if he did not know the alternative story of Prometheus' release from his torment to the one he recounts and interprets in the *De sapientia veterum*. In the *De sapientia veterum*,[126] Prometheus represents providence and gives fire to man, the most composite and so most powerful being. This gift helps the arts and sciences and causes men to be wise and forethoughtful, removing evils and misfortunes. But because of men's caution, the gift also causes them to be tormented by solicitude and inward cares. Bacon says that Prometheus represents man's tendency to be puffed up with the arts and much knowledge, a vice leading to the attempt to ravish the divine Minerva. Prometheus was saved, not by the gods, but by Herculean "fortitude and constancy of mind," which, as Virgil says, according to Bacon, can conquer fear, fate, and death when we know the "causes of all that is."[127] This wisdom is as the sun; that is, it is Vulcan's fire, the very gift that Prometheus gave to man.

Epimetheus represents the improvident, who take no care for the future but think only of pleasure and who therefore suffer distresses, difficulties, and calamities but are yet amused with "many empty hopes." He releases the mischiefs and calamities of Pandora, who represents pleasure and sensual appetite kindled, as it were, by the gift of fire. But Prometheus is the one who "applied himself with all haste to the invention of fire, which in all human necessities and business is the great minister of relief and help." According to the *De sapientia veterum*, Epimetheus and Prometheus are in fact the same, except that Prometheus can replace Epimethius' vain hopes with concrete success, so that with his powers he need not pay for crimes that need not merely be attempted. But if Prometheus becomes Epimetheus, then the Promethean conquest of many evils and misfortune, those arising from the conquest of the gods, is accompanied by distress, difficulty, and calamity that can only be salved, not conquered, by many empty hopes. If Pallas rather than Thetis helped Jupiter, then Prometheus is never unchained,[128] and the mischief produced by fire springs from the common desire of the very greatest of the great and the vulgar multitude. Moreover, the mischief produced by fire affects the minds, bodies, and fortunes of individuals and also kingdoms and commonwealths, upon whom it springs wars, civil disturbances, and tyrannies.

126. *De sapientia* 745–53.
127. Virgil, *Georgics* 2.490–92.
128. Had Thetis been Jupiter's ally, he would not have feared Prometheus' secret knowledge about Thetis. There would have been no way for Prometheus to escape by revealing the secret, and as Bacon says, Hercules did not release him.

Bacon comments that the fable of Prometheus has "not a few things beneath which have a wonderful correspondency with the mysteries of the Christian faith," especially Hercules' voyage to free the enchained rapist. He will say nothing of this, however, lest he "bring strange fire to the alter of the Lord." Prometheus represents the new proud science, which follows Christian charity in equating princes and peoples, the greatest of the great and the vulgar multitude. But such a science requires the management of political discontentments with the "artificial nourishing of hopes, and carrying men from hopes to hopes." However, unlike Machiavelli, Bacon knows that to manage hopes well one must appreciate moderation, *not* the claims of honor, which are, in fact, always modes of hopeful belief. One must know the limits of managed hopes. And as we see in the final Machiavelli essay, Bacon knows this truth because he knows the difference between hopeful or honorable belief and right opinion, a distinction he understood from the ancient utopian wisdom.

At the beginning of Essay 58, Bacon refers at once to Ecclesiastes and to Plato's *Meno* to establish his point that "the river of Lethe runneth as well above the ground as below," a point boldly establishing Machiavelli's more hidden Averroistic thesis that the world has no beginning.[129] At the same time that he refers to Solomon, who would disclose the vanity of all things, Bacon refers to Plato's argument that virtue is not teachable if in fact it is knowledge and that virtue is in fact not knowledge but right opinion, which depends not on knowledge but rather on correct belief.[130] These two references considered together disclose the difference between hopeful or charitable belief and the ancient utopian understanding of correct belief or right opinion. For the ancients, belief could be correct only when properly separated from knowledge, in particular, from the knowledge that convention is the product of an art. This is just the knowledge to which every honor-loving productive art is tempted, the knowledge implicit in Solomon's charitable wisdom. Christian charity trumpets Solomon's knowledge because it debunks as mere pretense the freedom of the noble or the virtuous from any needful art. But Christian charity combines this knowledge with belief in the possibility of a universal freedom for the resurrected body. Unlike right opinion, charitable belief combines belief and knowledge. For the sake of a universe and not for any patria, it liberates the political delusions of the productive arts from Socrates' criticism and from the restraints of conventional virtue. It is no surprise, then, that the uni-

---

129. Eccles. 1.9; Plato, *Meno* 81a10–e2.   130. *Meno* 87b2–94e2, 98e7–100c2.

verse of charitable believers is in fact a collection of ferociously warring sects. To calm sectarian fervor, the new knowledge directs charitable belief to new hopes for the conquest of nature. However, such a project combines knowledge and belief no less than does Christian faith and for the sake of the same kind of perfected body. Therefore, as Bacon says regarding the first distemper of learning, it is always necessary to "clothe and adorn the obscurity of philosophy itself with sensible and plausible elocution," which, even if it might be bad for philosophy, is important if such knowledge is to have use in "civil occasions" or "the like."[131]

For knowledge to calm Christian practice, it must be bad for political philosophy that tells of the difference between belief and knowledge. In fact, civil occasions and the severe inquisition of truth are never simply compatible. When philosophy is moderately adorned, it makes no pretense to combine truth and civil occasions. But when such adornment is taken to excess, as if truth and civil life are wholly compatible, Bacon says that it is as Pygmalion's frenzy and, as Hercules said of Adonis, "capable of no divineness."

As told by Ovid, the story of Pygmalion contrasts with the story of Orpheus, of which it is a part.[132] According to Ovid, Pygmalion indeed loved the beautiful statue, disgusted as he was by the Propoetides, but he prayed to Venus that he might have a wife like the ivory maid; his timidity and his modesty prevented him from requesting the ivory maiden itself for his wife. The goddess granted him what he was too moderate to request. The story of Pygmalion is part of the song of Orpheus, sung to the animals, birds, stones, and trees after Orpheus' failure to bring Euridice back from the netherworld. The moderate Pygmalion was granted life in his lifeless beloved, but Orpheus lost the grant of life from death because of his impatience. While Pygmalion built and loved the statue because of his resolve to remain a bachelor, the Thracian women silenced Orpheus' harmonizing music and destroyed him after he introduced the custom of homosexual love to Thrace. According to the *De sapientia veterum,* Bacon says that Orpheus "may pass by an easy metaphore for philosophy personified," for the story of Orpheus "seems meant for a representation of universal philosophy."[133] To be more precise, the music of Orpheus in the netherworld represents natural philosophy, which "proposes to itself, as its noblest work of all, nothing less than the restitution and renovation of things corruptible, and (what is in-

---

131. *Advancement* 284.   132. Ovid, *Metamorphoses* 10.1–11.66.
133. *De sapientia* 720–22.

deed the same thing in a lower degree) the conservation of bodies in the state in which they are, and the retardation of dissolution and putrefaction." This task can be accomplished, if at all, only by "due and exquisite attempering and adjustment of parts in nature, as by the harmony and perfect modulation of a lyre." And because this task is of all things the most difficult, it fails "from no cause more than from curious and premature meddling and impatience."

Bacon never asserts openly that the task is impossible. But in the *De sapientia veterum*, Bacon does say explicitly that the music of Orpheus on the Thracian hilltop represents the application of philosophy to civil affairs, "subsequent to the diligent trial and final frustration of the experiment of restoring the dead body to life." Moreover, he says that this harmonizing music must have its "periods and closes" so that perturbations, sedition, and wars silence the laws and return men to depraved conditions and barbarism until philosophy can rise up again "according to the appointed vicissitudes of things." We can now make sense of Bacon's opening reference to Pygmalion's frenzy. Orpheus and Pygmalion are *both* immoderate when they aspire to give to the body the perfection of a crafted image—when they aspire to the noblest work of all, the renovation of the corruptible body. And this immoderation is no clearer than when their aspiration is applied to civil affairs. That is, the worst "periods and closes" always follow the new knowledge because it measures mixed political bodies by the apparent simplicity of the individual human body. The new knowledge is itself a kind of fervent belief, being capable of no more divinity than any god who cares for man and who promises to give more than it is moderate for any mortal to desire.

The mastery of Promethean fire does not extinguish the need for managed hopes even as it serves and redirects the energies of Christian hopes; such mastery never outstrips the dangerous possibilities of distorted hopes. In the "Prometheus" of *De sapientia veterum*, Bacon comments that "the body of man is of all existing things the most mixed and the most organic," so that it is more powerful than any simple body.[134] But according to the new natural science, such a mixed body is but a conglomerate of simple bodies, a more powerful simple body, not a composite of body and soul or form as is any real human being or political body. The new science requires an orientation by ultimately simple bodies. For this reason, the memory of gentile deeds must be revived in order to show that such bodies are impossible, not because they might not some day be made incorruptible, but be-

---

134. Ibid., 745–53.

cause no body can be so simple, so free from the need to measure what is and is not its own, that its civil affairs might be perfectly just. In fact, such hopeful belief ultimately inflames civil affairs. To manage extravagant hopes, it is necessary to understand how very difficult it is to escape them. The secret vice of learning is that in the modern age, knowledge must exploit the charitable, and schismatic, belief in impossible bodies. To this point, Bacon has shown why Christian belief is more unmanageable than Machiavelli thought it and why the new learning has to answer to it. To conclude the treatise on Christian belief, we can turn to his discussion of the third disease of learning, where he shows that Christian belief is in fact as he has explained it.

[287–290]

The third vice or disease of learning, concerning "deceit or untruth," is the worst of all because it "doth destroy the essential form of knowledge." There are two forms of this vice: delight in deceiving, or imposture, and aptness to be deceived, or credulity. Although they may seem to be different, "the one seeming to proceed of cunning, and the other of simplicity," in fact they "do for the most part concur," for *percontatorem fugito, nam garrulus idem est.*[135] Just as an inquisitive man is a prattler, so too a credulous man is a deceiver, which is the same as in "fames," where those who believe rumors also augment them. This point "Tacitus wisely noteth, when he saith, *Fingunt simul creduntque.*"[136] There are two forms of "facility of credit," depending on whether its subject is belief of history (matters of fact) or of art and opinion. The first error can be seen in ecclesiastical history, which has too easily "received and registered reports and narrations of miracles wrought by martyrs, hermits, or monks of the desert, and other holy men, and their relics, shrines, chapels, and images."[137] These reports are believed for a time because of the "ignorance of the people, the superstitious simplicity of some, and the politic toleration of others." After the passage of time, however, "as the mist began to clear up," they came to be "esteemed but as old wives' fables, impostures of the clergy, illusions of spirits, and badges of antichrist, to the great scandal and detriment to religion."

In natural history the same lack of "choice and judgement" can be seen. This appears in the writings of Plinius, Cardanus, Albertus, and

---

135. "Shun a questioner, for such is a chatterer," Horace, *Epistulae* 1.18.69.
136. "The tale fabricated and believed at the same time," Tacitus, *Annales* 5.10.
137. The rest of this argument is omitted from the *De augmentis*.

"diverse of the Arabians," which are "fraught with much fabulous matter" and which discredit natural philosophy. In contrast to these writers, Bacon praises Aristotle, who mingled his natural history "sparingly with any vain or feigned matter" and put "all prodigious narrations" into one separate book, whereby he separated truth from "matter of doubtful credit" and yet preserved the latter. There are two forms of credulity in matters of art and opinion: when too much belief is attributed to the art, or when too much is attributed to certain authors in any art. Examples of the former are astrology, natural magic, and alchemy, but Bacon notes that while the ends of these sciences are noble, their means are "full of error and vanity," which their professors have tried to hide with "enigmatical writings" and "auricular traditions." Some right is due to alchemy, because it can be compared to the husbandman of Aesop's fable, who lied to his sons about gold he buried in his vineyard but whose lie inspired them to dig so much that they "had a great vintage the year following." So the search to make gold has brought to light many useful inventions and experiments "as well for the disclosing of nature as for the use of man's life." The sciences have received "infinite damage" from the extreme credit given to authors in sciences, for such credit "hath kept them low, at a stay without growth or advancement." In the mechanical arts the "first deviser comes shortest, and time addeth and perfecteth." But in the sciences, "the first author goeth furthest, and time leeseth and corrupteth." Whereas artillery, sailing, and printing were at first rude and then perfected, the philosophies and sciences of "Aristotle, Plato, Democritus, Hippocrates, Euclides and Archimedes are vigorous at first but are then degenerated and debased." The reason is that "in the former many wits and industries have contributed in one; and in the latter many wits and industries have been spent about the wit of some one." Water will rise no higher than its source, and so "knowledge derived from Aristotle, and exempted from liberty of examination, will not rise again higher than the knowledge of Aristotle." So although it is true that *oportet discentem credere*, this rule must be coupled with *oportet edoctum judicare*,[138] for "disciples do owe unto masters only a temporary belief and a suspension of their own judgement until they be fully instructed, and not an absolute resignation or perpetual captivity." To conclude his point, Bacon says

138. *Advancement* 290: "a man who is learning must be content to believe what he is told; when he has learned it he must exercise his judgement and see whether it is worthy of belief" (Spedding's trans.); cf. Aristotle, *De sophisticis elenchis*, trans. Forster, Loeb Classical Library (London: Heinemann, 1955), 164a20–65a3.

that we should let great authors have their due so as to let time, the "author of authors," have its due, "which is further and further to discover truth."

In discussing the third disease of "deceit and untruth," Bacon distinguishes between credulity regarding fact or history and credulity regarding art and opinion; fact or history includes both ecclesiastical and natural history and art and opinion include both the arts and the sciences. But close inspection shows the distinctions to be blurred rather than clear or distinct. Bacon's two examples of credulity in matters of fact are ecclesiastical history, on the one hand, and natural history on the other hand. To irresponsible ecclesiastical history corresponds the irresponsible natural history of the writers he mentions and "diverse of the Arabians," and while he gives no example of good ecclesiastical history, he praises Aristotle's book, which separates "prodigious narrations" from truth. But Aristotle's treatise on prodigious narrations, *De mirabilibus auscultationibus*, mixes indifferently natural oddities and divine miracles as if the question of miracles were simply reducible to the question of natural oddities.[139] Bacon says that scandalous credulity has caused wives' tales, impostures, and illusions to be taken as miracles, implying that time eventually sees the light of truth cast upon these frauds so as to dispel the mist that surrounds them when they are close, as if, implausibly, new impostures are more believable than old ones. But in the present context, and in the later context of the division of the sciences, Bacon offers no standard by which a person might judge the veracity of miracles.

In the division, ecclesiastical history is divided into the history of the Church or the times of the militant Church, the history of prophecy, and the history of providence.[140] The history of prophecy concerns "prophecy and the accomplishment," showing how "every prophecy of the Scripture" can be sorted with the event fulfilling the same" but with "that latitude which is agreeable and familiar unto divine prophesies," which is to remember that for their author "a thousand years are but as one day." Now, if this temporal horizon be considered, it is by no means clear that prophecies are simply exhausted by the scripture. The third part of ecclesiastical history, the history of providence, considers "that excellent correspondence which

139. *De mirabilibus auscultationibus*, trans. Hett, Loeb Classical Library (London: Heinemann, 1936), 836a6–18, 836b12–29, 840a27–b18.
140. *Advancement* 340–42.

is between God's revealed will and His secret will." It is made evident to man by "notable events and examples of God's judgements, chastisements, deliverances and blessings." It is hard to think of a prophecy that is not the miraculous foresight of a sign of divine providence, such as a divine judgment, chastisement, deliverance, or blessing. Therefore Bacon forces us to the conclusion that it is impossible to separate the history of prophecy from the history of providential signs. But in the absence of a standard to differentiate miraculous signs and natural oddity or fraud, these very signs are the "scandal and detriment to religion" because they can include imposture, illusions, and false signs of the "badges of antichrist."

Only in the *New Atlantis* does Bacon suggest a criterion by which the veracity of miracles might be judged, but there he blurs the distinction between history and art and opinion. In the *New Atlantis* a fellow of Salomon's House, a practitioner of the new natural science, pronounces the miracle of revelation to be a true divine miracle, or a "great sign," rather than something of the "works of nature, works of art, and impostures and illusions of all sorts."[141] But of course while natural science might distinguish between regular natural phenomena, on the one hand, and natural oddities and miracles on the other hand, it could not with certainty distinguish between natural oddities and miracles, both of which are by definition unique. And if an artful cause of an apparent natural oddity cannot be found, this does not prove that such a cause does not exist. The force of Bacon's argument is now clear. All divine signs must be interpreted according to variable opinion concerning nature and art, and there is no standard apart from this variable opinion that might differentiate "true" signs of miracles and art, imposture, or natural oddity. Whether a sign is true or false, then, depends not on knowledge and not upon belief in the end of days. Rather it depends upon contentious opinion that is driven, but never ordered, by such belief.

Bacon refers to Tacitus to argue that the more strongly one believes, the more easily one will augment rumors, at the beginning of the discussion adding to the defect of credulity the vice of "imposture" or "delight in deceiving." Tacitus' remark occurs in the context of describing a political conspiracy based on the imposture of Drusus by the son of Marcus Silanus. Although the conspiracy was foiled by Sabinus, it was based on a tale believed by its makers as soon as it was concocted.[142] According to the quotation and Bacon's argument so far, Christian faith requires contentious opinion that can serve

141. *New Atlantis* 137–39.   142. *Annales* 5.10.

possible imposture, and the longer the end of days is postponed, the greater must be the possibility of deception that can be used for political ambition and conspiracy. The Christian faith is like and unlike sects in general. It is like them because its belief in impossible bodies is always schismatic, so that one interpretation of signs always contends with others. And as the unrestrained vehicle for political ambition, each sect is driven to extinguish the other. But Christian faith is unlike sects in general because, in being free from the statesman and moral virtue, it so recalls beginnings as to strengthen itself by its very sectarian strife. Moreover, Christian memory is egalitarian, causing Christian conscience to be invisible. Christian strife would be more intractable than that of other religions even if the statesman could apply his art to it. And if time is the author of authors, as Bacon claims at the end of the discussion, then one can never really dispel the mists that protect ambitious impostors.

Regarding credulity in matters of art and opinion, Bacon's complaint is that both arts and authors are given too much belief. The arts of astrology, natural magic, and alchemy have noble ends, but their means are erroneous and vain. The ancient authors have been given too much credit, retarding progress in the sciences as can be seen by comparing the progress in the "arts mechanical" with stagnation in philosophy and science as propounded by Aristotle, Plato, Democritus, Hippocrates, Euclides, and Archimedes. In this light, Bacon has praise for alchemy because it performs the function described in Aesop's fable. Bacon refers to alchemy and this fable again in *Novum organum* 1.85, where he argues not that the mechanical arts progress while the sciences have languished but rather that admiration for the works man has long possessed causes human industry to flag. He says this phenomenon is not merely spontaneous, however, because such stifling satisfaction began "when the rational and dogmatical sciences began." But in the present context, Bacon says that Aristotle must be known with "liberty of examination" in the same breath that he recommends sayings based on Aristotle's distinction between arguments from accepted opinion and arguments from opinion to contradiction.

There is little doubt that the wisdom of the ancient authors did not lead to progress in the mechanical arts. By now we know very well why: the ancient utopian wisdom culminated in the critique of the productive arts. But these arts must now be freed not because Aristotle's wit imprisoned them but rather because Aristotle's wit was imprisoned by Christian hope, freeing the sophistical knowledge Aristotle attacked in the passages from *De sophisticis elenchis* to which

Bacon refers. Aristotle's incarceration led to credulity, imposture, and contention that the old wisdom could no longer control. As becomes clear in the division of the sciences, Bacon elevates the productive arts to the model for all of the arts and sciences, for the sake of freeing the productive arts for the conquest of nature and divine providence. But the problem with the arts is not their imprisonment by the rational and dogmatic sciences, especially those of Plato and Aristotle; rather it is their need to attend to the charitable orientation by the explicitly productive arts. This is nothing but the promise that Christian belief in impossible bodies can be borne out in this world rather than in another one. Man's orientation by the noblest end of science requires no less fervent belief and hope than that ineffectually constrained by Christian humility. The conquest of nature may for a time constrain fervent belief more effectively than Christian humility, but the new orientation by the productive arts neither obviates nor clarifies the sectarian history of providence, whose mists are a cloak for imposture and political conspiracy. As the productive arts become the model for all the arts and sciences, Aristotle will be as imprisoned by the new science as much as he was by Christian dogmatics. But Aristotle's moderate wit is no less relevant to human affairs for its having been made useless.

[5]

Learning's Peccant Humors: Innovation and the True Greatness of the Ancients

Bacon thinks that the ancient wisdom was imprisoned by Christian dogmatics and that, for reasons that are clear by now, this incarceration required freeing the productive arts and sciences to do the work of the statesman and moral virtue. In the modern Christian era, the constant progress of science will moderate charitable politics. But Bacon learned the danger and limits of such progress from the imprisoned ancient teaching, from the ancient utopians' critique of the productive arts. Of course Bacon could not make what he had learned from the ancients explicit, contradicting as it did his criticisms of the ancients on behalf of the productive arts and sciences. In the present section, Bacon demonstrates the true greatness of the ancient wisdom, applying that wisdom to show the limits of progress and innovation. For this reason he does not conclude his apology with the discussion of the three diseases of learning. Rather, before turning to the oblation at first promised to the king, Bacon turns to some other "rather peccant humors than formed diseases," which nevertheless are not "so secret and intrinsic but that they fall under a popular observation and traducement." To this discussion we now must turn.

[290–295]

The first peccant humor, or error, is the extreme affecting of either antiquity or novelty, showing that the "children of time" are like "the father" who devoured his children, because the lovers of antiquity envy everything new and the lovers of novelty seek to "deface" all that has come before. Bacon recommends the advice of the prophet: *State super vias antiquas, et videte quaenam sit via recta et bona, et ambulate in ea.*[1] Men should stand on antiquity so as to find the best way for progress, and "to speak truly," *Antiquitas saeculi juventus mundi.*[2] The second error, following from the first, is "a distrust that anything should be now to be found out, which the world should have missed and passed over so long time." In this the same objection is made to time that Lucian made to the gods, wondering why they had had so many children in the remote past but had not had any recently.[3] Likewise, men "doubt lest time is become past children and generation," and yet men's judgment is inconstant, for after doubting that new things might be done, they "wonder again that it was no sooner done" when new things are accomplished. This inconstancy is demonstrated by the doubt of Alexander's expedition to Asia and Livy's remark *nil aliud quàm bene ausus vana contemnere,*[4] by "Columbus in the western navigation," and, in intellectual matters, by the much more common experience of doubt and certainty that may be "seen in most of the propositions of Euclid." The third error is also tied to the second, and so, then, to the first. This is the belief that "of former opinions or sects, after variety and examination, the best hath still prevailed and suppressed the rest." On the basis of the conceit, men think that new searches must light on "somewhat formerly rejected," as if "the multitude, or the wisest for the multitude's sake, were not ready to give passage rather to that which is popular and superficial than to that which is substantial and profound." The truth about time is that it seems to be like a river "which carrieth down to us that which is light and blown up, and sinketh and drowneth that which is weighty and solid."

The fourth and fifth errors are related as the fifth "doth succeed" the fourth, and so both, like the fourth, are of "a diverse nature" from the rest. The fourth error is peremptory reduction of knowledge "into arts and methods," for to have a science too early "comprehended into exact methods" is to stunt its growth like a child grown too early to maturity. The fifth error is "that after the distribution of particular arts and sciences, men have abandoned universality, or *philosophia prima;* which cannot but cease and stop all progression." It is not possible, he says, to discover the "remote and deeper parts

1. "Stand ye in the old ways and see which is the good way, and walk therein" (Spedding's trans.), *Advancement* 290–91; Jer. 6.16.
2. "Ancient times are the world's youth." Cf. *Novum organum* 1.84.
3. Lactantius, *Divinae institutiones,* trans. M. McDonald (Washington: Catholic University Press, 1964), 1.16.
4. "It was but taking courage to despise vain apprehensions" (Spedding's trans.), Livy 9.17.

of any science" if one stands upon the level of "the same science" and does not ascend "to a higher science."

The sixth error springs from an "adoration of the mind and understanding" for the sake of which men withdraw from the "contemplation of nature and the observations of experience" only to become entangled in their "own reason and conceits." Such "intellectualists" are justly censured by Heraclitus, who condemned them for seeking truth in "their own little worlds and not in the great and common world,"[5] because they seek to divine and give oracles by their own spirits and are, then, "deservedly deluded." The seventh error "hath some connection" with the sixth and is that men have infected their meditations, opinions, and doctrines with some conceits they have "most admired" or some sciences that they have "most applied." From this error Plato mixed his philosophy with theology, Aristotle mixed his with logic, and the "second school of Plato, Proclus and the rest" mixed theirs with mathematics. Likewise, the alchemists grounded their philosophy on "a few experiments of the furnace," and Gilbertius made a philosophy "out of the observations of the loadstone." This defect was mocked by Cicero when he said *hic ab arte sua non recessit,* and so forth, of the musician who said the soul was but a harmony, a matter about which Aristotle spoke "seriously and wisely" when he said *qui respiciunt ad pauca de facili pronunciant.*[6]

The eighth error is "an impatience of doubt, and haste to assertion without due and mature suspension of judgement." The two ways of contemplation are like the "two ways of action" mentioned by the ancients: "the one plain and smooth in the beginning, and in the end impassable; the other rough and troublesome in the entrance, but after a while fair and even."[7] The same obtains in contemplation, where a person who starts with certainties ends in doubts and someone who begins with doubt ends in certainties.

The ninth error is "magistral and peremptory" delivery of knowledge, designed to be believed more than to be examined. This error is not to be disallowed "in compendious treatises for action," and in "the true handling of knowledge" men ought not to be "like Velleius

5. Sextus Empiricus, *Adversus dogmaticos*, trans. Bury, Loeb Classical Library (London: Heinemann, 1935), 1.133.

6. "He was constant to his own art" (Spedding's trans.), Cicero, *Tusculanae disputationes*, rec. Pohlenz (Leipzig: Teubner, 1965), 1.10.20; "they who take only a few points into account find it easy to pronounce" (Spedding's trans.), Aristotle, *De generatione et corruptione*, ed. H. Joachim (Oxford: Clarendon Press, 1962), 316a5–14.

7. Xenophon, *Memorabilia* 2.1.20; Hesiod, *Opera et dies* 287–92.

the Epicurean, *nil tam metuens, quàm ne dubitare aliqua de re videretur,*[8] or like Socrates, who ironically doubted "all things." Rather men ought to "propound things sincerely," with as much asseveration as warranted by their individual judgment.

The tenth error is not one but several, comprising the errors that spring from the mistaken scope men propound to themselves "whereunto they bend their endeavors," for men ought to make additions to their sciences rather than "to aspire to certain second prizes," such as interpretation or commentary, championing or defending, compiling or abridging.

The eleventh and final error is the greatest one of all. It is "the mistaking or misplacing of the last or furthest end of knowledge." Men have desired learning and knowledge because of "natural curiosity and inquisitive appetite," or for entertainment of the mind with "variety and delight," or for "ornament and reputation," sometimes for "victory of wit and contradiction," and mostly for "lucre and profession" rather than for giving account of their "gift of reason to the benefit and use of men." Men have sought in knowledge "a couch, whereupon to rest a searching and restless spirit," or a "terrace, for a wandering and variable mind to walk up and down with a fair prospect," or a "tower of state, for a proud mind to raise itself upon," or a "fort or commanding ground, for strife and contention," or a "shop, for profit and sale," rather than a "rich storehouse, for the glory of the creator and the relief of man's estate." Knowledge will be dignified and exalted, however, "if contemplation and action be more nearly and straitly conjoined and united together than they have been, a conjunction like unto that of the two highest planets, Saturn, the planet of rest and contemplation, and Jupiter, the planet of civil society and action."[9] By action Bacon does not mean lucre and profession, which "diverteth and interrupteth" the progress of knowledge like the golden ball thrown before Atalanta.[10] Bacon does not intend to do what was "spoken of Socrates: to call philosophy down from heaven to converse upon the earth; that is, to leave natural philosophy aside, and to apply knowledge only to manners and policy."[11] Just as heaven and earth both contribute to "the use and

8. "Who feared nothing so much as the seeming to be in doubt about anything" (Spedding's trans.), Cicero, *De natura deorum,* ed. Ax (Leipzig: Teubner, 1961), 1.8.18.
9. Macrobius, *Commentarii in somnium Scipionis,* ed. Willis (Leipzig: Teubner, 1963), 1.12.
10. Ovid, *Metamorphoses* 10.667.
11. Cicero, *Tusculanae disputationes* 5.4.10.

benefit of man," so from both philosophies the solid and the fruitful should be separated from whatever is "empty and void" so that knowledge will not be used as a courtesan, "for pleasure and vanity only," or as a "bond-woman, to acquire and gain to her master's use," but rather as a "spouse, for generation, fruit, and comfort."

To clarify this complicated list, the following schematic outline should be helpful (indentions indicate related propositions):

1. Affecting antiquity or novelty
   2. Distrust that anything new can be discovered
      3. Conceit that the best of former opinions and sects have prevailed and suppressed the rest
4. Peremptory reduction of knowledge into arts and methods
5. Abandonment of universality or *philosophia prima*
6. Adoration of the mind and understanding in place of the contemplation of nature and experience
   7. Infecting meditations, opinions, and doctrines with the most admired conceits or favorite sciences
8. Impatience of doubt and haste to assertion (the two ways of thought and action)
9. Magestral and peremptory delivery of knowledge
10. Errors in the scope men propound to themselves—seeking mere second prizes
11. The greatest error: mistaking and misplacing the furthest end of knowledge, which is the benefit and use of man, the glory of the Creator, the closer uniting of contemplation and rest with civil society and action.

Bacon's opening praise of antiquity is equivocal. Although he says that the learned should stand upon the ancients for the sake of "progression" or novelty, the reference to Jeremiah suggests that God prefers those who walk in old ways. The reference to Jeremiah proves to be the clue to understanding Bacon's equivocation, not so much by itself, but because it points us to the argument in Essay 24, "Of Innovations," where the reference occurs again. Just as with the discussion of Christian belief, Bacon supplements his discussion with arguments from one of the essays. And in this case, he leads us from the relevant essay to remarks in the *Novum organum*.

In the essay Bacon argues that time itself is an innovator and that new things, or innovations, are honorable.[12] In other words, at some time all old ways must have been honorable innovations. And Bacon

12. Jer. 16; *Essays* 433–34.

says that innovations are like the infancy of living creatures, at first "ill shapen." According to the essay, the honor afforded to innovation guarantees that infantile innovations will be improved, because the honor due to innovation cannot be secured by imitating originals but only by additional innovation. But also according to the essay, innovations are "forced motions," stronger at first but not continuing as strongly as when they began. Bacon says that the reason is obvious: innovations are the "children of time," and according to the present context, these children always imitate Cronos "the father," who castrated his father, devoured all but one of his children, and was overthrown by the survivor. According to the essay, innovation "cannot be content to add but it must deface," and as we can see, if time itself is the greatest innovator, then however honorable they may be, innovations are never free from the harshness of forced motions. If antiquity is the world's youth, then the course of time may have defaced it. It is not necessarily true, then, that this infancy is inferior to its novel replacement. We must follow this track by considering related arguments in the *Novum organum*.

According to the *Novum organum* 1.129, the "judgement of former ages" shows that "the introduction of famous discoveries appears to hold by far the first place among human actions." The authors of inventions were awarded divine honors, while "those who did good service in the state" were given only heroic honors. Bacon says that this judgment of antiquity was just, because while the benefits of discoveries are universal, political benefits are particular in respect to time and place, and whereas the former are the vehicles for blessings and harmless benefits, the latter are mostly accomplished with violence and confusion. Springing from the development of the arts, the difference between civilized and barbarous men is enough to justify the saying that "man is a god to man." Bacon says that this divine difference separates even the youngest ancients, who correctly judged the superiority of the mechanical or productive arts, from the moderns, who have been blessed with the arts of printing, gunpowder, and the magnet, which have changed the "whole face and state of things throughout the world in literature, warfare, and navigation" and which have led to "innumerable changes," so that these three mechanical discoveries exhibit greater power and influence in human affairs than any empire, sect, or state. Accordingly, human ambition is graded by the difference between politics and the productive arts, for the ambition to "establish and extend the power and dominion of the human race itself over the universe" is more wholesome and nobler than either local or even imperial political ambition. And again, the empire of man over things depends on the "arts and sciences" or, as must be clear by now, on progress in the sciences as they are ordered by the productive arts.

If such discovery is divine, says Bacon, how much higher must the honor be for the discovery of "that by which all things else shall be discovered with ease." By this time we know what it means for Bacon to say that his method for discovery deserves honor. In fact, we discussed this very passage in connection with Bacon's critique of Machiavelli's orientation by honor. Therefore, Bacon's next remark is less surprising than it must surely appear at first glance. After praising a universal method for artful innovation, Bacon says that, "to speak the whole truth," it is still the case that "beholding the light is itself a more excellent and a fairer thing than all the uses of it," so that the contemplation of things is "in itself more worthy than all the fruit of inventions." This remarkable assertion of the superiority of theory to practice accords with the ancient utopian critique of the productive arts, and it accords with what Bacon has just told us about the harshness of honorable innovation. In Essay 24, Bacon argues that, "if time of course alter things to the worse, and wisdom and counsel shall not alter them to the better," then "it were good that men in their innovations would follow the example of time itself; which indeed innovateth greatly, but quietly, and by degrees scarce to be perceived."[13] Bacon says that when an innovation is a surprise, it brings ill with good, and he says that while innovations in the productive arts are universal, innovation is scarcely possible in political affairs. At best, then, innovation must be slow and quiet, so that innovations in the arts and sciences do not cause the need for nearly impossible innovation in states. Bacon's discovery of "that by which all things shall be discovered with ease," his new method, will accomplish what new modes and orders, new foundings and repetitions of foundings, in political affairs cannot. It seems that careful progress or innovation will replace Machiavelli's political science.

However, as Bacon has said, *all* innovation is harsh, forced motion. Even quiet progress cannot but remind of beginnings that are harsh; it cannot but repeat such harsh beginnings the more it strives to remodel them as perfect. Despite the modern need to depend on progress in innovation, such progress is not self-managing. And the ultimate perspective from which progress might be managed is the ancient utopian knowledge of the superiority of theory to practice, a truth Bacon suddenly discloses from behind its enigmatical veil.

13. *Essays* 433.

Therefore, in *Novum organum* 1.129, Bacon says that the superiority of theory to practice is the "whole truth." But since the opinion of this superiority is more ancient than modern, it is not necessarily true that the "youngest ancients," who preferred the productive arts, were superior to the later ancients, the classical utopians, who did not. In the division of the sciences, Bacon shows enigmatically that he understands theory in just the same way as did the later ancients. And in the present context, he has lifted his enigmatical veil only very briefly: in *Novum organum* 1.84, he says that reverence for antiquity is a sign that Greek philosophy and contemplation are presently in "a bad condition." In *Novum organum* 1.71, he says that, while the earliest Greek philosophers deserve praise despite their desire for sects and applause, the later Greek philosophers, including Plato and Aristotle, were in fact no different from the sophists they scorned, except that they did not take money for battling over sects and heresies. The younger Greek philosophers were forgotten because the later ones appealed more "to the capacity and tastes" of the vulgar and because time is "like a river, bringing down to us things which are light and puffed up, but letting weighty matters sink."[14]

But as if to echo his bold remark about the superiority of theory to practice, Bacon shows that he knows his attack on Plato and Aristotle to be an unavoidable slander. He does this by way of a clever reference to Diogenes Laertius. When Bacon refers to Diogenes to say that the doctrines of the later ancients were, as "Dionysius not unaptly rallied Plato, 'the talk of idle old men to ignorant youths,' " he is silent about what he knows Diogenes says was Plato's perfectly apt response. According to Diogenes, Plato simply replied that Dionysius was a tyrant.[15] And we know that Bacon agreed with Plato that it is impossible to justify the ambition of the tyrant.

We can see the truth about time and innovation. Time is the author of authors and the progenitor of innovations, preserving the light and the contentious because it is in thrall to popular applause. But it is not the philosophy of Plato and Aristotle that is light and contentious; rather their teaching is the sole perspective from which it can be understood how far violent innovation can be managed. It is true, as Bacon says in *Novum organum* 1.77, that the "inundation of barbarians" into the learning of the Roman Empire caused the philosophy of Plato and Aristotle to be "floated on the waves of time," while the earlier philosophies sank. But the real reason is that Rome

14. Spedding's trans., *BW* IV, 72–73; *Advancement* 292.
15. *Novum organum* 1.71; Diogenes Laertius 3.18; cf. *Advancement* 287.

served the love of artful innovation or the belief that art alone can answer to human need. Ultimately, the ambition of the arts caused Rome to embrace the barbarians into the inclusive, universal empire of Jesus. The truth is that the contentious Christian sects were produced by the force of artful ambition, and they imprisoned the weighty, antisophistical teachings of Plato and Aristotle, who disclosed the limits of all artful ambition. The belief in progress, the hope that artful innovation will provide perfect justice, is not new. What is new is that moral virtue and the statesman can no longer supply the principles to manage the inescapable ambitions of the productive arts. Such is the real power of novelty, which in the modern age requires that progress in the arts be used against itself, we suspect, according to the principles of Machiavelli's political science. For reasons that have become clear, recourse to these principles cannot always restrain the danger of innovation, assuming as Machiavelli does the very charms of art and the new. Bacon cannot be bewitched by artful progress because he knows too well the teachings of the later ancients. They taught him just how important and precarious is the place of the statesman and the moral virtues in any order of productive arts. As Bacon makes clear in the sequel, the later ancients taught him how likely it was that their teachings would be imprisoned. It was no mere accident, and so Bacon explains that time itself is the author of innovation.

The second error or peccant humor is the assumption that time will have exhausted the things that can be discovered, so that it "is become past children and generation." Men erroneously make the same judgment that Lucian made about the gods and the same depreciation of what was thought impossible that Livy voiced of Alexander's expedition into Asia. But Bacon falsely identifies the speaker to whom the church father Lactantius refers, and is silent about the real context of Livy's remark. As Lactantius tells the story, Seneca preserved a lost fragment, not Lucian, and whereas Lactantius is concerned to demonstrate that the many pagan gods do not exist but that the one, true Christian God does, Lucian, who speaks about Jupiter's amours even with mortal women, is concerned to show that the gods depend on men for their very succor and prestige. Whereas Bacon speaks only of the restraint of the "law Pappia," Lactantius reports that Seneca wondered whether Jupiter did not fear to suffer at the hands of his own offspring as his father Saturn (Cronos) suffered at his.[16] Again, Livy is concerned not merely to belittle Alex-

16. Lactantius, 1.16; Lucian, *Juppiter Tragoedus,* trans. Harmon, Loeb Classical Library (London: Heinemann, 1968), 1–4.

ander's conquest of Darius but to show that Alexander's remarkable fame, based upon his being a remarkable general, was protected by his early death, ensuring that his successes were not marred by the misfortunes that Livy implies would have followed from the Persianizing degeneration he had already begun to suffer.[17]

These examples, meant to show that time itself is the model for well-managed innovation, actually concern the corruption of princes who think themselves gods because the gods depend on them. The truth is that time is always more or less like Cronos, and in imitating time, the silent innovations of the productive arts are never wholly free from the violent confusions of the highest political ambition. The reason is that faith in innovation is the transpolitical hope of the productive arts and that such hope is inseparable from political life. And from any imaginable beginning to any imaginable end, mankind is the political animal. The ambition to "establish and extend the power and dominion of the human race over the universe"[18] is never free from the harshness of innovation as such. Bacon tells us that the course of time is always against the moderation of this ambition. Both charity and its cousin innovation have time on their sides.

According to the *Novum organum*, the received sciences are defective because, "of the five and twenty centuries over which the memory and learning of men extends," barely six can be shown to have been conducive to the development of the sciences.[19] There are only three periods of learning—the times of the Greeks, the Romans, and the nations of Western Europe. The Arabs and the schoolmen need not be mentioned, because they have more "crushed the sciences with a multitude of treatises, than increased their weight." The problem is that, during these few ages, the "least part of diligence was given to natural philosophy," and natural philosophy "ought to be esteemed the great mother of the sciences, for all the arts and sciences, if torn from this root, though they may be polished and shaped and made fit for use, yet they will hardly grow."[20] With the advent of Christianity, most of the best wits applied themselves to theology, so that theology has preoccupied the third age, the more so because "about the same time both literature began to flourish and religious controversies to spring up." The Romans applied themselves to moral philosophy and public affairs, and among the Greeks, when Socrates drew philosophy down from the heavens, he was merely making moral philosophy even more fashionable than it was among the early Greeks,

17. Livy 9.17–18.   18. *Novum organum* 1.129.   19. Ibid., 1.78.
20. Ibid., 1.79.

or "the Seven Wise men (all except Thales)." As a result the unity of the sciences, grounded as it is in the great mother of natural philosophy, was missed, so that all the particular sciences, no more nourished by natural philosophy (neither brought to nor brought from natural philosophy), "altogether lack profoundness and merely glide along the surface and variety of things."[21] The ultimate reason is that the "true and lawful goal" of the sciences—"that human life be endowed with new discoveries and powers"[22]—has been misplaced.

But we know that the truth about time is otherwise. The ultimate superiority of contemplation to practice belies the superiority of works to the concern for morals and politics. If contemplation is superior to practice or works, then universal method measured by works can never simply comprehend moral philosophy, from which alone can spring knowledge of the superiority of contemplation to practice or works. The course of history is determined by time and innovation, but for Bacon, truth and the force of history are not the same. Therefore in the modern age the unity and order of the arts and sciences cannot be determined by the true superiority of contemplation to action. Rather it is determined by the innovating force of history, which reverses the relationship between theory and practice and which is the real source of Christian providence. Bacon explains this point in greater detail in his discussions of the fourth through the "greatest and last" errors, or peccant humors.

In discussing the fourth and fifth errors, Bacon warns that knowledge does not increase if it is too quickly reduced to arts and exact methods, even though when so reduced it may be polished and accommodated to action. When knowledge so reduced is distributed into particular arts and sciences, men have then abandoned universality or *philosophia prima*, first philosophy, but Bacon notes that the deeper parts of science will not be discovered without an ascent from particular to universal science. The precise determination of first philosophy is not discussed until the division of the sciences in book 2,[23] and it remains to be seen how Bacon's understanding of first philosophy differs from that of Aristotle. But before we know anything of the nature or status of first philosophy, we can discern a problem in the relationship of first philosophy to the rest of the particular arts and sciences. The division or distribution of the sciences presupposes the determination of their particular boundaries, and it is governed by Bacon's concern to see contemplation and action "more nearly and straitly conjoined" than they have been. But while Bacon

21. Ibid., 1.80.   22. Ibid., 1.81.   23. *Advancement* 346–49; Chap. 8 below.

says that action requires fixing the boundaries of the arts and sciences,[24] he here makes it clear that such fixed boundaries inhibit the progressive ascent from the sciences to philosophia prima.

The division of the sciences presupposes differences in the ways things are investigated or known, but the ascent to first philosophy does not necessarily presuppose differences in the kinds of things investigated or known. If such differences do not exist, then either every object differs completely from every other object, so that there are no kinds, or every object is the same, so that there are no kinds. In both cases difference would spring only from artful procedure, and there would be no necessary link between procedure and end or object. But if there are no differences between the arts and sciences except chosen differences of procedure, then the very division of sciences for the purpose of separating the deficient from the complete cannot be possible, not to mention separating the possible from the vain. Bacon's apology presupposes the possibility of dividing the sciences. Likewise, the division presupposes the separation of the objects of the various arts and sciences according to the differences of various, but not infinite, kinds. And no wonder, for if all the arts and sciences differ only in regard to procedure, and not in regard to any kind or end, then any science can accomplish any end, and the one who knows this truth can be the master of all the possible arts and sciences by mastering the art of imitation or disguise. Instructed by the later ancients' wisdom, Bacon knows that this is just what the dangerous sophists claimed to do. But even so, when he discusses first philosophy in the division of the sciences, Bacon argues that for natural philosophy to be the mother to which all the other arts and sciences return and from which they spring,[25] first philosophy must be understood in the light of a science of simple material bodies. Clearly, then, first philosophy does not recognize differences in the kinds of things investigated or known, and as such, the mother of the arts and sciences dissolves the differences between the various arts. It would thus seem that, according to Bacon, the force of history requires first philosophy to serve the deceptive arts of mere appearance.

In discussing the sixth and seventh errors, which belong together, Bacon comments that reverence for man's mind and understanding has led to contempt for nature and experience and that men who succumb to this reverence are justly censured by Heraclitus. As reported by the skeptic Sextus Empericus, Heraclitus notes that the

24. Ibid., 328.   25. Ibid., 349, cf. 365.

whole is so constituted that it is revealed only by what is apparent to all in common rather than by what is apparent to any single prudence *(phronēsis)*. Heraclitus' concern is with the world as it appears alike in common or private reason and prudence *(logos/phronēsis)*.[26] But Bacon has the Heraclitean identity of reason and prudence stand for his distinction between the mind's concern for mind on the one hand and the mind's concern for nature and experience on the other hand. According to Heraclitus, there is no theoretical consideration of nature that is not also a consideration of reason and prudence, or practice. Aristotle argues that this perfect equation of reason and all the parts of the whole, or between being and sensuous motion, means that everything is true, so that nothing can be said to limit the absolute certainty of every dogma.[27] Being familiar with this argument, Bacon knows that if the overthrow of dogmatic schools depends upon a new, Heraclitean doctrine, or the perfect unity of the arts and sciences in philosophia prima, it is not certain that dogmatism can be overcome. For with such a doctrine, skepticism and dogmatism are but the opposite sides of a single coin.

Therefore, despite the pressing need to return to the mother of the arts and sciences, Bacon says in discussing the seventh error that it is important *not* to intermingle the arts and sciences in a manner that is utterly untrue and improper. He says that Plato, Aristotle, the second school of Plato, the alchemists, and Gilbertius have all mistaken the separateness of the arts and sciences. But despite indicting Plato and Aristotle, Bacon immediately appeals implicitly to Plato and explicitly to Aristotle as men who speak seriously and wisely about the proper separateness of the arts and sciences. Bacon makes this point first by referring to Cicero's condemnation of meddling musicians and second by referring to Aristotle's comment about people who speak easily after taking only a few things into account. When he refers to Cicero's remark, taken from the *Tusculanae disputationes*, Bacon is silent about Cicero's remark in the very same context that clearly shows the difference between the meddling poet and Plato. According to Cicero, the poet had, from his own admired conceit, discovered something about the soul that is both valuable and long before explained by Plato.[28]

Bacon's more explicit reference is to Aristotle's *De generatione et corruptione*, and we need to pay close attention to the context of this remark. In the *De generatione et corruptione*, Aristotle discusses the no-

26. Sextus Empiricus, *Adversus dogmaticos* 1.133.
27. *Metaphysics* 1010a1–b1, 1012a25–28.
28. Cicero, *Tusculanae disputationes* 1.10.20.

tions of coming-into-being and alteration.[29] Aristotle says that the ground or beginning *(archē)* of all the problems is the question of whether the first existences *(prōtōn huparchontōn)* are or are not indivisible magnitudes and, if they are, whether they are bodies or planes, as is said in Plato's *Timaeus*. He says that it is more reasonable to assume that they are indivisible bodies rather than planes, however problematic even this view is, because in the latter view it is impossible to explain alteration and genesis. Plato's error was to have considered only how passing away is essential to things and to have considered coming-into-being only in regard to elements and not in every regard. For this reason he considered neither how flesh and bones come to be nor how change and growth are in things. Aristotle says that the only philosopher to have thought seriously about these matters is Democritus and that the reason we do not consider the facts of the matter is want of experience.

Aristotle criticizes Plato or rather the likely tale of a cosmogony that Plato has Timaeus rather than Socrates relate, in contrast to his partial praise of Democritus, whose view that thought is sensation and sensation is change led him to deny that anything is true. According to Aristotle, Plato abstracted from experience and the experience of concrete, physical life. Plato consequently failed to present a demonstrable cosmogony that would explain how form is always the form of some body that is, in one way or another, alive and in motion. Now, Socrates does abstract in this way in the *Republic*, and Socrates' persistent quest for the unity of the sciences does seem to prevent his embracing the fullness of practical experience because, as it leads him to equate virtue and knowledge, it erodes the differences between the several moral virtues. And yet in the *Republic*, Socrates admits that his abstract account of the unifying *eidos* of the Good may well be a fraud because it is not a direct account of it but only an imagelike relative of the Good.[30] And without a direct account, there is no way to measure the veracity of the account Socrates does provide. Moreover, if the image of the Good that Socrates presents is appropriate, then just as the sun, which grounds the possibility of sight, must blind the eye when viewed, so the invisible *eidos* of the Good, which is the ground of all knowing and generation,[31] must blind the mind when it itself is known.

Socrates' account of the direct, noetic comprehension of the Good

29. *De generatione et corruptione* 315a26–16a14.
30. *Republic* 500d4–8, 504d4–e3, 507a1–5; cf. *Protagoras* 329b1–34c6, 349d2–51b2, 360e6–62a4.
31. *Republic* 509b2–c11.

has the same effect as Democritus' and Heraclitus' attempts to ground knowing and generation on what is only visible and changeable. In both cases, the differences between things, including the moral virtues, become unintelligible. A desire for the perfect unity of the sciences, or for a perfect account of the forms of forms, is analogous to the desire to reduce all form to motion or change. In the former case, every individual thing is the same, being of only one kind, and in the latter case, every individual thing is absolutely unique, being the member of no unifying kind. And in both cases, any single art could claim to be the knowing master of every possible art. The former dissolves the difference between form and matter into the purely formal, which is lifeless. The latter dissolves the difference between form and matter into the purely material, which is formless. The former would produce a cosmology putting an end to all that moves, a deathly cosmology, and the latter would produce a cosmogony of the visible whole making it impossible to distinguish between what moving things ought to consume for the sake of their motion. Dogmatism grounded on the supposed knowledge of the form of forms is not different from skepticism grounded on the supposed knowledge of matter in motion; both views obliterate the mysterious difference between matter and form in their attempt to ground a perfectly self-sufficient art.

Human beings are moved to knowledge because their response to need generates the need for justice and the moral virtues, but these are not always consistent. For instance, courage and justice, and justice and even the highest common good, do not always cohere. The love of justice springs from the need for artful mastery, with both causing men to desire to gaze at the Good or to know that everything is matter in motion. That is, both cause men to aspire to the perfect unity of theory and practice, which requires either the perfect unity or the perfect heterogeneity of the arts and sciences. Socrates knows that the desires for knowledge and for moral certainty are linked, just as philosophy and the arts are linked,[32] because they all begin with the homely demands of the body. Both the arts and philosphy are possible only where there are needs, the most familiar of which are the manifold needs of the body.[33]

The desire for moral certainty springs from the desire to make the body perfectly known and self-sufficient rather than immediately familiar—both formal dogmatism and materialistic skepticism spring from an immoderate but truth-loving orientation by the body. Soc-

32. See Plato, *Apology* 22c9–e5.   33. See *Statesman* 271c8–72d4.

rates made his speech in response to his interlocutors' extreme demands to know perfect justice as an art, and Socrates' *moderate* justice led him to give what he warned was a botched account of the Good and to let Timaeus give a botched cosmogony of the visible whole. Knowing the humble origins of his interlocutors' zeal for demonstrable formal knowledge, Socrates did not simply attack such theoretical ambition. Rather he presented a moderately skeptical account of form and the knowable because, as it points the gaze beyond the somatic and the movable, it is more removed from the immediate source of the desire for mastery and moral certainty that unites formal dogmatism and dogmatic, materialistic skepticism. However much Aristotle attacks Plato's inability to account for beginnings, he himself, like Plato's Socrates, provides cosmology rather than cosmogony. And Aristotle's argument that the cosmos is eternal and uncreated is no more successful than Socrates' account of the Good in accounting for the possibility that anything with a formal difference from other things might move.[34]

Both Aristotle and Plato speak from the perspective of form or the knowable despite Aristotle's complaint about Plato's concern for the unity of the arts and sciences.[35] And yet Bacon suggests that Plato and Aristotle are serious and wise correctives to the unity threatening the heterogeneity of the arts and sciences. Bacon's subtle approval of both Plato and Aristotle shows him to have known how the later ancients steered a moderate course between dogmatism and skepticism and between the unity and the heterogeneity of the arts and sciences. Socrates looks conspicuously for the unity of the moral virtues, on the ground that virtue must be knowledge, and yet against both Protagoras and Meno he argues that virtue is not teachable, for the sake of the laws that do teach the right opinions upon which the separate or distinct virtues rest. Socrates knows that if virtue is knowledge, or if virtue is a single virtue-producing art, then, for example, courage and knowledge are the same. And if this is accepted as a maxim of action, then nothing remains of courage as it is experienced in practical life. Likewise Aristotle explains the separate, distinct moral virtues for the sake of demonstrating their mere likeness to the virtue of the one who would always rather know than opine.[36] Aristotle learned what Plato would have us learn from the problem of Socrates: to save the city from those who would aspire to the sophists' artful mastery requires the political philosopher to re-

34. *Physics* 250b11–52b6.
35. *Nic. Eth.* 1096a11–97a14.   36. Ibid., 1177a12–79a32.

semble the sophists in knowing how fragile the moral virtues are as separate and independent possibilities. Both know that the virtues are fashioned by a single art. But unlike the sophists, Aristotle's cautious and practical political science defends right opinion or the ignorance upon which the several independent moral virtues rest. And likewise Socrates' irony at least partially obscured his depreciation of the moral virtues, of which he himself was the single form or model.

The true greatness of the ancients was their knowledge of the human condition: all human practice strives for and requires either the perfect unity or the perfect heterogeneity of the arts and sciences. Either will put an end to the conflicts between moral ends. But this unity or heterogeneity would dissolve the very moral phenomena about which practice so urgently needs to know. The political philosopher cannot but strive for such perfect unity or heterogeneity, but his self-knowledge tells him that he should no more promise it to others than he should expect it for himself. The political philosopher's self-knowledge requires his experience of wonder at the concrete, practical fact of knowledge and error, which requires the mysterious combination of unity and heterogeneity in the arts and sciences. Moderation, rather than perfect moral certainty or justice, is the fleeting appearance of truth.

For both Plato and Aristotle, right opinion can only imitate the self-sufficiency to which those who desire moral certainty aspire. The political philosopher is the true model of knowing freedom. But his freedom is moderate, because his extreme love of truth bows to his inability to penetrate the mystery of knowledge. The danger to the city springs from the very truth-loving but dogmatic desire for moral certainty upon which the city, never knowing right opinion for what it is, depends. In the city this danger is always present in the claims made on behalf of the various productive arts, each of which would be self-sufficient and would so state a claim to rule. On the basis of such a claim, each art would be the principle unifying the whole order of the arts and sciences. But even as such a principle, each art takes its bearings by a particular need, so that the desire for masterful certainty would unify and yet presuppose a perfect separateness of the several arts, if only to protect itself from the sophists' deceptive imitation. Dogmatism in regard to form springs from the hope for perfect unity of the arts, and dogmatic, skeptical materialism springs from the hope for perfect separateness of the arts, preoccupied as such hope is with a single kind of motion. And again, both dangers spring from the productive arts' concern to perfect the body they serve.

For Plato and Aristotle, the several moral virtues moderate the several claims of the arts. The virtues transform the arts' claims to self-sufficiency into the forgetting of needs or the pretense to artlessness, thus directing them away from an orientation according to the particular need that the particular art must serve. Courage, for example, is the moral virtue of the art of war, but as grounded on knowledge it points to philosophy, and as grounded on right opinion it points to a philosophylike freedom from the love of life and the need for victory. The same could be said for liberality as the moral virtue of the arts of money making. Even though for Plato moral virtue and right opinion collapse more easily into philosophy than they do for Aristotle, both Plato and Aristotle show that moral virtue is the partial but never perfect oblivion of the harsh, immediate need for every art to proceed as if its grounds and methods were absolutely, or certainly, secure. Bacon signals his subtle agreement with Plato and Aristotle when he warns in the eighth error against "impatience of doubt and haste to assertion."[37] He speaks here in general, not in regard either to the unity or to the heterogeneity of the arts. We are reminded not to forget the tension between perfect unity and perfect heterogeneity, a tension generated by productive arts as they hastily, that is, practically, go on their way.

When Bacon calls for the closer conjoining of theory and practice, he is well aware that the conjunction serves the delusions of the productive arts. Christian providence requires nothing different from what any city as an order of arts requires. To the extent that such providence demands a new harmony of theory and practice, it imitates the productive arts' desire for moral certainty at the same time that it undermines right opinion, which moderates the dangerous delusion of artful self-sufficiency. According to Christian dogma, men are never free from need because, as sinners, they are never free from their needy bodies, and no virtuous deed can ever embody a kind of artless freedom from need. Such a conventional or imitative freedom would be the prideful imitation of the gods, for whom the arts are never essential. But the denial of such conventional self-sufficiency leads to an exclusive orientation by the arts serving the body, so that Christian providence promises the resurrection of a body perfected by perfect arts, that is, of a body that has no limiting needs. Christian charity elevates the knowing certainty of the productive arts, or the arts as they honorably proceed on their way, over the forgetful, habitual self-sufficiency of virtuous deeds. Charitable humility in

37. *Advancement* 293.

fact liberates the impossible claim to merely useful self-sufficiency. We begin to learn how Christian conscience, like any art, can be money love and why money love is always tyrannical when it is perfected. If we are to understand why and how far theory and practice may in fact be conjoined, we have to understand the political wisdom and the true greatness of the later ancients, who themselves understood the difference between the virtuous and the honorable.

Bacon confirms his agreement with the later ancients when he argues that the two ways of contemplation are like the "two ways of action commonly spoken of by the ancients." He says that in contemplation, to begin with certainty is to end in doubt, and vice versa, and this is the same as action that, if it begins smoothly ends with impasse, whereas, conversely, if it begins with difficulty, it will end in smooth faring. The saying to which Bacon refers in this remark is reported in the conversation between Socrates and Aristippus as recounted by Xenophon.[38] There Socrates speaks not simply of ease and trouble but of the noble and virtue, which are first difficult and then easy, and vice, which is simply easy. For Socrates, the superiority of virtue to vice springs from the superiority of the good and the noble to the bad, not from the order of the easy and the hard. According to Bacon, then, when the ways of contemplation and action are taken to be alike, virtue is superior to vice because virtue is easier than vice. As contemplation and practice become alike, virtue and vice differ only as differing ways to the objects to which any deed is directed, so that virtue and vice differ as do art and chance.

In the context to which Bacon refers, Socrates and Aristippus discuss the possibility of freedom from politics and the possibility of freedom within political life. Against Aristippus Socrates denies the former and affirms the latter.[39] According to Hesiod, to whom Socrates refers in refuting Aristippus, virtue and difficulty go together because of the gods' displeasure at Prometheus' gift of fire.[40] But Socrates mentions only the difficulty and not the eventual ease of virtue, as does the poet. Bacon sides with Hesiod because Bacon speaks as if virtue might become easy, but what he really says is that, as virtue is equated with the ease to which an art pretends, the very distinction between virtue and vice disappears. And who claims to work with greater ease than the inspired poet? Bacon warns against impatience of doubt because he knows that the desire for the unity of theory and practice reflects the dangerous desire for practical, artful

38. *Memorabilia* 2.1.20.   39. Ibid., 2.1.7–20.
40. Hesiod, *Opera et dies* 40–295.

certainty. This certainty oscillates between the elements of practical dogmatism: the perfect integrity of the several, separate arts or the perfect unity of the several arts. Both of these practical dogmas culminate in Aristippus' view that political life is never free, and they spring just as much from Aristippus' dream that men can be free, by way of art, from the dependences of political life. The Christian orientation by the body and the arts repeats the error of Aristippus, which is that in debunking the possibility of virtuous political freedom, it looks beyond the scope of political life only to be open to the harshest political necessities.

As Bacon points out in discussing the ninth error, in treatises designed for practice one need not be ingenuous and faithful. But in the true handling of knowledge, one must avoid the fear of any doubt, as did Velleius the Epicurean, and one must also avoid the ironical doubting of all things. Of course Socrates did not doubt everything because he never doubted his own ignorance. His irony showed that to know this one thing is to know many other things. As Cicero reports Velleius' habitual, Epicurean certainty, it concerns his view that the gods are passionless and motionless and not to be feared and that the world was never in need of an artificer.[41] In *Novum organum* 1.67, Bacon identifies the lovers of certainty as the "philosophy of Aristotle" and then "the older sophists, Protagoras, Hippias and the rest," in mockery of whom Plato's school "introduced *Acatalepsia,* at first in jest and irony." The later New Academy, however, "made a dogma of it and held it as a tenet." Those who see no need for a divine artificer appear as the dogmatic lovers of certainty; they are the same as the sophists, who are by their special art the masters of every art. But we know that, according to the *Novum organum,* what begins with doubt can become dogmatic tenet; that is, doubt can be dogmatic. Furthermore, we know that Bacon's indictment of Aristotle's dogmatism is a slander necessitated by Aristotle's having been imprisoned by the dogmatic Christian schools. Of course Aristotle did doubt the need for a divine artificer, but he did not doubt the possibility of divinity, and Bacon knows that this sets Aristotle apart from the lovers of doubt and the lovers of certainty.

Bacon knows that the true difference between Aristotle and Plato, on the one hand, and the schools or Christian charity, on the other hand, is the difference between doubt that is open to the divine and certainty about the divine that is open to the hope for artful self-making. Moderate practice requires both Socratic, knowing doubt and

41. Cicero, *De natura deorum* 1.8.18–20.56.

Aristotelian, doubting certainty, both of which can appear neither ingenuous nor faithful. But practice according to Christian providence requires a standard of certainty that opens the way for replacing ingenuous, faithful, and certain belief in God's grace with belief in the certain progress of the productive arts. The new learning can take the place of Christian hope only because the latter is so much open to the former.

In discussing the tenth error, Bacon depreciates the individual who defends the old and praises people who always strive to add to what has come before. As we know, time measured by both the old and the new end of days proceeds by innovation, always taking after the nature and malice of the father who would devour his children. The saving grace of the Christian world is its own openness to the new science of nature. Devotion to progress—to the reformation of knowledge toward its "furthest end," the union of contemplation and action—can dampen the violence of the sects. But Bacon is himself no convinced progressive because he knows that, like its Christian roots, the new knowledge looks past the conventional constraints of man's political life only to be speechless in the face of the harshest political practice. As he shows in the next chapter, our modern devotion to progress is required by the charitable god, whose malicious vanity always sets the order of the arts in motion.

[6]

The True Dignity of Knowledge:
The Call and the Way of the Creator God

By the end of his apology, we know that Bacon is moved to his new project for the productive arts by the historical force of Christian charity. But an interesting question still remains. On the one hand, the Christian doctrine of sin takes the conventional imitation of freedom to be a vainglorious imitation of God, who needs no art. But on the other hand, Bacon has earlier suggested that art and reason are drawn together by knowledge only when men are open to the gods. Christian doctrine explains this by the account of original sin. But why did God create a being free to sin? According to Bacon's argument so far, the divine gift of the arts to man is really a divine revenge. It seems that Christianity, more than any other religion, knows this fact. But Christianity seems to ignore the possibility that such revenge may be unjust.

In the present chapter, Bacon shows that Christianity more vividly than any other religion reveals that human beings are always in the grip of a malicious god who created them in his own image. But Christianity does not understand this fact, thinking God to be one who could never be angry at the frustration of some need or art. This failure of self-understanding is not to be quickly despised. But it does explain how Christianity is charitable as Bacon understands it. In the present chapter, Bacon explains Christian scripture to show that its God is quite different from what Christian doctrine takes it to be. Bacon has yet to "weigh justly" the dignity of knowledge. To do so, he will not speak for himself. Rather he will let "divine and human testimonies" speak on behalf of knowledge. They show that God unjustly requires men to need justice that can never be perfect.

The Divine Testimony: Bacon's Divine Justice

[295–301]

Bacon first seeks the dignity of knowledge in the "arch-type or first platform," the attributes and the acts of God. God's knowledge is different from merely human learning because learning must be acquired, whereas "knowledge in God is original" and is better called wisdom, as the scriptures do.[1] The creation shows two of God's virtues: power and wisdom. God's power is demonstrated in the making of matter and his wisdom is shown "in disposing the beauty of the form." Matter was created in a moment, but the order and disposition of matter into form took six days, a difference God chose to "put upon the works of power and the works of wisdom." Accordingly the Bible reports that in his work of power God did not say "let there be heaven and earth," as it is said of the "works following," that is, of the works of wisdom, but rather God "made heaven and earth," so that the works of power carry the "style of manufacture" and the works of wisdom carry the style of a "law, decree, or counsel."[2]

After saying that the testimony of angels shows that knowledge and illumination are placed above office and domination,[3] Bacon moves from "spirits and intellectual forms" to "sensible and material forms," a reconsideration of God's testimony by reference not to God's agency but to its products. The "first form" created was light, "which hath a relation and correspondence in nature and corporeal things, to knowledge in spirits and incorporeal things." Likewise, in distributing days, the day on which God "did rest and contemplate his own works was blessed above all the days wherein he did effect and accomplish them." After the creation was finished, the Bible tells us that man was placed in the garden to work, but his work "could be no other than work of contemplation, that is, when the end of the work is but for exercise and experiment, not for necessity." Since at first man did not have to sweat, his work must have been for "delight in experiment" rather than "labor for the use." Therefore man's first acts in the garden were the two "summary parts of knowledge," the "view of creatures" and "the imposition of names."[4] The knowledge that caused the fall was not the "natural knowledge of creations" but the "moral knowledge of good and evil," as Bacon says he "has touched before."[5] The supposition was, says Bacon, that God's command-

---

1. See Prov. 8.22–31.   2. Gen. 1–2.9.
3. Dionysius, *De caelesti hierarchia*, ed. P. Hendrix (Leiden: Brill, 1959), 6–9.
4. Gen. 2.18–20.   5. *Advancement* 264–65; Chap. 2 at nn. 1–8 above.

ments or prohibitions were not the causes of good and evil but that they had "other beginnings, which man aspired to know, to the end to make a total defection from God, and to depend wholly upon himself." After the fall, the image of the "two states, the contemplative state and the active state," is presented in "the two persons of Abel and Cain" and in the "two simplest and most primitive trades of life," the shepherd, who because of his leisure represents the contemplative life, and the husbandman, who represents the active life.[6] Here again, the dignity of knowledge is demonstrated by the fact that "the favor and election of God went to the shepherd, and not to the tiller of the ground." That is, God's favor was shown to the contemplative rather than to the active life. In the age before the flood, the Bible mentions and honors the "name of inventors and authors of works in music and works in metal," and in the age after the flood, the first judgment of God upon human ambition was "the confusion of tongues; whereby the open trade and intercourse of learning and knowledge was chiefly imbarred."[7]

Bacon then descends to "Moses the lawgiver," said by the Bible to have been "seen in all of the learning of the Egyptians,"[8] which was one of the most ancient learnings, as Plato teaches when he has the Egyptian priest refer to the perpetual infancy of the Greeks, who have "no knowledge of antiquity and no antiquity of knowledge."[9] Moses' ceremonial law has been "travelled profitably" for natural and moral sense by some of the most learned rabbis, as in the "law of leprosy,"[10] from which one sees a principle of nature and another sees a "position of moral philosophy." In this and in many other places in that law is to be found "much aspersion of philosophy" in addition to the theological sense. The same thing can be said for the book of Job, which is "swelling with natural philosophy."[11]

In the person of King Solomon we can see the gift of learning, both in his "petition and in God's assent thereto," preferred "before all other terrene and temporal felicity."[12] God's grant to Solomon enabled him to write about divine and moral philosophy and also to write his natural history. Solomon claimed none of the glories in which he excelled except the "glory of inquisition of truth," for he said that "the glory of God is to conceal a thing, but the glory of the king is to find it out,"[13] as if God delighted in the hiding and as if there were no greater honor for a king than to be "God's playfellows in

6. Gen. 4.2.   7. Gen. 4.21, 22, 11.1–9.   8. *Acts* 7.22.
9. *Timaeus* 22b3–23d1.   10. Lev. 13.12, 13.
11. Job 9.9, 10.10, 26.7, 26.13, 28.1, 2, 38.31.   12. 1 Kings 3.5–14.
13. Prov. 25.2.

that game, considering the great commandment of wits and means, whereby nothing needeth to be hidden from them."

God's preference for learning did not abate with the coming of Jesus, who showed his power to subdue ignorance before he subdued nature by miracles,[14] and the coming of the Holy Spirit "was chiefly figured and expressed in the similitude and gift of tongues, which are but *vehicula scientiae.*"[15] In order to declare his "immediate working" more evidently and to "abase all human wisdom and knowledge," God first used unlearned persons for planting the faith. These persons had only the knowledge supplied by inspiration. But no sooner had he done so than "in the next vicissitude and succession" he sent his truth to the world "waited on with other learnings as with servants or handmaids," so that we see most of the New Testament written by the learned Paul. Many ancient bishops were well versed in heathen learning, so much so that Julianus' prohibiting of the admission of Christians to schools was considered worse for the faith than the blood persecutions practiced by his predecessors. Gregory's attempt to "obliterate and extinguish the memory of heathen antiquity and authors" came to be censured, and in fact, the Christian Church saved the relics of heathen learning from the Scythians and the Saracens.

In recent times, when God called the Roman Church to account for its degeneracy and abuses, it was ordained by divine providence that there should be a renovation "of all other knowledges," and even "on the other side" the Jesuits have "much quickened and strengthened the state of learning."[16] To conclude the divine proofs, Bacon says that there are two duties and services that philosophy and learning must perform for faith and religion "beside ornament and illustration." First, they are "an effectual inducement to the exaltation of the glory of God." The Psalms and other scriptures invite us to "consider and magnify" God's works,[17] and we injure God's majesty if we should only rest in contemplating "the exterior of them as they first offer themselves to our senses." Second, they help and preserve against "unbelief and error." The Savior says "you err, not knowing the scriptures, nor the power of God,"[18] showing that God lays before us two books that must be studied if we are to be preserved from error. The first is scripture, revealing God's will, and the second is the creatures, expressing God's power. The book of God's power is a key to the book of his will, not just for conceiving the true sense

14. Luke 2.46.   15. "Carriers of knowledge" (Spedding's trans.), Acts. 2.
16. This argument is omitted from the *De augmentis.*   17. Ps. 19, 104.
18. Matt. 22.29.

of the scriptures, but for "opening our belief" in "drawing us into a due meditation of the omnipotency of God, which is chiefly signed and graven upon his works."

At the outset of the discussion, Bacon says that the creation of God's formal creatures represents his wisdom, while the creation of matter represents his power. Furthermore, Bacon says that the superiority of God's wisdom to his power is shown by the fact that while the creation of matter was instantaneous, the creation of forms took six days. But if so, then it is surely perplexing that God's wisdom is not like knowledge because his wisdom is "original and not acquired," so that it has no beginning, middle, and end, which stands to reason because God's omnipotence would seem to rule out his ever being in need of learning, experience, or time. We wonder how divine wisdom can be superior to divine power if such wisdom takes time just as knowledge does. If God's will is wise, then it may be, as Bacon in fact says, that God's inferior power is the key to knowledge of his will. This suspicion is fortified when we note that at the end of the discussion Bacon flatly contradicts his opening distinction by saying that the book of God's formal creatures, nature, expresses God's power and that the scripture reveals God's will. Wisdom and power are interchangeable, and the wise and powerful creation of the articulated world, as opposed to the merely powerful, instantaneous creation of matter ex nihilo, takes time. Furthermore, the articulated world is governed by divine will, which also takes time: the scripture demonstrates divine providence occurring over time, which makes perfect sense if God's will is also wise.

How should we understand Bacon's opening distinction between divine power, wisdom, and will and his subsequent denial of this distinction? By confusing them, Bacon suggests that divine power, wisdom, and providence are somehow the same, and again, we wonder how it is possible, if they are the same, that they can be shown as a whole as having a beginning, a middle, and an end. We get a clue to Bacon's intention here if we think about the nature of divine providence. Surely providence is temporal because mortal men need God's providence, not because God needs it. But then, how can providence, will, and wisdom be akin to divine power, especially if God's wisdom is superior to his power because wisdom takes time? We are asked to believe that such wisdom is both temporal and yet "original and not acquired."

We pursue an answer to our question by noting that, according to

Bacon, God's works of power carry the style of manufacture, while his works of wisdom carry the style of law, decree, or counsel, again demonstrating God's view of the superiority of wisdom to power. According to Bacon, this distinction is shown by the Bible's report that God did not say "let there be" when he created the matter of heaven and earth but that he "made" them and that he did use such language when he created the works following heaven and earth. In fact, however, the Bible is not so clear about this distinction. God is said to have *created* heaven and earth and is said to have said "let there be light," so that light occurred with no reference to making except for the separating of the light from the darkness. But God is also said to have *made* the firmament and to have separated the waters above and below after saying "let there be a firmament," so that the upper waters became the heavens.[19] God not only said "let there be" as he did with the light, which appeared spontaneously, with no mention of making, but, it is said, also made the firmament, which separated the heaven. Again, while God said "let there be," and then it "was so" for dry land, vegetation, plants and fruits, and lights in the firmament, he is said to have made the sun and the moon. God said to let the waters bring forth the living creatures, and yet he made the sea monsters and every living creature that moves in the water and in the air, with the same being true for the kinds of creatures of the earth.[20] Of man God said let *us* make, and God is then said to have made man in his image.[21] Of course in Genesis 2, the order of the creation is revised, and God is said to have *fashioned* man from the dust and also the beasts of the fields and the air to serve as man's helpers before creating woman, while he is said to have caused the edible trees to sprout from the earth.[22]

Close inspection shows that the Bible does not support Bacon's distinction between divine power and divine wisdom. Bacon tells us subtly that in contradicting this very distinction himself, he comes to agree with the teaching of the scripture. But why, then, does he present the distinction in the first place? The answer is to be found in the distinction's implied difference between the "style of manufac-

---

19. Gen. 1.1–8. In Genesis 1.1 God's creating is expressed by *bara*, which means "to create" and also "to fashion or shape"; in Genesis 1.7, God is said to have *made* (*ya'ash*) the firmament. As Bacon explains it, the Bible does not support his distinction, and as becomes clear, Bacon knows perfectly well that it does not.

20. Gen. 1.9–18, 20–25. For the sun and the moon, the Bible says that God *made* (*ya'ash*) them; for the sea monsters, etc., the Bible says that God created or fashioned (*bara*) them.

21. Ibid., 1.26, 27 (*bara*).

22. Ibid., 2.7, 9, 19. Here the verb is *yatzar*, "to form or to fashion as a potter."

ture," on the one hand, and "style of law, decree or counsel," on the other hand. The former is like needy art, and the latter is like free reason. Bacon's argument is that, according to the Bible, wisdom and power must but cannot be the same: were they not the same, God would have no omnipotent will that appears in the providence governing his creatures; but insofar as they are the same, the whole of God's creation has the style of a manufacture, which is never instantaneous and which has the need for a beginning, a middle, and an end. The implication is that, for a wise creator god, such wisdom is like the knowledge of any art, and such a god must have needed to create what he created, even if his creation of matter were ex nihilo. For Bacon, the proper subject of God's works reveals his will to have been constrained. Consequently the books of creatures and the scriptures reveal providence to be the necessity governing all artifice or manufacture. Somehow God and men are equal in the course of providence, and it remains to be seen what problem or need causes providence to move as it does.

According to Bacon's comment on the distribution of days, God does not just demonstrate the superiority of knowledge to manufacture and rule; quite specifically, his being and acts demonstrate the superiority of "rest and contemplation" to his own works. This superiority, Bacon says, is demonstrated after the creation by the work God gave to man in the garden, contemplation ungoverned by any necessity, and even after the fall, God showed his preference for the contemplative Cain to the active Abel. But Bacon cannot possibly be right on this point, forcing us to conclude that, if we accept his terms, in Cain's eyes God's preference for Abel must have been for Abel's mere pretense to rest and contemplation. Bacon correctly tells us that after man's attempt to fathom the "moral knowledge of good and evil," all men were condemned to "labor for the use." Cain would seem to be correct to have despised one who was honored for being what no postlapsarian man could be and to have despised the one who conferred such honor. And God's judgment of Cain would seem to have recognized the justice of Cain's anger, because not only was Cain spared the whole of the punishment, which was too great for him to bear, so that he was spared the payment of full recompense for his brother's life, but his offspring were those who, before the flood, were the inventors and authors of music and works in metal who have been honored in the "holy records."[23]

Bacon's bizarre use of the example of Cain and Abel focuses our

23. Ibid., 4.11–23.

attention on the condition of man before the fall. The Bible is not clear about the time of man's appearance in relation to the rest of the creation; as has been noted, according to Genesis 2, man was created before the beautiful and edible fruits and before the beasts of field and air and woman were created to be his helpers. The story of Genesis 1 would appear to be more plausible, accounting as it does for the creatures of the waters. But even so there is really no difference between Genesis 1 and 2. According to Genesis 2, the creation of all the things after man was caused in part by man's need for help, and the appearance of woman and ultimately, then, the knowledge of good and evil and the fall can be traced to an original human neediness. And according to Genesis 1, man was compelled by some need from the beginning, for however much there was no "reluctation of the creature and sweat of the brow" in the garden, even in Genesis 1 man was faced with the task of subduing and filling the earth.[24] But according to Bacon, man in the garden was simply at rest at the very first and did not rest as respite from work, however easy it may have been. The astounding core of Bacon's argument about the Christian God now comes clearly to light. A creator God must have been at the very first compelled by some need to create, and according to the Bible itself, the reason for man's fall must have been divine jealousy that could not suffer the creation of a being whose leisure was superior to His own neediness. If God did not at first create a being who was his superior, then he created one equally as needy: the story of the fall reflects the fact that any divine creation has the character of divine rancor, not divine justice.

Bacon says that it was not the contemplative knowledge of God's creatures that led to man's fall but the knowledge of good and evil showing the source of good and evil to be not God's commandments and prohibitions but rather "other beginnings." Bacon does not say what they are, but the only possible beginning in *Bacon's* garden could be the knowledge of the superiority of restful contemplation over needful, constrained manufacture. God's commandments and prohibitions sprang from his knowledge of the difference between his own evil neediness and the good contemplative rest of the man he either made or was unable to make. God's commandments and prohibitions were not the source of good and evil, because man's very superiority or likeness to God was the source of good and evil. Man's knowledge of good and evil revealed God's neediness, and God's

---

24. Ibid., 1.28; 2.15, 18–23.

commandments were the result, not the origin, of God's evil circumstances. The first divine commandments and prohibitions reflect the divine jealousy that caused God to make man into a constrained maker like himself. And with this act, God either repeated creating man in his own image, or else he completed the creating of man in his own image, which was botched before the fall. But in either case God made man into a constrained maker who then rivaled his own artifice, as is shown by Bacon's examples of the flood and man's activities before and after the flood, when life was devoted respectively to inventing and building.

If the creator God of the scriptures exists, then according to Bacon's blasphemous account of divine testimony itself his first home must have been the human polity as an order of productive arts. And if he does not exist, then the belief in such a god is latent in the very structure of political life itself. The rancorous creator god causes men to need art, in turn causing men to need to moderate their consequent love of justice. But the creator god does not teach moderation. Human life is always in the grip of such a god, but the more its true nature is known by political philosophy, the less such a god can be heeded and the more likely moderation is to temper the love of justice. Christian doctrine did not know this god's true nature, because the Christian notion of sin obscured the possibility of conventional human freedom, thereby obscuring the creator god's inferior neediness. When the doctrine of sin became universal, it became easier to hear the creator god's call and more difficult to heed the critique of the productive arts. The consequence is that in the modern Christian age, the new science of nature must secularize, and so manage, the hope for a perfected body that has been projected to another world to which conscience is the only guide. But now we know the deepest reason why such secularizing is possible: the charitable creator god has always stalked the earth, not the silent heavens.

To see Bacon's point, we have to note his remark that in the next vicissitude after the era of the unlearned and inspired apostles came the learned apostle Paul. He says that Paul's learning was continued by the ancient bishops but was interrupted by the prohibitions of Julian and then "the emulation and jealousy of Gregory." They were then followed by the Christian rescue of learning from the barbarians and a new revival of learning caused by the reformation of the Church of Rome. Bacon concludes by saying that learning inclines men to exalt the glory of God and ministers against "unbelief and error," for Jesus said that "you err, not knowing the scriptures, nor the power of God."[25]

Now, Bacon refers to the same words of Jesus again in *Novum organum* 1.89, which treats the relationship between the study of natural philosophy and "superstition, and the blind and immoderate zeal of religion." There Bacon says that the latter has always been the adversary of the former, in the times of the Greeks as well as among the "ancient fathers of the Christian Church." He says that as things now are the investigation of nature is made "hard and perilous" by the schoolmen, who in trying to fashion theology into an art incorporated more than was fit the contentious and thorny philosophy of Aristotle into "the body of religion." The same effect has been wrought by thinkers who have tried to deduce the principles of religion from those of philosophy, and such mixtures have excluded all change for the better in philosophy. Some simple divines close off philosophy altogether, thinking that the Bible must prohibit it; others have done the same, thinking that ignorance makes God's agency more evident or fearing assaults on religion in general or on its authority in particular. These latter two fears "savor utterly of carnal wisdom and so they spring from the subtle belief that men doubt and mistrust religion in their hearts." Bacon then says that philosophy is a cure for superstition and a nourishment for faith. Jesus did not err when he said that men err in failing to know the scripture and the power of God, which coupled knowledge of God's will and knowledge of his power. It is not surprising, he says, that the growth of natural philosophy has been checked when religion, "the thing which has most power over men's minds," has been directed against knowledge by the "simpleness and incautious zeal of certain persons."

According to the earlier discussion of the vicissitudes of things, the Church was rent by schismatic discord from its very beginning, and the sectarian revival of ancient wisdom served less to revive it than to imprison it within the confines of Christian charity.[26] Here Bacon signals his final disagreement with Machiavelli's opinion regarding Gregory, for the ferocious Christian sects, whose money-loving subversion of the difference between the noble and the vulgar has imprisoned the moral teachings of the ancients, have completed Gregory's jealous work and have evened the score with the apostate Julian. In imitating Jesus' charity, the new knowledge forsakes the wisdom of the later ancients in order to make man as good a maker as the creator god. But just for this reason, men do not become truly di-

25. Matt. 22.29.   26. Chap. 4 at nn. 104–11 above.

vine. And the further reason why they do not is made clear from closer attention to the words of Jesus to which Bacon refers.

When Jesus spoke of knowledge of divine will and power, it was to refute the Sadducees, who said that there is no possible resurrection of the body.[27] They asked Jesus how the matter of the wife of seven dead brothers would be treated after the resurrection, for according to Moses, each must have married the woman after the death of a brother, so that at the time of the resurrection she would have to be wife to them all. Jesus responds that in the resurrection they are like angels in heaven, who neither marry nor are given in marriage. But heaven is not the home of the creator god; the worldly city is, and in this world, men cannot be as angels, mere voices or bodiless bodies. From the example of Jesus we learn a lesson about the new project of science: the hope for the perfect freedom of a universal body is always the dream of some particular body that is never free from the tension between what is common and what is irreducibly one's own, such as a faithful spouse. As it is modeled on such a hope, the new knowledge actually promises the freedom the later ancients understood to be the delusion of the tyrant. Despite its power to manage turbulent sects, the new knowledge at once demands justification for the tyrant and cannot produce the economy of desire upon which such a justification might be made. Jesus' secret orientation by carnal desire is the vehicle for the revenge of the demiurge. This is the Olympian who will *always* reside in the human polity and who, in revenge for his forced service to the city, calls men to become cosmic gods by the impossible means of imitating his own, merely artful nature.

The Human Proofs:
Sin, Virtue, and Machiavelli's Immoral Justice

Having explained how the renewed attempt to justify the tyrant is a divine necessity, Bacon closes his apology by letting the words of others—"human testimonies"—show his new attempt to be just. In the present section, he explains how the Christian notion of sin came to be prominent and universal in the world, so that his task before the rancorous creator god was more difficult than anything Socrates experienced. The doctrine of human sin was the real legacy of the

27. Matt. 22.23–33.

honorable Roman conquest of the Jews and the rest of the world. In this world, Bacon is less free than Socrates was in his. With these points established, Bacon can show that his new project is just, even though it grants more to Machiavelli's immoral science of political honor than it does to the true wisdom of the later ancients.

[301–309]

The field of human proofs of knowledge's dignity is so large that Bacon will have to choose from among them rather than to "embrace the variety of them." The first proof is the testimony of human honor. Among the heathen, the highest honor was to "obtain to a veneration and adoration as a God." This fruit is denied to the Christian, of course, but since he now treats human proofs, Bacon can consider what the "Greeks call *apotheosis,* and the Latins *relatio inter divos,*"[28] the supreme honor awarded to men, especially when "given not by a formal decree or an act of state, as was used among the Roman emperors, but by an inward assent and belief." This honor was so high that it had "a degree or middle term," so that there were between human and divine honors "honours heroical." Men of great political or "civil merit" were accorded the middle honor, as we see with Hercules, Theseus, Minos, Romulus, "and the like," but "inventors and authors of new arts, endowments, and commodities towards man's life" were "consecrated amongst the gods themselves," as were Ares, Bacchus, Mercurius, Apollo, and others. This honor was just, because the political men benefited only an age or a nation, but the latter kind of men gave permanent and universal benefits, like "heaven," and while the former benefits were mixed with "strife and perturbation," the latter were like the "divine presence," coming in "*aura leni,* without noise or agitation."[29]

The second proof is related to the first because the merit attributed to political benefactors is not "much inferior" to that accorded the inventors. The reason is that suppressing the "inconveniences which grow from man to man" is not much inferior to "relieving the necessities which arise from nature." Plato's statement about philosophers and kings[30] is verified to the extent that "under learned princes and governors there have been ever the best times." If kings are illuminated by learning, then whatever their imperfections in passions and customs, they receive the counsel of religion,

28. Herodian, trans. Whittaker, Loeb Classical Library (London: Heinemann, 1969), 4.2.
29. 1 Kings 19.12.   30. *Republic* 473c11–e5.

policy, and morality that preserves them and "refrains" them from errors and excesses. The same can be said for senators and counselors. Bacon still keeps to the "law of brevity," so he demonstrates his point by referring to "eminent and selected examples." These examples appear in the Roman age from the death of Domitian until the reign of Commodus. This age of six princes was the "most happy and flourishing of the Roman Empire," which was "prefigured" in the dream that Domitian had the night before he was killed.[31] Although the matter is vulgar and so thought fitter for "declamation" than for inclusion in the present treatise, yet because it is pertinent, *neque semper arcum tendit Apollo,*[32] and because it would be "too naked and cursory" simply to name them, Bacon does not omit them altogether.

The first was Nerva, of whom Tacitus said *Postquam divus Nerva res olim insociabiles miscuisset, imperium et libertatem.*[33] The last act of Nerva's short reign was a letter to his son that was a token to Nerva's learning, for he complained from "some inward discontent at the ingratitude of the times" by using a verse of Homer's.[34] Trajan was not learned, but as Jesus said that "he that receiveth a prophet in the name of a prophet, shall have a prophet's reward,"[35] so he deserves to be placed among the learned princes, for Nerva was a great admirer and benefactor of learning. Legend tells that Gregory the Great, who hated "all heathen excellency," prayed for the deliverance of Trajan's soul from hell on account of Trajan's moral virtues. The prayer was granted "with a caveat that he should make no more such petitions." During Trajan's reign the persecution of Christians "received intermission," as we are told by Pliny Secundus, a learned man advanced by Trajan.[36]

Adrian was the "most curious man that lived and the most universal inquirer," so much so that it was counted a defect that he wanted

---

31. Suetonius, *De vita Caesarum*, rec. Ihm (Leipzig: Teubner, 1967); *Domitianus* 23.

32. "And Apollo does not keep his bow always bent" (Spedding's trans.), *Advancement* 303; Horace, *Odes* 2.10.

33. "He united and reconciled two things that used not to go together—government and liberty" (Spedding's trans.), Tacitus, *Agricola* 3. The next five paragraphs, *Advancement* 303–7, are condensed to one paragraph in the *De augmentis*, *De aug.* 471–72. All of the references to Christianity and the Bible are omitted, as is the discussion of Queen Elizabeth. The important argument to follow, nn. 68–83, could not be drawn from the *De augmentis*.

34. Dio Cassius, trans. Cary, Loeb Classical Library (London: Heinemann, 1970), 68; *Telis, Phoebe, tuis lacrymas ulciscere nostras* ("O Phoebus, with thy shafts avenge these tears"; Spedding's trans.), Homer, *Iliad* 1.42.

35. Matt. 10.41. 36. Pliny (the Younger), *Epistulae* 10.97.

to comprehend everything and did not reserve the worthiest things for himself.[37] He became like Philip of Macedon, who was well answered by the musician with whom he argued: "God forbid, Sir, that your fortune should be so bad, as to know these things better than I."[38] Adrian's curiosity contributed to the peace of the Church in those days. His veneration of Jesus as a "wonder or novelty" to be "matched with Appollonius" in his gallery allayed the bitter hatred of the times against the Christians, so that the Church was in peace.[39] He did not match Trajan's justice or glory in arms, but he exceeded him in "deserving of the weal of the subject" because of his monuments and buildings, which were so numerous that Constantine the Great called him *Parietaria*.[40] Although his buildings were more for glory than for use, he went all through the Roman Empire rebuilding and renewing, "policing cities with new ordinances and constitutions, and granting new franchises and incorporations, so that his whole time was a very restoration of all the lapses and decays of former times."

Having the patience and subtle wit of a schoolman, Antoninus Pius was dubbed *cymini sector*.[41] He would "enter into the least and most exact difference of causes," which was the fruit of the "tranquility and serenity of his mind." He was noted as the man of the purest goodness that ever reigned or lived, and he "likewise approached a degree nearer unto Christianity" and became, as Agrippa said to Paul, "half a Christian."[42] He admired Christian religion and law and not only ceased persecution but even advanced Christians. Then the first *Divi fratres* ruled, Lucius Commodus Verus, son to Aelius Verus, who would "call the poet Martial his Virgil,"[43] and Marcus Aurelius, who survived and outshone his brother and who was called "the philosopher." Marcus excelled "all the rest" in the perfection of the royal virtues just he did in learning, as shown in Julianus' *Caesares*, where Silenus was unable to "carp at him."[44] The virtue of Marcus "continued with that of his predecessor" and made the name Antoninus so sacred that even though it had been dishonored by Commodus, Caracalla, and Heliogabalus, when Alexander Severus refused it, the

37. Dio Cassius 69.3, 11.

38. Plutarch, *Quomodo adulator ab amico internoscatur* 67f–68a; *Quaestionum convivialium* 634c.

39. *Historiae*, Lampridius, *Alexander Severus* 29.

40. Sextus Aurelius Victor, *Epitome de Caesaribus*, ed. Pichlmayr (Leipzig: Teubner, 1966), 41.13.

41. "A cutter of cumin seed," Dio Cassius 70.3.   42. Acts. 26.28.

43. *Historiae*, Spartianus, *Aelius* 5.

44. Julianus, *Symposium*, trans. Wright, Loeb Classical Library (London: Heinemann, 1913), 334.

Senate said *Quomodo Augustus, sic et Antoninus.*[45] During this emperor's time the Church was "for the most part" in peace, so that in these six princes we see the "blessed effects" of learning "painted forth in the greatest table of the world."

Because learning does not only affect "civil merit and moral virtue" and the arts of peace and peaceful government, the third proof can be seen in the good effect of learning on "martial and military virtue and prowess," as shown by the examples of Alexander the Great and Caesar the Dictator. Alexander was taught by Aristotle and was attended by Callisthenes and other learned persons. His esteem for learning is shown in three examples, his envy toward Achilles, his judgment of the cabinet of Darius, and his letter berating Aristotle for publishing the "secrets or mysteries of philosophy," in which he told Aristotle that he "esteemed it more to excel other men in learning and knowledge than in power and empire."[46] To demonstrate his point, Bacon mentions several of Alexander's wise speeches, concluding his praise of Alexander by remarking that just as some say hyperbolically that all the sciences might be found in Virgil, so too the "footprints and footsteps of learning" are to be found in the speeches reported of Alexander, the admiration of whom considered as Aristotle's scholar rather than as Alexander has carried him too far.[47]

Julius Caesar's learning is witnessed by his *De analogia,* in which he labored to make *vox ad placitum* become *vox ad licitum* and to reduce custom of speech to congruity of speech and took "the picture of words from the life of reason."[48] His reformation of the calendar was a monument to his power and his learning, and in his *Anti-Cato* we see that he aspired to victory in wit as well as in war.[49] In his *Apophthegms* we see that he thought it better to collect the wise words of others than to have every word of his own made into an oracle, as vain princes

45. "Let the name of Antonius be as the name of Augustus" (Spedding's trans.), *Advancement* 306; *Historiae,* Lampridius, *Alexander Severus* 10.

46. Plutarch, *Alexander* 8, 15, 26, 7; Pliny (the Elder), *Naturalis historiae,* ed. Ian and Mayhoff (Leipzig: Teubner, 1967), 7.29.

47. Plutarch, *Alexander* 14, 15, 22, 28, 29, 31, 47, 53, 74; Seneca, *De beneficiis,* trans. Basore, Loeb Classical Library (London: Heinemann, 1935), 5.4.4; Plutarch, *Quomodo adulator ab amico internoscatur* 65f; *Ep.* 59.12; Homer, *Iliad* 5.340; Plutarch, *Regum et imperatorum apophthegmata* Alexander 180d–e; Plutarch, *Caesar* 11; *Apology Concerning the Earl of Essex,* in *BW* VI, 249–50. The sentence referring to Virgil is omitted from the *De augmentis.*

48. Suetonius, *Divus Iulius* 56; Quintilianus, *Institutio oratoria* 1.7.34–35; Cicero, *Brutus* 72; Aulus Gellius, *Noctes Atticae* 1.10, 9.14, 19.8.

49. Suetonius, *Divus Iulius* 11, 56; Aulus Gellius, *Noctes Atticae* 4.16; Plutarch, *Caesar* 54; *Att.* 12.40, 41, 13.50.

do, and yet if Bacon were to enumerate his speeches as he did with Alexander, they would be seen to be what Solomon noted when he said that the words of the wise are like goads and nails.[50] Bacon mentions three of Caesar's admirable speeches. The first was the one in which Caesar appeased mutiny in his army with but one word, addressing his soldiers as *quirites* rather than *milites*. The second was when Caesar objected to the weak and poor cry evoked by his being saluted as king, putting it off "in a kind of jest, as if they had mistaken his surname, *Non Rex sum, sed Caesar*." This speech unseriously refused the name, signified his confidence and magnanimity, and yet was "of great allurement towards his own purpose," as if the state "did strive with him but for a name." The third speech was the one to Metellus when Caesar took possession of Rome and, more precisely, Rome's treasury. This speech combined the greatest terror and the greatest clemency that could proceed out of the mouth of man.[51] To conclude, Bacon notes that Caesar knew his own perfection in learning, as was shown by his scoff that because Sulla was poor at letters he did not know how to dictate, so that it was not, as some said, strange that he should have resigned his dictatorship.[52] After mentioning the example of Xenophon, Bacon moves to the fourth human proof.[53]

The fourth proof is found in "moral and private virtue," which is aided by liberal studies, as demonstrated by remarks of Ovid, Koheleth, Seneca, Epictetus, and Virgil.[54] The unlearned man does not know what it is to "call himself to account" or to have the pleasure of feeling himself better than he was the day before.[55] In general, *veritas* and *bonitas* are "but the seal and the print," and the "clouds of error descend in the storms of passions and perturbation." The fifth proof is to be found in "matter of power and commandment." Just as free political rule is superior to tyranny because command over wills is superior to command over deeds, so the "commandment of knowledge is even higher than that over will," because it is over the highest parts of man: reason, belief, and understanding, which give law to the will itself.

50. Suetonius, *Divus Iulius* 56; *Fam.* 9.16; *Eccles.* 12.11.
51. Suetonius, *Divus Iulius* 70, 79; Plutarch, *Caesar* 35.
52. Suetonius, *Divus Iulius* 77.
53. Xenophon, *Anabasis* 2.1.12, 2.5.37–42; Plutarch, *Agesilaus* 15.
54. Ovid, *Ex ponto* 2.9.47; *Eccles.* 1.9; Plutarch, *Agesilaus* 15.6; Seneca, *Naturalis quaestiones*, ed. Gercke (Leipzig: Teubner, 1970), 1; *Praefatio* 10; Epictetus, *Encheiridion*, rec. Schenkl (Leipzig: Teubner, 1898), 3, 26; Virgil, *Georgics* 2.490; *Eccles.* 12.23.
55. Plato, *Alcibiades I* 132c7–34b5; Xenophon, *Memorabilia* 1.6.4–11.

The sixth proof is found in fortune and advancement. Learning aids not only states and commonwealths but also particular persons. As was noted long ago, Homer gave more men their living than did Sulla, Caesar, or Augustus, and it is hard to say whether arms or learning has "advanced greater numbers."[56] And with regard to sovereignty, if "arms or descent have carried away the kingdom, yet learning hath carried the priesthood, which ever hath been in some competition with empire."

The seventh proof is from pleasure and delight. The pleasure and delight of learning "surpasseth all other in nature." In all other pleasures "there is satiety," showing them to be but "deceits of pleasure"; but knowledge has no satiety, for "satisfaction and appetite are perpetually interchangeable," so that it is good in itself, and the pleasure Lucretius described of seeing the perturbations of men from the fortification of the certainty of truth is of no "small efficacy or contentment."[57]

The eighth proof is found in what man's nature most aspires to: "immortality or continuance." Monuments of wit and learning are more durable than monuments of power or the hands, for Homer's verses have lasted for twenty-five hundred years, while palaces, temples, castles, and cities have decayed or been demolished. No clear pictures of Cyrus, Alexander, Caesar or great kings remain, but the images of men's wits do. And they are not even images, because they "generate still." Even materialistic philosophers have believed that understanding, but not the affections, might remain after death. And we who know by divine revelation that the purified affections and the "body changed" can have immortality "do disclaim in these rudiments of the senses." Even so, Bacon does not think it possible for him to reverse Aesop's cock's preference for the barleycorn over the gem, Midas' preference of Pan, or plenty, over the Muses, Paris' preference for beauty and love over wisdom and power, Agrippina, who preferred "empire with condition never so detestable," or Ulysses, who in preferring an old woman to immortality is a "figure of those which prefer custom and habit before all excellency."[58] These will continue as they have been, but "so will that also continue whereupon learning hath ever relied, and which faileth not: *Justifi-*

56. Plutarch, *Regum et imperatorum apophthegmata* 175c.
57. *Advancement* 317; Lucretius, *De rerum natura* 2.1–10.
58. *Phaedrus*, trans. Perry, Loeb Classical Library (London: Heinemann, 1965), 3.12; Ovid, *Metamorphoses* 9.153; Euripides, *Troiades* 924; Tacitus, *Annales* 14.9; Homer *Odyssey* 5.218; Plutarch, *Gryllus* 985d–86b; Cicero, *De oratore* 1.44.

cata est sapientia a filiis suis."[59] And with this final caveat, Bacon ends the first part of this treatise for the king.

According to Bacon, the highest human honor is to be venerated as a god. But Bacon is careful to distinguish between veneration springing from "inward assent and belief" and that springing from political convention, as was the case in the customary deification of the Roman emperors. He quotes Herodian, who says that the Romans deified only those emperors who left children as successors,[60] although Bacon says nothing of this fact. But actually, says Bacon, the deified Romans could not really have been divine, regardless of the cause, because they were merely men of civil merit and in this regard were just like Hercules, Theseus, Minos, and Romulus. Such men's benefits are fraught with strife and perturbation, he says, as if to remind us of the conflict between Theseus and Minos, and this shows them to have been inferior to the truly divine benefactors, the inventors and authors of new arts.

But the list of inventors and authors of new arts is silent about the troubles that accompanied the very first such benefactor, Prometheus, who Bacon said in the earlier discussion of innovation represented divine providence. Moreover, the list mentions no men, but only gods who were never men. We know, of course, that the divine gift of the arts is always a part of the human polity. In this vein, then, we note that Bacon begins the human proofs by reminding of the argument of the preceding discussion: the human and the divine can never by wholly severed or differentiated, and the more they are likened according to the model of a god who could give the gift of art, the more such a gift is a strife-ridden and turbulent blessing. And as we should expect, this turbulence is associated with the confusion of universal and merely political benefits, a confusion that occurs with the very largess of the gods.

After reminding us of the necessity that moves his apology, Bacon refers to the speeches in which Socrates works out his critique of the city as an order of productive arts. Bacon uses Socrates' famous statement in the *Republic* about the concurrence of philosophy and monarchy in order to claim that the best times are when "learned princes and governors" rule. We have to think briefly about Bacon's reference, because Bacon follows this laudatory reference to the later

---

59. "Wisdom is justified by her children" (Spedding's trans.), Matt. 11.19.
60. See n. 28 above.

ancient with a list of Roman emperors that is the gateway to his last extended discussion of Machiavelli.

Now, according to everything we know so far, the happy concurrence of learning and rule can, in Bacon's time, be accomplished only by way of progress in the productive arts. We cannot but be reminded, therefore, that Socrates built his city in speech according to Thrasymachos' understanding of art, or the precise definition of art as simply unerring.[61] Thrasymachos demands that justice be the same as the perfectly self-sufficient art of securing only what is one's own, taken to be all of the things that might satisfy the needs of the body, as if any body might be self-sufficient as long as it is, by way of a perfect and universal art, the strongest. In building the best—or the most just—city on the model of the arts as unerring and self-sufficient, Socrates satisfies Thrasymachos' tyrannical desire to be the source of justice, or to imitate the self-sufficiency of the city as such, apart from any of its parts.[62] But satisfying Thrasymachos proves ultimately to be impossible, because no body can be self-sufficient; no body, or soul for that matter, can be the same as the pretense of the city as a whole.

Therefore, two contradictory principles of justice govern the fashioning of the best city and are treated as if they were the same: the principle that one should perfect one's art and the principle that one ought to mind one's own business. They are not compatible if some art requires minding others' or everyone's business, which proves to be the case for the philosopher kings who mind the artisans' business and for the artisans who practice the only common art, or the art that runs through the order of the arts as a whole, the money-making art.[63] As the city is modeled on the productive arts, there is no standard for distinguishing between what is and what is not one's own. The communism that is introduced to solve this dilemma founders on the same shoal, for just as no body can be perfectly self-sufficient because there is no single, self-sufficient art of perfecting the body, so there is no body that can share everything, or have nothing that is peculiarly its own. This appears in the dilemma of the place of women in war, which springs from the differences not between male and female souls but rather between male and female bodies.[64]

The best city's impossibility is grounded at least in part on Thrasymachos' delusional hopes for the productive arts. His hopes rep-

61. *Republic* 473c11–e5, 340d1–41a4, 370a7–c5, 433a1–34d1.
62. Ibid., 336c–37d2, 338c1–3, 348c11–12.
63. Ibid., 334b3–6, 345e5–52a3, 519c8–21b10, 540e5–41b5.
64. Ibid., 449a1–57c2, 462c10–e2, 466c6–72b2.

resent the moving principles of any city and ultimately the moving principle of any art within a city. In representing the city, Thrasymachos is like the tyrant and the sophists because he sees the productive art, and hence the willful origins, behind every conventional virtue or claim to be free from the need to make and acquire for the sake of the body. The city built on his knowledge is impossible because it cannot be just, but forgetting its justice, his city would be nothing but a perfect warrior who fights for the sake of money makers who do not. In the *Republic,* the tyrant's desire to be the city itself is comical as well as impossible: in his desire for perfect self-sufficiency, the tyrant becomes the servant of merchants. And the possibility that the political philosopher might accomplish the highest political hope is just as comical. Socrates' founding of the city according to the desire for perfect justice is itself a laughing matter: Socrates founds the city as if he had already founded it and as if he were able to coerce bodies that he could not coerce. And this culminates in the request that he and the speech-loving interlocutors expel all the citizens (more than ten years old) of an existing city so that philosophy and monarchy might coincide.[65] Not only might such a founding be harsh and unjust; it is fantastic to think that one who always questions the just and the good would be willing, not to say able, to accomplish it.

If we consider Bacon's present discussion carefully, we see that he imitates Socrates' comical procedure. To demonstrate the supposed harmony between learning and rule, Bacon presents the long discussion of learned Roman emperors before describing Caesar, the armed founder of the empire in which their learning was supposed to have reigned. And as we see below, the armed founders, Caesar and Alexander, were no true friends of learning. The question arises, then, whether Bacon intends to be comical as he knows that Socrates was. Just as Socrates said that necessity forced him to be brief,[66] Bacon tells us that in discussing the human proofs he cannot give the full range of proofs their due and is forced to choose among them. And in presenting his list of learned rulers, he says that by giving only a few select examples he obeys the "law of brevity." Some necessity must prevent him from making a comprehensive speech to "weigh justly" the true dignity of learning. This necessity requires that the comprehensive account be presented by way of Bacon's "enigmatical method," the same one once used by the ancients. Moreover, we know that the necessity governing Bacon's attempt to be just and to ex-

65. Ibid., 540e5–41b5.   66. Ibid., 435c9–37a10, 506d2–507a5.

plain his justice is the same as that governing Socrates' presentation of the perfectly just city: the demand for an account of justice commanded by the hope for a perfected order of the arts. Bacon can hardly flaunt his agreement with Socrates' critique of the productive arts when Christian charity requires a new project for the artful mastery of nature.

Charity limits Bacon's ability to be comical. Socrates could make Thrasymachos blush, because despite his tyrannical bluster, Thrasymachos respected conventional virtue and propriety.[67] To respect such propriety makes it possible to see the comedy of attempts to be free from it—where there is no capacity for shame, there is no capacity for laughter. In fact, Thrasymachos was playful at the very beginning of his encounter with Socrates, so that he was within Socrates' comical grasp from the beginning. Socrates could no more rule for perfect justice than a tyrant could. But Socrates could reasonably try to dampen a tyrant's desire for perfect justice with playful speeches. And he could do this because conventional virtue and propriety had not been wholly debunked by the doctrine of human sin. Unlike Thrasymachos and Socrates, Jesus never laughed. In the sinful Christian world, there is in men no natural or conventional excellence by which to measure shame. When all men as sinners are ashamed, none really is, and without real shame there can be no sense of the comic or ridiculous. In the present era, all men are moved more powerfully than in Socrates' time by deadly serious hopes.[68] The necessity governing Bacon's task is more powerful than what Socrates faced, giving Bacon no choice but to attend to the all too serious hopes of the productive arts. Therefore the justice of Bacon's project must be judged by an easier measure than should be applied to Socrates or to any of the later ancients.

Although Bacon has been moved by necessity to brevity, he does not simply speak at some length about the six princes who ruled during Rome's happiest age. In addition to the six, he also mentions Domitian, Commodus, Caracalla, Heliogabalus, and Alexander Severus, and speaks at length about seven princes, not six, because he includes Lucius Commodus Verus along with Marcus Aurelius. This subtle violation of the law of brevity is important, for it alerts us to the importance of the list of emperors, which turns our attention to Bacon's last discussion of Machiavelli. It is important, therefore, to

67. Ibid., 350c12–d8.
68. See Sir Thomas More, *Dialogue of Comfort against Tribulation*, ed. L. Miles (Bloomington: Indiana University Press, 1965), 1.13; Friedrich Nietzsche, *The Will to Power*, trans. W. Kaufman and R. J. Hollingdale (New York: Vintage, 1968), 187.

examine Bacon's discussion of the list of learned rulers with great care.

According to Bacon, Trajan deserves to be counted as learned, even though he was not, because he greatly admired and benefited learning. But by this standard, Domitian should count as a learned prince, for according to Suetonius, to whom Bacon refers in mentioning Domitian's dream the night before his murder, Domitian was a lavish builder and supporter of libraries and affected before his reign a love of poetry, even though at the beginning of his rule he neglected liberal studies. Suetonius also comments that at the beginning of his rule Domitian was good regarding justice and public morality, perhaps, we are led to suspect, because he ignored liberal studies. It was only as his rule progressed that he turned first to cruelty and later to avarice.[69] By the force of this example learning would appear to be no guarantee against cruelty and avarice, whereby we are led to wonder whether the attempt to combine learning, justice, and rule is in fact impossible. Domitian was just, while he ignored liberal studies, and then cruel and avaricious because he was not calmed by liberal studies. The effect of learning may have to precede the accession of the learned to political power, but in the light of Socrates' comical procedure in the *Republic*, it is by no means clear that such an order is possible. The question of Domitian points to the problem of just and possible beginnings.

Bacon refers to Domitian's dream again in the Essay "Of Prophecies," where it is one of a list of prophecies "that have been of certain memory, and from hidden causes."[70] Every prophecy in the essay's list foretells the violence of a founding or a tyrant or the violence that befalls either a tyrant or a founder, with two exceptions: the prophecy regarding Vespasian's or Jesus' rise from out of the East, and the prophecy about the fall of England made nugatory by the enlargement of England into Britain. Bacon says of the former prophecy that Tacitus "expounds it of Vespasian" however much it may have been meant "of our Savior." But Tacitus does not say only that it referred to Vespasian. Rather, according to Tacitus the prophecy was said to have sprung from the superstitious interpretation of omens witnessed by the Jews besieged in Jerusalem. The vulgar of the Jews did not interpret the omens as fearful, because

---

69. Suetonius, *Domitianus* 2, 3, 8–11, 20.

70. *Essays* 463–65; 1 Sam. 28.19; Virgil, *Aeneid* 3.97; Seneca, *Medea*, ed. C. Costa (Oxford: Clarendon Press, 1973), 2.374–78; Herodotus 3.124; Plutarch, *Alexander* 2; Appian, *Bella civilia*, trans. White, Loeb Classical Library (London: Heinemann, 1912), 4.134; Suetonius, *Galba* 4; *Domitianus* 23; Aristophanes, *Equites* 195–210.

they believed them to confirm the prophecy in their priestly writings that men originating from Judea would possess the world. Although the prophecy had in reality pointed to Vespasian, the common Jews, according to human ambition, interpreted the signs in their own favor" and could not be dissuaded even by adversity.[71] According to Tacitus, the prophecy he describes was not just a prophecy from the East but a prophecy based on on the priestly writings of the Jews, that is, the prophecies of the Old Testament. At the end of the essay Bacon comments that all of the prophecies he has mentioned "ought to be despised" and ought "to serve but for winter talk by the fireside." They ought to be despised as far as belief is concerned but not with regard to their efficacy as they are "spread and published," by which they have accomplished "much mischief."

According to the argument of the essay, Bacon is bolder than Tacitus because Bacon despises not a prophecy of the enemies of Rome but the prophecy foretelling the empire of Jesus. Yet he does not disagree with Tacitus' estimation of the power of the prophecy that, according to Tacitus, so appealed to the universal ambitions of the vulgar Jews that they believed them of themselves rather than of Vespasian, even in the face of adversity. Now, Vespasian was the emperor who began the dynasty built on the power of the legions outside of Rome, culminating in the rule of Domitian and followed by the adopted princes whom Bacon celebrates as the learned rulers of Rome's happiest age. Nerva combined empire and liberty, as Bacon says, but as Tacitus notes, this measure did not immediately cure the listlessness that men first hated under Domitian and soon came to love because of its charm.[72] According to Dio Cassius, to whom Bacon refers in the quote from Homer, Nerva was not discontent with the times; rather he asked Trajan to use Apollo's avenging shafts for his having been held in contempt for his old age by the mutinous praetorian guards, upon whom he depended and whom he was unable to prevent from executing anyone they wished.[73] Nerva's revenge was to turn the softness of the slavish, vulgar many against the martial ferocity of the praetorians. It was executed by Trajan, and as Bacon presents them, Trajan is the first of the golden princes to have given "intermission" to the persecution of the Christians.

Bacon likens Trajan to the learned and to prophets, with Pope Gregory being said to have loved and esteemed Trajan's moral virtues, while he, Gregory, had the most extreme "envy" or malice to-

71. Tacitus, *Historiae* 5.13; cf. Suetonius, *Divus Vespasianus* 4.   72. *Agricola* 2–4.
73. Dio Cassius, trans. Cary, Loeb Classical Library (London: Heinemann, 1968–70), 68.3–4.

ward all "heathen excellency." Bacon says that Pliny tells us of Trajan's having halted the persecution of Christians, but in fact, Pliny reports no such thing. Rather, in correspondence between Pliny and Trajan, Pliny explains that he has sought Trajan's counsel in the matter of prosecuting Christians with threats of capital punishment to be inflicted on those who will not recant, curse Jesus, and worship the gods and the image of Trajan. Pliny seeks the emperor's counsel on the matter because the contagious superstition of Christianity has spread from cities to rural districts and towns, endangering people of all ages, ranks, and sexes. Trajan objects to anonymous accusations, for the sake of the spirit of the age, but he approves Pliny's harsh punishments.[74] In this light, it is hard to see how Gregory, who hated heathen excellency, found it possible to love Trajan's virtue and pray for his soul.

To see how Gregory could have loved Trajan, we have to pursue Bacon's description of the learned Roman princes. After speaking of Trajan, Bacon refers to Dio Cassius in reporting that Adrian was curious and a universal inquirer. But actually, Dio says that Adrian's curiosity led him to use divinations and incantations and to build a city and set up statues in honor of Antinous, either out of love for him or because he had sacrificed his life for the achievement of Adrian's goals. According to Dio, Adrian wanted to comprehend everything because of his excessive love of honor *(philotimia)*, which then caused his insensibility to the difference between public and private and between the greatest and the slightest of things.[75] Bacon says that God used Adrian's curiosity to bring peace to the Christian Church, because he venerated Jesus as a wonder or novelty, but Lampridius, who tells the story to which Bacon refers, says that it was not Adrian who so venerated Jesus but Alexander Severus.[76] According to Dio, Adrian's excessive love of honor led him to want to surpass everyone in everything, causing the downfall and the destruction of many even though he had sworn that he would not emulate Trajan, who had killed many great men both at the beginning and at the end of his reign, for which he was almost denied deification.[77]

The truth of Bacon's references is that Trajan did not avenge Nerva, because Trajan killed many of the great, and that Adrian's excessive love of honor led him to ignore the difference between what does and what does not deserve honor. Bacon presents this love of honor

---

74. Pliny (the younger), *Epistulae* 10.96, 97. 75. Dio Cassius 69.3, 11.
76. *Historiae*, Lampridius, *Alexander Severus* 29. 77. Dio Cassius 69.1–2.

as love of learning, assimilating it to the Christian sympathy of Alexander Severus, a prince *not* of the golden years, who was, as we already know, killed by his soldiers because of their contempt for his dependence on his mother. As Bacon reports the progress of Nerva's revenge against the soldiers, the emperors became more dependent upon the army, killed the very men whom Nerva would have spared, became indifferent to the distinctions between noble and base because of the excessive love of honor, and began to venerate Jesus in a fashion that in fact outlived the golden age of which Bacon speaks. Regarding the latter, it is important to note that, according to Lampridius, Adrian considered building a temple to Jesus and instead had temples with no image built in every city, and Alexander decided not to build a temple to Jesus because his augurers told him that, if he did, all men would become Christians and the other temples would be abandoned.[78] Adrian did not refrain from building, then, but in fact built temples dedicated to no particular god, which means that they could be taken to have been built for Jesus as much as any other god.

To continue, we notice that Bacon refers again to Dio in his discussion of Antoninus Pius, called *cymini sector* by those who would tax any virtue. Bacon says that Antoninus' love of inquiry "unto the least and most exact differences of causes" sprang from the same "tranquility and serenity of mind" that caused him to "approach a degree near unto Christianity." The importance of Bacon's further reference to Dio becomes clear when we see that, according to Dio, Adrian was almost denied deification by the Senate on account of certain murders of great men. Antoninus ensured Adrian's deification only by manipulating the Senate's respect for himself and fear of the soldiers. Antoninus, then, was like his predecessors in his dependence on the soldiers and his attention to the small as well as the great; only with the ascension of this one, who was half a Christian, did the killing of the eminent men came to an end, for Dio says that Antoninus refused to punish men who had been accused.[79] According to Acts 26, where the comment of Agrippa occurs, Paul's appeal to Caesar prevented his being freed by Agrippa and caused him to be sent to Rome, where he remained quite unconstrained.[80] It is clear, then, that Nerva's revenge was discharged not by Trajan, or by any of Trajan's successors, who never became simply free from the sol-

78. *Historiae*, Lampridius, *Alexander Severus* 43; see Spartianus, *De vita Hadriani* 13.
79. Dio Cassius 70.1–3.   80. Acts 26, 28.30.

diers. Rather his revenge was executed by the spokesman of the prophet whom Trajan was said to resemble, a prophet whose fate it was to suffer and extraordinary execution.

We can understand now why Gregory loved Trajan. It was because of Trajan's unwitting likeness to the prophet who conquered without arms and honor and who was not dependent upon soldiers. And we can see now that, in loving Trajan, Gregory simply appreciated the gift of the emperors who, in loving honor, came only to be vengeful toward the soldiers upon whom their martial love of honor made them dependent. Roman honor became dependent vengefulness, and it was just one step from this development to the burden of sin that only an unarmed prophet could expiate. And only with this last step could the soldiers be said to have been overcome—so that the subjugated Jews prove to have been correct about which men starting from Judea would possess the world.

Such a conclusion is born out by the discussion of the two divine brothers who succeeded the near-Christian Antoninus. Despite Lucius Commodus Verus' extraordinary corruption and lack of any natural ability in liberal studies,[81] Bacon considers him together with Marcus, so that the six golden princes are in fact seven; he considers him as one of the divine brothers along with Marcus, who excelled all the rest both in learning and royal virtues, as is attested by Julianus in his *Caesares*. Bacon says that in the *Caesares* Silenus could say nothing against Marcus except for a "glance" at his patience toward his wife. But closer inspection shows that Silenus does have something more to say, for he berated Marcus for his inability to rear his own son, as if his virtue based on philosophy were not teachable and as if the laws of succession could not guarantee such education.[82] In the contest enacted in the *Caesares*, the two emperors who were the most opposite were the last to speak. Marcus wished that he might imitate the gods and so have as few needs as possible and benefit as many people as possible. Constantine wished to amass great wealth or the means to satisfy needs. When each of the contestants chose the guardian and guide with whom he would live in the future, Marcus chose Zeus and Cronos, while Constantine chose Pleasure, Incontinence, and Jesus, who cried to all that he would cleanse any seducer, murderer, or anyone who is sacrilegious and infamous. According to the *Caesares*, imperial ambition rose from Caesar's desire to be first and to conquer the world to Octavian's interest in govern-

81. *Historiae*, Spartianus, *Aelius* 5; *Historiae*, Julius Capitolinus, *Vita Marci Antonini philosophi* 15.
82. Julianus, *Symposium* 334.

ing well and then from Trajan's desire to conquer the world to Marcus' concern with imitating the gods and being wholly free from need. Then the empire culminated in the rule of Constantine, who was nothing but a pleasure-loving money maker, who placed himself and his entire empire at the feet of Jesus, who would forgive every sin.[83]

Bacon's just brevity presents the "greatest table of the blessed effects of learning in sovereignty" so as to give an account of Nerva's revenge, which was incomplete because it did not master the turbulent soldiers. In Nerva's failed revenge, Bacon sees the transformation of Roman honor and the victory of sin-forgiving Jesus over the soldiers. The truth of praiseworthy honor is that, in the hands of the masterful Romans, it became the burden of universal sin. Roman honor spread the shameless doctrine of human sin, and in a shameless world it is impossible for conventional virtue to moderate the call of the rancorous creator god. Rome's imperial ambition resulted in the victory of the jealous Gregory. As we know from the whole of Bacon's argument so far, the consequence of this victory was the liberation of the money-loving passions. To own up to one's original sinfulness is, of course, to admit to being in the grasp of harsh needs that are common to all men. And lucre is the medium by which such commensurable needs can be ordered so that they can be met by the productive arts. Like Thrasymachos' perfectly artful city, the Christian city is an order of money makers protected by arms. But unlike Plato's austere guardians, the new arms are born by money-loving but conscientious Christian soldiers.

Bacon's account of the "greatest table" illustrates the coincidence of learning and rule as Plato presents it in the *Republic*. And now by comparing these two accounts we can see how much harsher is the necessity compelling Bacon's speeches than that which Socrates had to face. The necessity governing Bacon's speech springs from a tendency latent in any city insofar as it is an order of the arts ruled by forgetful convention. Every city is open to a vengeful demiurge, but in Bacon's time this god has discovered the universal hopes of Christian charity that were spread by the vengeful decay of Roman honor. Thrasymachos was playful precisely because of his irrepressible respect for conventional propriety, so that he and Socrates could become friends.[84] But there is no such thing as Christian comedy or playfulness, because Christian charity explodes the dignity of every conventional virtue and perhaps because its forgivingness is in fact so close to Nerva's desire for revenge.

83. Ibid., 329d–36c.   84. *Republic* 349a4–10, 498c9–d1.

Unlike Socrates, then, Bacon is not free to become the founder of a comical city in speech that might charm or bedazzle the tyrant. Rather Bacon must direct the learned to be benefactors who aim at the perfection of the productive arts and "endowments and commodities towards man's life." But this statement means that the learned must embrace rather than charm the tyrant, because the new founding attends precisely to the delusion that moves the tyrant's passionate and dangerous love of physical self-sufficiency. The "greatest table" is a vulgar matter because it shows that the learned must address the tyrant as he is in his likeness to the many rather than as he is in his likeness to the philosopher. Only the new revenge against nature can moderate the harsh political legacy of Nerva's revenge.

Bacon shows that he understands Plato's comic procedure in presenting the philosopher as an unarmed founder, but unlike Plato, Bacon does not proceed in the manner of Socrates' jests. Rather justice requires Bacon to recommend a serious, new founding that promises as much as the tyrannical Thrasymachos demanded from Socrates. Bacon knows that, for simple justice, learning must now minister to the tyrant's impossible hopes. Moreover, he also knows that it is impossible for learning to justify the tyrant. Precisely for this reason, it is ultimately not possible for learning now to charm and soothe the tyrant or to moderate the tyrannical possibilities that are latent in the hope for a perfectly just new founding. Consequently, when he turns to the armed founders, whose compatibility with the learned founder he has presupposed, he shows them precisely *not* to be well disposed toward learning.

Bacon says the three works of Alexander's love of learning are his envy of Achilles at having been praised by Homer, his wise judgment concerning the cabinet of Darius, and the letter to Aristotle regarding the publication of his secret works. Now, if in alluding to the cabinet of Darius Bacon refers to Pliny's version of the story, then Alexander wished to put Homer's works in the famous box, but if he refers to Plutarch's version, then his intention was to put in the box the *Iliad,* but one of Homer's works, which of course preserves the deeds of the martial Achilles—the butt of much of Plato's humor in the *Republic*—rather than the deeds of the wise and steadfast Odysseus.[85] As Bacon tells it, it is not clear whether Alexander preferred Achilles to Homer and Odysseus. But it is obvious from the remaining quotations regarding Alexander that Bacon knew Plutarch's account as well as he knew Pliny's.

85. Pliny (the Elder), *Naturalis historia* 7.29; Plutarch, *Alexander* 26.1–3.

That Bacon presents Plutarch's Alexander, who admired Homer only because he admired Achilles, is evident from Bacon's reference to Plutarch in the matter of Alexander's letter to Aristotle.[86] According to Plutarch, Alexander was angry because the publication of Aristotle's works made the truth no longer his own, as if the truth as such could ever be one's own. If Alexander wished to excel all other men in learning and knowledge more than in power and empire, he did so only because he took the objects of learning and knowledge, or the truth, to be just like the objects of power and empire, so that the truth is like any nation that can be taken as a prize for one's own. True converse between the armed founder Alexander and the political philosopher, a friendship that is simply free from calumny, is impossible, because according to Alexander any further discovery of truth can always be a loss as well as a gain.[87] And in the rest of the texts to which Bacon refers, Alexander appears as the avaricious tyrant who thought himself a god and who turned against philosophy, while Caesar appears as the thieving tyrant who overthrew the Roman constitution and whose learning made him superior to Sulla in being able to rule speeches and limit learning.[88]

In turning knowledge from the defense of private life and moral virtue to merely useful and common ends, the learned are forced to bend to the tyrant, who, even if he might be charmed by it, could never become converted to political philosophy. Neither Socrates nor Aristotle thought they might single-handedly transform the tyrant; only the secret effects of right opinion could restrain the tyrant sufficiently to have him hear the political philosopher's music. But in the modern age, the task of learning must be much more ambitious, even though the tyrant is no more tractable to moderate justice than in a prior age. Just as Socrates was not bested by the tyrant, neither was Aristotle. But the same cannot be said for Cicero and the age bequeathed by Cicero's Rome. Bacon knows that although the new learning must look beyond the boundaries of Machiavelli's insufficient political science, the new knowledge itself will have limits on its power to calm the passions it so boldly sets free. And when these limits are met, as Bacon knows someday they will be, there will be no alternative to the Machiavellian science of honor, which will itself be no more sufficient as part of the scientific project than it was before.

86. *Alexander* 7.4–9.   87. Aristotle, *Nic. Eth.* 1156a6–57b5.
88. Plutarch, *Alexander* 14, 15, 22, 28, 30, 47, 52–54, 74; Seneca, *De beneficiis* 5.4.4; Plutarch, *Quomodo adulator ab amico internoscatur* 65f; *Regum et imperatorum apophthegmata* 180d–e; *Phocion* 29; *Ep.* 59.12; Suetonius, *Divus Iulius* 56, 70, 76–79; Cicero, *Brutus* 75.262; Plutarch, *Caesar* 3, 4, 35, 54, 60; *Fam.* 9.16; Eccles. 12.11.

Bacon's apology ends on this pessimistic note, and appropriately, then, he closes with a dark comment that turns our attention once more to Machiavelli. Regarding learning and the vulgar many, Bacon does not expect men to prefer the gem before the barleycorn, as is related in Phaedrus' tale, or to reject the judgments of Midas, Paris, the vile Agrippina, and Ulysses (Odysseus).

Bacon refers to the gem and the barleycorn in Essay 13, in the context of agreeing with Machiavelli's bold claim that Christian charity had "given up good men in prey to those that are tyrannical and unjust."[89] To moderate this danger of charity, Bacon warns against treating all men as equals, as if charitable egalitarianism were the true source of the victory of the tyrannical and unjust. And in fact, Machiavelli's having mistaken the power of this egalitarianism blinded him to the tyrannical ferocity not just of the tyrants but of the charitably good men themselves. In fact, because the Christian soldier takes the gem and the barleycorn to be the same, Christian conscience is as ferocious as it is invisible. For this reason, the new science must imitate the transpolitical hopes of Christianity in order to moderate its political effects. Therefore men have no choice but to choose, as Midas did, for plenty rather than for the Muses, or as Paris did, who according to Euripedes chose not between beauty and wisdom and power, as Bacon says, but between two kinds of power.[90] That is, men have no choice but to choose plenty by way of the choice of power over beauty and wisdom. The new empire, like the old, will never be wholly free from the likes of the vile Agrippina and the even viler Nero because it has such scorn for Ulysses.

In Essay 8, Bacon refers to Ulysses as one who possessed a grave nature grounded in custom.[91] The essay treats marriage and single life, and there it is obvious that the "old woman" (Bacon says in Latin *vetulam*) whom Ulysses preferred to immortality was his trusting wife. According to the essay, Ulysses set greater store by grave obedience to custom than by any faithless hopes. In Plutarch's *Gryllus*, to which Bacon refers in commenting on Ulysses, a sophistical Greek pig blames Ulysses, the Greek human being, for his virtue that depends on the artful service of custom.[92] The Ulysses that Bacon condemns here is the Ulysses as taken by the sophistical pig, that is, the Ulysses whose artful service of custom exposed his artless fidelity to his faithful wife to be sham virtue. The new popular judgment cannot be awed by Ulysses' virtue because it sees only his art, and because the new

---

89. *Essays* 403–5. 90. Euripides, *Troiades* 914–65.
91. *Essays* 391–92. 92. Plutarch, *Gryllus* 986f–88f.

learning is likewise as indifferent to virtue as it is attentive to art, wisdom will be "justified by her children," as Bacon says in concluding book 1. But these children include gluttons and tyrants as possibilities that replace the son of man.[93] However much the new knowledge may save body and soul as no divine promise can, it is not more successful in producing the "body changed."

In the course of presenting his self-justifying human proofs, Bacon rank orders artful and civil merit, includes "moral and private virtue," gives a list of Roman emperors, and makes it clear that Caesar was not an emperor: like Alexander, Caesar founded an empire with the force of arms, but he did not rule this empire by legitimate succession. We now know for sure, then, that when Bacon referred earlier to a list of emperors, he referred to Machiavelli's list that included Caesar as an emperor, rather than to his own, which does not. In *Discourses* 1.10, where Machiavelli gives his list of emperors, he does so as a part of a discussion of the kinds of men who are praised and blamed, a discussion that considers the possible justification of the tyrant-founder. By turning our attention to Machiavelli at the very end of the human proofs, Bacon invites a final comparison of his ranking and list with Machiavelli's.

In *Discourses* 1.10, the highest of all praised men are the founders of religions, the next are founders of republics or kingdoms, followed by commanders of armies and those who extend their kingdom or republic, literary men of different kinds and different degrees of excellence, who are separate but who may be "joined" to the martial men, and, finally, all others who are praised according to their arts and professions. The blamed are the destroyers of religion, the overturners of republics and kingdoms, the enemies of virtue, the enemies of letters, and the enemies of the arts that bring utility and honor to human generation. Machiavelli mentions the blame of the enemies of virtue, but he does not mention the praise of the virtuous. The omission is not surprising in the light of *Discourses* 1.9, where Machiavelli excuses the founder's making himself alone by fratricide and the murder of associates. The reputation of the founder depends on his being a successful founder, whose wisely contrived crimes serve the common good, understood as the selfish ways that "men use the prince's authority for themselves."[94] Machiavelli's founder must have the courage for extraordinary homicides, and while vice can always be blamed, such an understanding of virtue cannot be praised. Indeed it could not, because such virtue serves solitary

93. Matt. 11.18–19.   94. Mansfield, 64.

striving. In fact, the deeds are excused by their being solitary, by their being denied praise, and by the selfish common good produced in their image.

In *Discourses* 1.10, after he distinguishes between Caesar's tyranny and Romulus' nontyranny, Machiavelli blurs the distinction between them. Romulus' crimes led to a reordering, as opposed to Caesar's destruction of a corrupt city. But this comment comes after the list of emperors, which will show a prince in a republic that good emperors deserved more praise than bad ones in the "twenty-six emperors from Caesar to Maximinus." Of course Caesar was not an emperor. In fact, he was alone precisely to the degree that he was not; that is, precisely to the degree that he neither depended on nor took from successors. For Machiavelli to excuse the founder requires excuse for the tyrant, because both presume the absolute coincidence of the common good and their own good insofar as they are alone. The tyrant believes that he might be alone and never need those who are his successors; he would be one who never needs to take, and these hopes lead him to his extraordinary deeds of appropriation. Obviously, founders such as Caesar, and even such as Romulus, are not simply alone, because they can be praised and blamed, as *Machiavelli* does in *Discourses* 1.9 and 1.10. To be alone truly requires being the bestower of praise and blame, not being the receiver of them. At the end of *Discourses* 1.10, Machiavelli says that he will give some excuse to princes who could not order a city without falling from rank but no excuse to princes who found a city and keep their rank *(grado)*. In the context, to have rank is yet to be praised and blamed; even princes who become as alone as princes might be can never wholly be alone, so that they cannot be wholly excused, while a prince who keeps his rank and is wholly dependent on successors and others from whom he might take can have no excuse. Caesar's attempt to govern praise and blame by his successors' restraint on free speech simply demonstrated his dependence on praise and the wiles that give freedom to the writer even when there is no freedom.[95]

The tyrant's solitude would excuse if it were possible. Machiavelli's perfect solitude as a writer, presumably above and beyond all excuse, makes use of the tyrant for whom there can be no complete excuse or justification and yet who must be a founder for the sake of any excuse. But Machiavelli's harsh use of the founder-tyrant can be completely justified or excused insofar as it is virtually alone, or virtually beyond all excuse, because Machiavelli claims that what is

95. Ibid., 68–69; cf. Plutarch, *Alexander* 7.

wholly good for himself, his perfect solitude as a founder-writer, can be perfectly harmonized with the common good of every patria. For Machiavelli, there is no imitation of the gods that is not ultimately commensurable with the common, *political* good. That is, Machiavelli teaches that the tyrant is perfected in the perfected, godlike writer, so that for himself and for others, virtue is worthless because virtue presumes the disharmony between public and private goods as well as a conventional freedom from the needs causing the tyrant to take or the writer to write.

Both Socrates and Jesus wrote nothing, but Jesus' unabating seriousness shows his difference from Socrates and from the later ancients who could always turn a joke. Machiavelli follows Jesus because Machiavelli understands the godlike solitude of the writer to be like the god who could not but create, however much he is the original source of praise and blame. For Machiavelli, all godliness is praised as much as it is revered. In this he learns from Christianity, whose God is, for the sake of the body's salvation, more often praised than held in awe.

Now, Bacon's criticism of Machiavelli is that his imitation of Christianity failed properly to conquer the Christian empire because it was unable to restrain the Christian passions. Bacon's new learning replaces Machiavelli's political science because it proceeds beyond Machiavelli's political horizons and because the new patria of science will aim to replace any patria that might be merely one's own. But as with Machiavelli's political science, for the new learning there is never a godlike freedom from need: not even God is so free. Bacon knows that his project for the new learning *is* Machiavellian; despite its ambition it can never escape the horizon of political life: it necessitates the use of the tyrant, requiring excuse for the tyrant when in fact no such excuse is possible. In this important sense, Bacon is politically more realistic than Machiavelli, who thought that, in using the tyrant, the writer can be a perfected tyrant whose private good is wholly or justifiably compatible with the common good. As such a writer-tyrant, Machiavelli thought he could be both the source of praise and blame and praiseworthy by reference to the common good. But by this account there is simply no kind of gracious or beautiful human freedom.

Bacon's list of praiseworthy men differs from Machiavelli's in several decisive respects. Bacon lists the inventors of the arts in the place Machiavelli gives to the founders of religion. The founder of the method that discovers for all replaces the founder of the religion of the God who used his Son for the sake of all. Likewise, Machiavelli joins the literary men to the commanders of armies and extenders of countries, so that Machiavelli is a conquerer armed with a pen rather than the sword. But Bacon presents those who found and who continue an empire to show, in the manner of Plato, that there can never be an order to be traced to an unarmed founding. Bacon's knowledge of the ancient utopian teaching is far more sober than Machiavelli's utterly unrealistic "realism." For all of his remarkable art, Machiavelli is crafty but never puckish. To be open to the moderating effects of comedy, a person must know that while there are limits to the human good, that good has some measure of grace or beauty. Machiavelli sees only shameless need, and so unlike Bacon and the later ancients, he cannot see the comedy of impossible foundings. Consequently, he cannot see the danger in trying to master them.

Machiavelli is silent about the praiseworthiness of moral virtue, but moral virtue is included in Bacon's ranking. While Machiavelli's ranking is wholly concerned with political possibilities, Bacon's demonstrates the tension between honorable political possibilities, the freedom of the mind's activity, and praiseworthy moral virtue that stands in between. Moreover, Bacon's list of emperors differs from Machiavelli's list to which Bacon has earlier referred. Unlike the first such list, which proved to be Machiavelli's, the present one contains no references to the tyrannical Severus, the best "almost" of the first list.[96] And Machiavelli's list of *Prince* 19 begins where Bacon's list ends, with Marcus "the philosopher," for while Commodus is the last of the six mentioned in the list, Bacon does not discuss Commodus except to say that he, along with Caracalla and Heliogabalus, dishonored the name of Antoninus. Even though Bacon's list ultimately looks beyond the six princes he mentions, and even though it presents the account of Nerva's revenge, his list still distinguishes between tyrants and lawful or just rulers.

Bacon knows that the politics of the new learning is inevitably Machiavellian, even though it can never accomplish Machiavelli's vain purpose. But Bacon is not for this reason a Machiavellian. For Bacon, Machiavelli did not see how far history had changed the world, so that Machiavelli did not see just how and why virtue had come to be worthless in human affairs. To see this properly, a person had to appreciate the utopian teaching of the later ancients, to see the grace of moral virtue and the charms of political philosophy and comedy. For Bacon, such knowledge is necessary to detect the ultimate limits

96. *Advancement* 303–6; *Essays* 447–78.

of the new learning, which never outstrips what Bacon in book 2 says is Machiavelli's *immoral* orientation by honor. Bacon has no dangerous illusions about perfect justice wielded by a pen or by any art, for that matter. Therefore he can tell how and when necessity pushes the modern age to strain the bounds of any possible justice. Historical necessity governs Bacon's project, but Bacon is not blinded by the cure he prescribes for modern times. Bacon is just because he knows the difference between the necessary and the good. He tells us justly that urgent necessity, not the truth, requires the urgent reordering of the arts and sciences.

PART II

THE DIVISION OF
THE SCIENCES

[7]

Introduction: Divine Revenge and the Problem of Political Science

After presenting his apology, Bacon's principal task is to describe the arts and sciences as they must be pursued in his charitable age. In such an age, the furthest end of knowledge is the unity of theory and practice, and the highest human ambition the discovery of the means to this unity: a single method or means for the discovery of all things and the commanding of nature in action.[1] These ends govern the division of the sciences and the account of the sufficient and insufficient arts and sciences. But from the apology, we know this agenda to be more urgent than true because it does not accord with the relationship between theory and practice as understood by the later ancients. For them contemplation could not be applied to practice so as to satisfy every practical need. Rather the freedom of contemplation could be the model for practical virtue, which could be neither as free nor as self-knowing as its model and which could be moderate but not perfectly just. Furthermore, contemplation could never demonstrate a perfect unity or perfect heterogeneity of the arts and sciences, either one of which would satisfy the desire for practical mastery and moral certainty. Bacon's new method aspires to such unity: it aspires to satisfy the most extravagant hopes to meet, rather than to be conventionally free from, every political need. But Bacon knows these ends to be as impossible and dangerous as it is necessary in our age to pursue them. Therefore the "enigmatical" discussion continues the themes of the apology with an eye to the relations of the arts and sciences.

1. *Instauration* 20–21, 24, 115.

If contemplation and action cannot be compatible as the modern project requires, then it cannot be true that a single method for conquering nature can disclose a comprehensive science of the whole of man and nature. More particularly, it is not possible that such a method could disclose the human good and the proper relations between those who claim to know and to minister to the human good. In fact, from what we know so far, the pursuit of such a method is a charitable hope whose effect on the human good can be judged only in the light of the wisdom of the later ancients. As it turns out, the real question posed by the pursuit of a new universal method is not its ways and means but its effect on the relations between the traditional claimants to the knowledge men need for practical life: the poets, the rhetoricians, the sophists, and the statesman—the very individuals in any city who, along with the other productive artisans, must be ordered by any political science. In the division of the sciences, we discover Bacon's comprehensive understanding of the causes of the whole, we see his account of the human good, and we see his detailed understanding of the later ancients' political science.

[329–343]

Because the division of the sciences is governed by historical necessity, it is fitting that Bacon begins with the sciences that exercise the memory: the several varieties of history. Bacon divides history into natural, civil, ecclesiastical, and literary. Whereas the first three are extant, the fourth, he says, is deficient. The reason is that no one has propounded to himself the task of describing the "general state of learning from age to age" as has been done for nature, the civil state, and ecclesiastical works. A just story of learning is wanting. The use of such a history is not for curiosity or satisfaction; rather it will serve to make men "wise in the use and administration of learning." Bacon likens such a history, the absence of which is similar to the statues of Polyphemus with his eye out, to ecclesiastical history, for just as a divine would become wise not by reading St. Augustine or St. Ambrose but rather by a reading of ecclesiastical history, so the learned would become wise by the reading of such a literary history.

Natural history is divided into three parts, the history of creatures, the history of marvels, and the history of the arts, of which the first is extant and perfect but the latter two are so poorly handled as to be considered deficient. There is no history of natural irregularities, but such a work has been honored by the precedent of Aristotle's *De mirabilibus*. The important use of such a history is to satisfy "curious and vain wits," to "correct the partiality of axioms and opinions," and to open the passage that exists from the wonders of nature to the wonders of art.

The history of art, or the history of "nature wrought or mechanical," has been ignored because of the long-held prejudice against "matters mechanical." This arrogance was justly derided in Plato, where Socrates mocked Hippias' offense at Socrates' homely examples of the beautiful.[2] The truth is that the highest examples do not give the best information, as can be seen from the example of "the philosopher's distracted fall into the well."[3] Aristotle notes well, says Bacon, "that the nature of every thing is best seen in his smallest proportions."[4] Just as he investigated the commonwealth by studying first the family "and the simple conjugations of man and wife, parent and child, master and servant, which are in every cottage," so too "the nature of this great city of the world and the policy thereof must first be sought in mean concordances and small portions." The history of art is the most important for natural philosophy that will be "operative to the endowment and benefit of man's life," because it will suggest the unity of the arts and will give true illumination of nature's causes. Just as Proteus did not change shape until forced to do so,[5] so the "passages and variations of nature" appear fully when nature is denied liberty by the "trials and vexations of art."

Civil history is divided into three kinds, likened to "three kinds of pictures or images": unfinished, perfect, and defaced. Memorials are unfinished histories, antiquities are defaced histories, and perfect histories are, of course, perfect. "Just and perfect history" is itself divided into three kinds: chronicles, representing times; lives, representing persons; and narrations or relations, representing actions. Although chronicles are the "most complete and absolute kind of history and hath most estimation and glory," the history of lives is more useful, and narrations are truer and more sincere. The reason is that the history of times represents "the magnitude of actions and the public faces and deportments of persons" and ignores the "smaller passages and motions of men and matters." But just as God often hangs the greatest weight on the smallest wires, the history of times focuses on the "pomp of business" rather than on "the true and inward reports thereof." Well-written histories of lives, mixing great and small, public and private, "must of necessity contain a more true,

2. *Greater Hippias* 291a3–4.
3. Plato, *Theaetetus* 174a4; Diogenes Laertius 1.34.
4. Aristotle, *Physics* 184a10–b14; *Politics* 1252a24–53a39.
5. Virgil, *Georgics* 4.387–414.

native and likely representation." Likewise, the narrations of actions like the Peloponnesian War, the expedition of Cyrus Minor, and the conspiracy of Cataline must be "more purely and exactly true" than histories of times, because they deal with events that can be comprehended by a writer. The same is less true for the history of times because a writer meets up with "many blanks and spaces," to be filled with the author's own wit and conjecture.

Bacon says that the history of times has been distributed by God's providence, for God "ordained and illustrated two exemplar states of the world" for arms, learning, moral virtue, policy, and laws. The histories of Greece and Rome stand midway between "antiquities of the world" and modern histories. Bacon then speaks of the deficiencies of the various kinds of history. The antiquities of the world cannot help being deficient, for antiquity is like fame, *caput inter nubila condit,* her head is muffled from our sight.[6] The history of the "exemplar states" is "extant and sufficient," but Bacon wishes that the history of Greece from Theseus to Philopoemen, when Greece was overtaken by Rome, and that the history of Rome from Romulus to Justinian, who was truly said to be *ultimus Romanorem,*[7] might be set into a "perfect course." To provide such a "course" would be simply to supply and continue the texts of Thucydides, Xenophon, Livy, Polybius, Sallust, Caesar, Appianius, Tacitus, and Herodian. But doing so would be a matter of "magnificence," which is not required, or "of supererogation."

Bacon says that modern histories are for the most part mediocre. He will not speak of foreign countries because he will not be *curiosus in aliena republica,*[8] but he cannot fail to show the king the unworthiness of the history of England and the recent history of Scotland. It would be an honor if the island of Great Britain "as it is now joined in monarchy for the ages to come, so were joined in one history for the times passed; after the manner of the sacred history, which draweth down the story of the Ten Tribes and of the Two Tribes as twins together." Should such a history prove to be inexact in relation to the greatness of its subject, there is "an excellent period of a much smaller compass of time," the period from the uniting of the roses to the uniting of the kingdoms. In this period of time there have been the "best varieties" of any equal number of successions in any hereditary monarchy, culminating in "this most happy and glorious event"

6. Virgil, *Aeneid* 4.177, "with her head hidden among the clouds."
7. Tacitus, *Annales* 4.34; Suetonius, *Tiberius* 61.
8. Cicero, *De officiis,* ed. Atzert (Leipzig: Teubner, 1963), 1.34.125; Tacitus, *Historiae* 1.1.

when Britain, separated from the rest of the world,[9] is "united in it-self." The oracle of rest given to Aeneas, *antiquam exquirite matrem*,[10] has now been fulfilled by England and Scotland, which, like "massive bodies," underwent "prelusive changes and varieties" before settling in the king and his "generations."

Bacon finds it strange that the virtue of present times has been so little esteemed that the history of lives has not been more frequent. Even though most states are monarchies, and there are few sover-eign princes or absolute commanders, yet there are still many wor-thy persons who might be the objects of such histories. In present times, Bacon complains, such good fame has been allowed to lay waste, so that its cultivation is deficient. Another part of history must not be forgotten. They are Tacitus' annals, treating "manners of estate," and journals, treating "acts and accidents of a meaner nature." Like-wise, Bacon says that he is not ignorant of a kind of writing used by "grave and wise men," "ruminated history" or "political discourse and observation" on, but separate from, a "scattered history of those ac-tions" thought worthy of memory. These books are more policy than history and so should be discussed "at that place,"[11] because "mix-tures are things irregular, whereof no man can define." History is one thing, which represents events "together with the counsels," but the observations and conclusions based on history should be left "to the liberty and faculty of every man's judgement."

Another kind of mixed history is the history of cosmography, composed of natural history, civil history, and mathematics. In pre-sent times this part of learning has "obtained most proficience" of all other parts. Although the ancients had knowledge of the antipodes, as, according to Bacon, can be seen from a statement by Virgil,[12] this knowledge consisted in "demonstration and not in fact, and if by travel, it requireth the voyage but of half the globe." But "to circle the earth, as the heavenly bodies do" is something possible only in recent times. Therefore recent times "may justly bear in their word" not only "further yet," in precedence of the ancient "no further," and "imit-able thunder" in place of the ancient "inimitable thunder," as we see in the lines from Virgil's *Aeneid*,[13] but also "imitable heaven" because of the voyages "after the manner of the heavens" around the earth. The example of such voyages will "plant an expectation" of the ad-vancement of all the sciences, because it may seem that "they are or-

9. Virgil, *Eclogues* 1.64–66.
10. "Seek out your ancient mother" (Spedding's trans.), Virgil, *Aeneid* 3.96.
11. *Advancement* 453; see Chap. 11 below.   12. Virgil, *Georgics* 1.250.
13. *Aeneid* 6.590.

dained by God to be coevals, that is, to meet in one age." It is "already performed in great part," so that the learning of latter times does not "much give place" to the "former two periods or returns of learning, the one of the Grecians, the other of the Romans."

Ecclesiastical history is divided exactly as is civil history. Considered by itself, ecclesiastical history is divided into history of the Church, history of prophecy, and history of providence. The first describes the history of the militant Church, whether it be "in persecution, in remove, or in peace," and it is not deficient. The history of prophecy has two related parts, prophecy and accomplishment. This kind of history is deficient, with future work in it requiring the sorting of every prophecy in the scripture with the event that fulfills it, "throughout the ages of the world." This task when accomplished will confirm the faith and will illuminate the Church concerning unfulfilled prophecies. But with divine prophecies and "their author," a "thousand years are but as one day,"[14] so that prophecies are not fulfilled at once but rather proceed "through many ages," although the "height or fulness of them may refer to some one age." The history of providence contains the "excellent correspondence which is between God's revealed and His secret will." Even though God's secret will is "so obscured as for the most part it is not legible to the natural man," at times it pleases God to write it "in such text and capital letters" so that merely sensuous persons, who "hasten by God's judgements" and who do not think about them, are "nevertheless in their passage and race urged to discern it." This work, Bacon says, is not deficient.

Bacon concludes his division of history by noting that other parts of learning are "appendices to history." All the "exterior proceedings" of man are words and deeds, and history preserves the memory of deeds and words as "inducements and passages to deeds." Bacon regrets the loss of Caesar's book of apophthegms, for just as Caesar's history and the extant apophthegms have excelled all others, so Bacon supposes that Caesar's collection in his book would have excelled the collections of others. With this remark, he concludes the division of history.

In order to conform to the just law of brevity, we begin by noting that two forms of history appear to comprehend the whole of learning. The first is literary history, making the learned wise in the "use

14. Ps. 90.4; 2 Pet. 3.8.

234

and the administration of learning." The other is the history of cosmography, recounting man's imitation of the heavenly bodies in the circumnavigation of the earth. It is obvious that this circumnavigation depends upon the productive arts, in particular on the navigational arts, with regard to which recent times are superior to divinely ordained exemplary times. Modern times have surpassed the limited hopes of the ancients, who, according to Bacon's reference to Virgil, were limited by the knowledge of punishments inflicted upon those who imitate the artful power of the gods.[15] The model of such conquest will assure that all the sciences will flower at the same time. Obviously the wisdom for organizing and setting the sciences on this path will be contained in a literary history or in an account of the "general state of learning from age to age." Such a history must have its one eye focused on the productive arts, and it is again obvious that the crucial literary history is contained in the present treatise that Bacon gives to the king. The core of the present treatise is, therefore, the history of the mechanical arts, which is the most "radical and fundamental" for a "natural philosophy that is not speculative" but is rather "operative to the endowment and benefit of man's life." This finding should not by now be surprising. And neither should it be astounding that Bacon likens a perfected literary history to the sighted but single-eyed Polyphemus, who was, we must remember, the terrible and cannibalistic Cyclops. But the precise importance of this remark for the hoped-for future unity of all the sciences is spelled out in Bacon's discussion of the history of the mechanical or productive arts.

Bacon says that the history of the mechanical arts is deficient because it has been thought base to "descend to inquiry or meditation upon matters mechanical." Bacon praises Plato for having Socrates deride Hippias' sophistical and "vain and supercilious" arrogance, and he refers to the example of Thales to demonstrate his claim that the highest things are not the most secure. Likewise, as proof that small and mean things better disclose great things than conversely, Bacon refers to Aristotle's procedure in the *Politics*, which demonstrates the principle that "the nature of every thing is best seen in his smallest proportions." For this reason Aristotle inquired into the nature of the "commonwealth" first in the "family . . . which are in every cottage." In the same fashion, says Bacon, "the nature of this great city of the world and the policy thereof" must be discovered by way of the smallest parts.

15. *Aeneid* 6.585–94.

Now, Bacon says that Socrates mentioned the fair virgin as a homely example of the beautiful as such. But this is not in fact what transpired in the *Hippias Major*. Actually Hippias, not Socrates, gave the example of the fair virgin. Because he loved whole speeches, the tools his sophistical art uses to encompass both private and public matters, Hippias accused Socrates of not seeing wholes and of always cutting things into parts.[16] In the course of the dialogue we learn that it is really the sophist who forgets the difference between the whole and the parts of the city and in fact knows nothing of the perplexing relations of dependence between wholes and parts. In the *Apology*, Socrates said that the artisans are the only members of the city to have any knowledge of the noble things, as opposed to the politicians and the poets. But Socrates ultimately belittled even the artisans' knowledge, and he certainly denied being a sophist.[17] In contrast, when he presents Plato as the proponent of the productive arts, Bacon assimilates Socrates' partial praise for the productive arts to the sophistical claim that by artfully reducing all wholes to parts one can master all things private and public. To demonstrate his point about Socrates, Bacon mentions Socrates' reference to Thales in the *Theaetetus*. But in doing so, Bacon shows that he has falsified his account of Socrates' estimation of the productive arts.

In that dialogue, Socrates tells the geometrician Theodorus that the philosophers are superior to the rhetoricians and poets and only appear to be inferior, as did Thales, because they do not attend to *any* practical affairs. But Bacon knows that this contrary view is also incorrect, because he earlier acknowledged that Socrates brought philosophy down from the heavens, and so in the next breath he refers to Aristotle's discussion of the elements of the city in book 1 of the *Politics*. There Aristotle shows that Thales used his theoretical knowledge of the heavens to acquire enormous wealth.[18] In referring to Aristotle's method of beginning with simple elements and conjugations, Bacon claims that, according to Aristotle, to discover the nature of parts is to discover the nature of the whole of the city. But Bacon does not report that Aristotle's intention in discussing the parts of the city is to establish the essential differences between the kinds of rule corresponding to the parts of the city and the city as a whole. Nor does Bacon point out that, according to Aristotle, the limits of wealth getting, so well perfected by Thales, are determined by the

16. *Hippias Major* 281b5–d2, 287c8–80d3, 301b2–4e9.
17. *Apology* 22c9–d2. 18. Cf. *Theaetetus* 174a4–b6; *Politics* 1259a5–23.

236

inferiority of the affairs of the household, which attend to mere life, to the affairs of the city, which attend to the noble or the good life.[19]

Bacon knows perfectly well that for Plato and Aristotle, philosophy must chart a course between absorption in and abstraction from the affairs of the practical world. And insofar as the philosopher is the model for the statesman, the wisdom of those who deny a tension between theory and practice—the early ancients like Thales, the sophists, the rhetoricians, and the poets—is revealed to be dangerous because it elevates the several claims of the productive arts. Bacon knows with the later ancients that the wealth-getting art belongs to the household. He knows that the new knowledge elevates the wealth-getting art by way of the mechanical arts. He must therefore also know that the new knowledge takes the household to be the model of all other forms of rule, and he must know with Aristotle that such a model is the same as the Cyclops' savage rule.[20]

Bacon opens the division of the sciences by telling us that the mechanical arts are the core of the perfect unity of the arts and sciences. But he actually tells us that his task is to demonstrate the proper place of the productive arts in the whole of the arts and sciences and that this task is bounded by the traditional question regarding the order and place within the city of those who claim that their knowledge is indispensable for political life. While all these claims are always heard, not all are plausible, and some are dangerous because they mistake the differences between the parts of any city and between the different kinds of authority. And the new knowledge must be subjected to the traditional critique, no matter how urgent and timely the new knowledge might be. The most important issues raised by Bacon's literary history concern the standards and principles by which civil history must be judged.

When he discusses the deficiencies of the three kinds of civil history, Bacon says that the defective antiquities of the world precede the first of the two kinds of the history of times, the extant and perfect histories of the exemplary states of Greece and Rome. Despite the extant perfection of these histories, Bacon says that he wishes for a "perfect course of history" from the beginning of Greece to the time when its affairs were "drowned and extinguished in the affairs of Rome" and from the beginning of Rome to the time of Justinian, who was the *ultimus Romanorum*, a term borrowed from Tacitus and

19. *Politics* 1252a7–23, 1252b27–34, 1256b26–58b8.
20. Ibid., 1252b15–27; Homer, *Odyssey* 9.114–15.

Suetonius. According to Tacitus, the last of the Romans was Cassius, and for so describing him a writer was accused by the friends of Sejanus. According to Suetonius, the last Romans were Brutus and Cassius, and for so describing them a historian was killed by Tiberius.[21] For Tacitus and Suetonius, the last of the Romans were the last defenders of the free republic; but for Bacon, the last Roman was Justinian, the emperor who closed the Athenian schools of philosophy and who built the famous Christian church at Constantinople.

A perfected course of the history of Greece and Rome had not been presented because the full history of Roman conquest had not been told. Such a civil history would recount the conquest of the "exemplar state" of Greece by Rome and the conquest of Rome by the very people of the East whom the Romans conquered. Furthermore, such a civil history would demonstrate modern civil history to be the legacy of the Romans' conquest by Christian love. Therefore the civil history of the world is determined by a people about whose history Bacon is silent until he speaks about modern history and the modern history that completes the Delian oracle's charge to Aeneas—the history of Great Britain. Of course if Great Britain is to seek out an ancient mother, as the oracle charged to Aeneas, then the unity of Great Britain is the model for the unity of the whole world.[22] And Bacon explicitly says that the civil history of Great Britain completes the incomplete story of the "Ten Tribes" and the "Two Tribes as Twins together"; that is, it completes the legacy of the Jews.

Regarding such a history, Bacon borrows a phrase from Tacitus and from Cicero's *De officiis* to support his claim that he will not speak about foreign states. Cicero refers to foreigners and resident aliens who reside in Rome, and Tacitus refers to the disappearance of truthful history after the battle of Actium because of men's considering politics to be foreign to them, because of the desire to flatter, or because of the hatred of their masters. The ultimate source of these developments was that, after the battle, freedom required that political power be concentrated in one man's hands.[23] For Tacitus, truthful history is the history of how imperial ambition put an end to truthful history. It is obvious that Bacon's treatise is itself the comprehensive civil history of Greece, Rome, and the modern world, and if Tacitus demonstrates the reason for his attention to imperial Great Britain, then Bacon's civil history is a truthful account of an end of truthful history. Bacon's literary history of the arts and sciences is a

21. Tacitus, *Annales* 4.34; Suetonius, *Tiberius* 61.   22. *Aeneid* 3.90–98.
23. *De officiis* 1.34; Tacitus, *Historiae* 1.1.

truthful and truthless—that is, enigmatical—civil history of the charitable legacy of the Jews. Because of this legacy, Bacon cannot openly tell the truth about the fundamental principles of civil history or give an open account of the parts of any well-ordered political life.

Modern times require a new approach to the ordering of civil life. In particular they require that the noble statesman be belittled by the promises of the practical arts and, especially, the arts of acquisition. For this reason, while the history of times is most complete and "hath most estimation and glory," the history of lives is superior "in profit and use." Bacon says that the history of lives might compound the great and the small and the public and the private, and for this reason the history of lives, being the "true and inward" resorts of the outward pomp of the history of times, is a "more true, native, and lively presentation" than the history of times. The truth about private matters is the more useful history because it exposes the private resorts behind public pomp and glory. One certainly need not extoll public pomp and glory, because in present times, men must be wisely led to hope for the private acquisition that was the secret object of Roman honor. Modern civil wisdom cannot attend to the difference between private need and any public virtue or self-sufficiency. Consequently it cannot attend to the differences between different kinds of authority, and especially to the difference between the household and any free city. Bacon begins his division of the sciences by showing that modern civil wisdom eliminates the political philosopher and statesman from their rightful places in the city. And as we see in the sequel, he concludes the introduction by questioning the places of the poets and the rhetoricians.

[343–346]

Poetry is the art that exercises the imagination. Bacon says that it is restrained with respect to the measure of words, but in "all other respects" it is extremely licensed. It exercises the imagination because, "being not tied to the laws of matter," it can match what nature has severed and vice versa, so as to make "unlawful matches and divorces of things" or, as Horace says, *pictoribus atque poetis, etc.*[24] Poesy is to be taken in two senses, with respect to words, in which case it belongs not by itself but as part of the knowledge that pertains to understanding and reason and in particular to rhetoric, or with respect to matter, in which case it is a kind of history, "feigned his-

---

24. "Painters and poets have always been allowed to take what liberties they would" (Spedding's trans.), Horace, *Ars poetica* 9.

tory." The purpose of feigned history is to satisfy the mind when nature cannot. When the acts and events of "true history" do not satisfy the mind, then poesy makes up for the lack. Poesy supplies deficient greatness and merit of virtue and vice, in accordance with retribution and revealed providence, and rareness and variation. For this reason, poesy serves magnanimity, morality and delectation, for which reason it is always thought to be somewhat divine. Poesy raises the mind, "submitting the shows of things to the desires of the mind," whereas reason submits the mind to nature.

Poesy admits of two divisions, one according to the divisions of history, of which it is a parallel kind, and the other according to a division into narrative, representative, and allusive. Narrative poesy is the imitation of history, representative poesy is an "image of actions as if they were present," and allusive, or parabolical, poesy is a narration used "to express some special purpose or conceit." Parabolical poesy was used more in ancient times because it was necessary to express any "point of reason" that went beyond vulgar understanding. Even now, however, parabolical poesy is useful "because reason cannot be so sensible, nor examples so fit." There is another use of parabolical poesy, for an opposite purpose to the one just mentioned. This other kind is used to "retire and obscure" what is taught, as when the secrets of religion, policy, or philosophy are related in fables. Such use of fables is authorized in "divine poesy," and it was well used by the heathens, as can be seen from the fable of the giants, expounded by Virgil. This fable shows that, when princes and monarchs have put down rebels, the malignity of people, "which is the mother of rebellion," generates "libels and slanders and taxations of the state."[25] Likewise, the fable of Jupiter, Pallas, and Briareus shows that monarchies do not need to fear limitation of their power by "mighty subjects" as long as they "keep the hearts of their people." And the fable that Achilles was brought up by Chiron the Centaur has been "expounded ingeniously but corruptly by Machiavelli," who said that princes must know how to imitate the "lion in violence and the fox in guile" as much as "man in virtue and justice."[26]

In such poesy it is more likely that the fable came first and the exposition was then devised than that the "moral was first, and thereupon the fable framed." Chrysippus was simply vain to ascribe the doctrines of the Stoics to the "fictions of the ancient poets."[27] Ba-

25. Virgil, *Aeneid* 4.178–80.   26. *P* 18; Homer, *Iliad*, 11.832.
27. Cicero, *De natura deorum* 1.15.

con will express no opinion as to whether "all the fables and fictions of the poets were but pleasure and not figure." Of the poets now extant, even Homer, who was taken to be scripture by the "later schools of the Grecians," cannot be said to have had such an "inwardness in his own meaning" in his fables. What they meant originally is hard to say, however, because he did not invent most of them. Bacon reports no deficiency in poesy, because since it springs from "the lust of the earth, without a formal seed, it hath sprung up and spread abroad more than any other kind." But since he is eager to give the poets their "due," Bacon remarks that "for the expressing of affections, passions, corruptions, and customs, we are beholding to poets more than to the philosophers' works; and for wit and eloquence, not much less than to orators' harangues."

Bacon argues that poetry is the work of the free imagination because it is not "tied to the laws of matter." To attest the commonly accepted liberty of the poets, who make unlawful matches and divorces of things, he quotes Horace. But actually, Horace's remarks do not exactly correspond to Bacon's use of them. According to Horace, the poets take license to connect and disconnect the things of nature, but Horace says that such license must not extend to the matching of the savage and the tame, the serpent and the bird, or the lamb and the tiger.[28] Bacon says that poetry serves magnanimity, morality, and delectation but only by submitting the appearances of things to the "desires of the mind" and that in this respect it differs from reason, which submits the mind to the nature of things. But in the *Novum organum*, Bacon says that reason bows to nature in order to conquer her. Moreover, in discussing the history of the arts, Bacon noted that, when nature is vexed by art, she is likened to Proteus, who, when "straitened and held fast," would change his shape.[29] What Bacon here calls the feigned liberty of poetry is actually promised by the scientific mastery of nature, and so the liberty of poetry is actually to be found in the new method for discovering all things.

But even so, in the present context Bacon says that poetry is not free from history, memory, or reason that submits to the nature of things. We are forced to wonder, then, just how poetry serves magnanimity, morality, and delectation if, when it occurs as the possibility of science, it sees past the essential differences of nature and the difference between need and virtue. The answer is discerned from

28. *Ars poetica* 9–14.   29. *Novum organum* 1.129.

Bacon's appeal to Horace, who, in the context to which Bacon refers, says that poetic liberty—the liberty we know to be possible only by way of science—is like a sick man's dreams. This is the poetry whose wisdom about peoples and princes includes the "corrupt" wisdom taught by Machiavelli in chapters 18 and 19 of *The Prince*, a wisdom counseling the prince to act as both beast and man and not to act according to virtue but only to appear to do so.

As soon as he criticizes such masterful poetry, Bacon makes it clear that it is not the poets as such who possess wisdom but rather those who can properly interpret the poets. Therefore Chrysippus was wrong to think that the poets knew anything about the origins and nature of the gods.[30] But Bacon does admit that in regard to affections, passions, corruptions, and customs the poets know more than do the philosophers and vie with the orators for excellence in wit and eloquence. On the one hand, the poets as such know nothing about the origin of the gods, but on the other hand they vie with the philosophers and the rhetoricians in regard to knowledge of the human things. Now, these poets are the very ones the later ancients said have a place in the city only as long as they serve moral virtue and, properly chastened, do not aspire to the comprehensive questioning of philosophy. While these proper poets cause the necessary gods to speak, according to the poets' knowledge of affections, passions, corruptions, and customs, they can be "scripture" only insofar as they are ignorant of their own status as causes of these gods. They are ignorant of the source of the gods in the very productive arts of which they, more than any other of the separate arts, claim to be the most exalted. But in the present age, such proper poets who do not interpret themselves have no place in political life. There is room only for the poets who know and discover all things: in Bensalem there is no philosopher or statesman, nor a proper poet, one who speaks through right opinion that he has not made. However, in Bensalem there are various imitative arts, and the inhabitants sing hymns and prayers.[31]

Bacon says that the masterful poets who submit the appearances of things to the "desires of the mind" are themselves divine and can only be those who actively serve the scientific conquest of nature. These new poets eschew right opinion and speak for themselves; only they have a cognitive place in the new, charitable city. This exclusive place does not accord with the political teaching of the later ancients,

30. Cicero, *De natura deorum* 1.15.39.
31. *New Atlantis* 161–64, 166; Plato, *Republic* 377b5–83b7, 595a1–608b10.

for whom the artisans, the poets, the rhetoricians, the statesman, and even the sophists were all necessary parts of any city, which had to be balanced in accordance with political philosophy for the city to be its moderate best. If any of these claims attempted to stand alone, apart from the model of the philosopher, the city would be tempted to its immoderate worst. According to a later remark, Bacon says that reason and imagination come together not just in the new poetry but also in the art of rhetoric.[32] The rule of the new poetry—or the rule of the new science—is the same as unbridled rule by rhetoricians. To understand just what this statement means, we have to understand the later ancients' political science that prescribed correct relations between the philosophic openness to the whole, the statesman, the poets, and the sophists and rhetoricians. These are the parts of any political order, and in the modern age the relations between them have become disturbed. These new relations are explained in the division of the sciences that exercise man's reason.

32. *Advancement* 382, 409; Chap. 9 at nn. 1–2, 33–39 below. In the *De augmentis* Bacon distinguishes between style or words and matter in the broad division. In the *De augmentis,* all of the allusions in the discussion of poetry are omitted, and the chapter is enlarged with discussions of the fables of Pan, Perseus, and Dionysus.

[8]

Natural Philosophy: First
Philosophy, Politics, and Wonder

[346–366]

Bacon begins the division of the sciences corresponding to reason by making a broad division according to sources. Some knowledge descends from above from divine revelation, and some springs from beneath, from the light of nature, which consists of the mind's notions and the reports of the senses. Knowledge that men receive from teaching is "cumulative and not original." According to the two originals, knowledge is divided into divinity and philosophy, and in philosophy man's thought penetrates to three objects: God, nature, or himself. All things are stamped with a "triple character," the power of God, the difference of nature, and the use of man.[1] The parts of knowledge are not just like lines that "meet in one angle"; rather they are "like the branches of a tree that meet in a stem, which hath a dimension and quantity of entireness and continuance, before it come to discontinue and break itself into arms and boughs." Therefore, before Bacon enters the former distribution, he says that it is good "to erect and constitute one universal science, by the name of *Philosophia Prima*, primitive or summary philosophy, as the main and common way."

Bacon first says that he is doubtful as to whether this science is deficient. He sees a mixture of natural theology, logic, the principles of natural philosophy, and the part of it treating the soul, but this mixture is bombast rather than anything solid. By "first philosophy," Ba-

1. This sentence, which is important in itself and for the later discussion of the good, is omitted in the *De augmentis*. See Chap. 10 at nn. 1–15 below. He refers to human art but not to man's use. *De aug.* 496.

con means simply that it should be a "receptacle for all such profit-able observations and axioms as fall not within the compass of any of the special parts of philosophy or science" but are "more common and of a higher stage." There are many such axioms and observa-tions uniting justice and mathematics, mathematics and logic, and natural philosophy and natural theology, as Bacon shows by quoting Ovid and Ecclesiastes regarding the eternal quantum of nature, and government, religion, and nature, as Bacon shows by referring to Machiavelli's argument that the way to establish and preserve gov-ernments is to reduce them *ad principia*.[2] Nature and government, music and affection, justice and rhetoric, music and the behavior of light on water, as shown by reference to Virgil,[3] and organs of sense and organs of reflections all can be seen reduced one to the other as "the same footsteps of nature, treading or printing upon several subjects or matters." After discussing first philosophy in greater de-tail, Bacon abandons his initial hesitancy and justly reports it to be deficient according to his understanding of the science.

When he proceeds next to natural theology, Bacon calls first phi-losophy the "common parent" of the "former" division of philoso-phy into divine, natural, and human. First philosophy is "like unto Berecynthia, which had so much heavenly issue."[4] After the account of natural theology, Bacon divides natural science or "theory" (he no longer speaks of "natural philosophy") into physic and metaphysic. Straightway Bacon is concerned to note that he uses the word "me-taphysic" differently from the manner in which it is received. In this and in other instances, even though his understanding differs from ancient views, he is careful to "keep the ancient terms." He wishes to be clear, but he wishes to depart from the ancients as little as possi-ble. Bacon marvels at Aristotle, who wished to "confound and extin-guish all ancient wisdom," which served him well for "glory" and at-tracting disciples. Regarding human truth it happens that *veni in nomine Patris, nec recipitis me; si quis venerit in nomine suo, eum recipietis.*[5] But this comment was made of the deceiving Antichrist, so that we can see that coming in one's own name with no regard for "antiquity or paternity" is not a good sign of truth, even though it might guar-antee being received. It seems to Bacon that Aristotle "learned that humor of his scholar" so that, "as Alexander would conquer all na-tions, so Aristotle would conquer all opinions." By doing this, Aris-

2. Ovid, *Metamorphoses* 15.165; Eccles. 3.14; *D* 3.1.
3. Virgil, *Aeneid* 7.9. 4. Ibid., 6.777–87.
5. "I have come in my Father's name, and ye receive me not; if one come in his own name, him ye will receive" (Spedding's trans.), John 5.43.

totle may receive from those of "bitter disposition" a title like that which Lucan gave to Alexander: a fortunate robber of learning.[6] For Bacon, who wishes to maintain a "sociable intercourse between antiquity and proficience," it seems best to "keep way with antiquity *usque ad aras*,"[7] and so to keep the ancient terms, even though the "uses and definition" may be changed. This practice accords with the "moderate proceeding in civil government," where although there is some change, as Tacitus says, *eadem magistratuum vocabula*.[8]

Bacon intends first philosophy and metaphysics to be two different things, although hitherto they have been considered to be the same. The first is the common parent of all knowledge while the latter is a branch of natural science. First philosophy contains the principles and axioms common to the "several sciences" and the inquiry "touching the operation of the relative and adventive characters of the essences as quantity, similitude, diversity, possibility, and the like" as they have efficacy in nature and not in logic. Bacon has separated natural theology from the rest of the sciences, so that it cannot be confused with metaphysics. The question is, then: what remains as the province of metaphysics?

Bacon preserves enough of the "conceit of antiquity" to say that physics considers what is "inherent in matter" and therefore transitory and that metaphysics treats what is "abstracted and fixed." Physics treats what supposes in nature "a being and moving" and metaphysics treats what supposes "further in nature a reason, understanding, and platform." Just as natural philosophy is divided into inquiry regarding causes and the production of effects, so the inquiry into causes is divided, "according to the received and sound division of causes," into physics, treating material and efficient causes, and metaphysics, treating formal and final causes. Physics is midway between natural history and metaphysics for three reasons: because natural history describes "the variety of things," because physics describes the "variable or respective causes," and because metaphysics describes the "fixed and constant causes." As an example of a cause according to physics, Bacon quotes Virgil with respect to fire, wax, and clay,[9] and then he divides physics into its three parts regarding

6. *De bello civili* 10.20.
7. "As far as may be without violating higher obligations" (Spedding's trans.), Plutarch, *De vitioso pudore* 531d; *Praecepta gerendae reipublicae* 808b; Aulus Gellius, *Noctes Atticae* 1.3.20–21.
8. "The officials went by the same or old names," Tacitus, *Annales* 1.3.
9. *Eclogues* 8.80.

nature united or collected or nature diffused and collected. The three doctrines are the contexture and configurations of things, the principles and originals of things, and the variety and particularity of things, with the latter part but a gloss to natural history. These have been labored and are not deficient, but Bacon will not comment on whether they have been treated perfectly or truly.

Of metaphysics, Bacon says that the assignation of formal cause to metaphysics might seem useless because of the "inveterate opinion" that men cannot know "essential forms or true differences." Bacon agrees with this opinion insofar as it implies that "the invention of forms is of all other parts of knowledge the worthiest to be sought," but it is an ill discoverer who says there is no land because he can only see the sea. Plato was correct to say that forms are the true object of knowledge,[10] but he erred in thinking them to be "absolutely abstracted from matter." This mistake led him to turn his opinion "upon theology," which infected all of his natural philosophy. However, if man keeps an eye on "action, operation, and the use of knowledge, he may advise and take notice what are the forms, the disclosives whereof are fruitful and important to the state of man."

Bacon says that the forms of substances are "so perplexed that they are not to be enquired." Just as the "gross forms of the sounds which make words are infinite," the form of the sounds that make "simple letters" is easily comprehensible. Therefore, to inquire about the forms of things as substances, or as they appear, such as lion, oak, gold, water, or air, is vain. But this is not so for the forms of voluntary motion, vegetation, color, and "all other natures and qualities, which like an alphabet are not many, and of which the essences (upheld by matter) of all creatures do consist." To inquire of the "true forms of these" is the metaphysics of which he speaks. There is one exception to the inscrutability of apparent form, or substances—man, of whom it is said *Formavit hominem de limo terrae, et spiravit in faciem, ejus spiraculum vitae* and not, as it is said of all other creatures, *Producant aquae, producat terra.*[11] Physics takes account of these "same natures," but only with regard to the material and efficient causes of them, not with regard to form. The intermixture of air and water is the cause of whiteness in snow or froth, but this is the cause or carrier, the efficient cause of the form of whiteness. This part of metaphysics

10. *Republic* 596a5–98d5.
11. "He formed man of the dust of the ground, and breathed into his nostrils the breath of life" (Spedding's trans.), Gen. 2.7; "let the waters bring forth, let the earth bring forth" (Spedding's trans.), Gen. 1.20, 24; *Advancement* 355.

(formal cause) is deficient because forms could never be found by the received "course of invention" whereby men fly too quickly and too far from particulars.

But the use of this part of metaphysics is the "most excellent" because the "duty and virtue of knowledge" is to "abridge the infinity of individual experience" as much as is consistent with truth and to remedy the complaint *vita brevis, ars longa* by uniting the notions and conceptions of sciences. Bacon says that knowledge is as a pyramid, so that natural history is the basis of natural philosophy, "the stage next the basis is physic," and "the stage next the verticle point is metaphysic." The "verticle point" is the work that God performed from the beginning to the end, or the "summary law of nature,"[12] and of this, says Bacon, it is not known whether human inquiry "can attain unto it." These are the three true stages of knowledge; to the depraved they are like the giants' hills of which Virgil speaks,[13] but for those who revere God, they are like the "Sancte, sancte, sancte,[14] holy in the description or dilation of his works, holy in the connection or concatenation of them, and holy in the union of them in a perpetual and uniform law." Plato and Parmenides were correct to speculate that all things "by scale did ascend to unity."[15] The worthiest—and simplest—knowledge, metaphysics, considers the "simple forms or differences of things," which are few and out of which springs "all this variety."

The second respect in which this part of metaphysics is most excellent is that it gives to man's power the "greatest liberty and possibility of works and effects." Physics is constrained by the "ordinary flexous courses of nature," but to "sapience, which was anciently said to be *rerum divinarum et humanarum scientia*,"[16] there is "ever choice of means." Physical causes show new invention in similar material, but "whosoever knoweth any form, knoweth the utmost possibility of superinducing that nature upon any variety of matter, and so is less restrained in operation, either to the basis of the matter or the condition of the effect." This knowledge is like that described by Solomon, not liable to "particularity or chance."[17]

Bacon says that the other part of metaphysics, the inquiry of final cause, is not omitted but that it has been misplaced. This misorder-

---

12. Eccles. 3.11.   13. *Georgics* 1.281–82.   14. Rev. 4.8.
15. Plato, *Parmenides* 165e2–66c5.
16. "The knowledge of things human and divine" (Spedding's trans.), *Advancement* 357; Cicero, *De officiis* 1.43; *Tusculanae disputationes* 4.26; *De finibus*, rec. Schiche (Leipzig: Teubner, 1961), 2.12.37.
17. Prov. 4.12.

ing causes it to be deficient, or at least "improficient," however. Treating final causes "mixed with the rest of physical inquiries" has impeded the study of "real and physical" causes. Such mixing was done by Plato, Aristotle, Galen, and others, and Bacon notes that teleological accounts of cause are "well enquired and collected in metaphysic" but that they do not belong in physics.[18] Concern for such accounts of cause has led to the search for "physical causes" to be neglected. Democritus and others who supposed no "mind or reason in the frame of things, but [who] attributed the form thereof able to maintain itself to infinite essays or proofs of nature,"[19] seem to Bacon to be better than Plato and Aristotle in regard to the "particularities of physical causes," into which both Plato and Aristotle mixed final causes. Bacon does not condemn these final causes as worthless or not true. Rather if their "precincts and borders be kept," men are wrong to think that there is "an enmity or repugnancy between them." The knowledge of final cause does not "impugn" the physical causes; rather, both are "true and compatible," with "the one declaring an intention, the other a consequence only." This does not "question or derogate from divine providence"; rather it exalts it. Just as in civil actions the greater "politique" is the one who can make other men the "instruments of his will and yet never acquaint them with his purpose," so that they do what is expected and yet do not know it, so too the wisdom of God is more admirable "when nature intendeth one thing and providence draw forth another" than if God "had communicated to particular creatures and motions the character and impressions of his providence."

When Bacon pronounces his division of natural philosophy to be complete, he says that, if he has departed from ancient and received doctrines so as to "move contradiction," his aim has not been to dissent or to contend. If his division is true, then the voice of nature will consent, as Bacon says is implied by a line from Virgil's *Eclogues*.[20] Bacon prefers to come as Alexander Borgia said the French came to Naples, with chalk rather than weapons in their hands. He will come with chalk to "mark up those minds which are capable to lodge and harbor it" rather than to come "with pugnacity and contention."

There is another division of natural philosophy, "according to the report of the inquiry," which is the report of assertions or doubts. Registering doubts is useful for preventing error and for inviting in-

---

18. He gives examples taken from Xenophon, *Memorabilia* 1.4.2–8.
19. Cicero, *Tusculanae disputationes* 1.11; Diogenes Laertius 9.44–45.
20. *Eclogues* 10.8.

vestigations, but doubts must not be simply preserved as doubts, as we see done by lawyers and scholars; rather they are to be made into certainties. To a proper calendar of doubts should be appended a calendar of popular errors, chiefly in natural history, so that knowledge is not debased by them. As for "doubts in general or in total," Bacon takes them to be the differences of opinions that give rise to the different sects, schools, and philosophies of such as Empedocles, Pythagoras, Democritus, Parmenides, and "the rest." Bacon will not be like the Ottoman-like Aristotle, who sought to kill his brethren, so he thinks it will be useful for those who seek truth to see the various opinions about nature. One should not expect exact truth from them, because the "ordinary face and view of experience is many times satisfied by several theories and philosophies." To find the real truth requires another "severity and attention." Just as Aristotle said that children will at first call every woman mother and will later distinguish truly, so experience will first call every philosophy mother and will then come to "discern the true mother."[21] It is useful to see all the various opinions because each one sees something more clearly than the others. Therefore some collection should be made of the ancient philosophies, and this work is deficient.

Bacon warns, however, that such a collection must be done carefully, each philosophy by itself and not "faggoted up together" as one sees with Plutarch. The harmony of a philosophy in itself gives light and belief, not when it is "singled and broken." When Bacon reads of Nero or Claudius in Tacitus, he does not find it strange as he does when he reads of them "gathered into titles and bundles" by Suetonius, where they seem monstrous and incredible. The same thing applies, says Bacon, to the report of philosophies. He does not exclude such an account of later philosophies, and with this, Bacon says that he has finished treating "two of the three beams of man's knowledge," the one referred to nature and the one referred to God, which "cannot be reported truly because of the inequality of the medium."

We recall from the previous chapter that we need to know what it means for the new unity of the sciences to be the same as exclusive political rule by those who practice the art of rhetoric. We might equally well say that, in the new charitable age, political life has no place for the statesman whose moral virtue is modeled on the free-

21. *Physics* 184a10–b14.

dom of the political philosopher who practices no productive art. For the later ancients, the political philosopher is turned from any needy productive art by wonder at the whole comprised of radically different and yet related phenomena: needy human experience, the self-sufficient, the divine, and the eternal. In loving or needing wisdom about the whole, the political philosopher, unlike the mathematician and the physicist, never forgets the differences between the parts of the whole that prevent ever converting wonder, which no truly divine being could experience, into absolute wisdom about *all* things divine and human. The philosopher's wonder leads to the greatest possible human freedom, which is the model the statesman uses to direct the productive arts' contempt for wonder to forgetful moral virtue.

It is obvious by now that for Bacon it is no longer just for the mind's perfection to begin and end in wonder; rather the furthest end of knowledge is the conquest of nature for human purpose. Instead of wondering at the tension between theory and practice, the lover of wisdom must look forward to their unity. But all along we have seen that, for Bacon, what is just is not necessarily true, and if so, then the conquest of wonder cannot really be possible. And in turn if this is so, then in the present age the political philosopher is denied his own necessary place in political life. The absence of political philosophy discloses the limits of the modern project, known in detail when we know what it means for rhetoric to replace political philosophy. But before Bacon can demonstrate this, he must show that wonder can never be outstripped and that the causes of the whole can never be wholly understood. He does so in his discussion of how, in the present age, the highest and most universal knowledge of causes differs from such knowledge as it was understood by the later ancients.

We begin by noting that Bacon speaks openly about his own way of proceeding. In discussing metaphysics, Bacon says that he will use the ancients' terms and that he wishes to depart from them as little as possible. But he is clear that he intends to change their use and definition. He makes no attempt at all to hide the difference in his views from those of the ancients: in this context he repeats his earlier slander against Aristotle by saying that Aristotle was like Alexander in wanting to conquer all opinions and that Aristotle was not really different from the sophists because, although he took no financial reward, he "set up and battled for philosophical sects and controversies." This attack on Aristotle is not new; we encountered it earlier, in the apology, where it was revealed to have been nothing but a necessary slander. We are certainly led to wonder, then, just what it means for Bacon to say that he will keep company with antiquity *usque ad aras*.

Bacon takes the phrase from two sources. According to Plutarch, it means that Bacon will not praise insincerely and will stay within the bounds of law, equity, and advantage. According to Gellius, it means that Bacon will depart from the straight path to help a friend but not so far as to violate an oath to the gods.[22] But which one is it? From the apology, we should think it the latter, so that Bacon might both praise and blame insincerely in order not to defy the power of the gods, that is, to bow before the necessity of divine revenge. This meaning is indeed confirmed in the present context by Bacon's claim to be using the moderate civil proceeding described by Tacitus. Now, what Bacon knows but does not say is that, according to Tacitus, this proceeding was the use of republican terminology in the reign of Augustus, which simply disguised a wholly reversed world in which no trace of pure Roman character could be found.[23] If Bacon speaks in the way described by Tacitus, then he speaks insincerely in a world that has been wholly changed: in order to bow justly to divine revenge, Bacon praises and blames insincerely, so that his obvious attack on Aristotle and his obvious praise of the new knowledge is not sincere. In the present context Bacon obviously does not dissemble his disagreement with the ancients—it is not dissembling to have called Aristotle a tyrannical sophist. Rather because of divine necessity, one that has wholly changed the world, he dissembles his agreement with them, just as he had done in the course of his apology.

Bacon's first departure from ancient terminology is to change the meaning of first philosophy. In the present context, first philosophy is the unifying base of the sciences; it is "primitive" or "summary" rather than the highest of the sciences. First philosophy is a common parent of the sciences who is like Berecynthia, the mother who sent Aeneas on his trek that led to the first of two universal empires. In this sense, surely, first philosophy is not the perplexing theoretical science of final cause and what is like final cause in every formal cause, as it was for Aristotle and Plato.[24] It is not the science that begins and ends in wonder before the divine and being. But Bacon distinguishes between first philosophy, which is nothing like the first philosophy of Aristotle, and metaphysics, at least a part of which *is* the same as the later ancients' first philosophy. And the difference and

---

22. *Praecepta gerendae reipublicae* 808b; *Noctes Atticae* 1.3.10–21.
23. *Annales* 1.3–4.   24. See Aristotle, *Metaphysics* 1026a7–32.

tension between them discloses that despite the ambitions of the new knowledge, moderating wonder can never be overcome.

As Bacon describes the new first philosophy, it abstracts from the forms of things as they are both visible and intelligible or as they are experienced as different. In this regard, first philosophy is the study not of eternal forms but of bodies in motion. As such it comprehends and unites justice, mathematics, logic, natural philosophy, government, religion, music, affection, and rhetoric. The end of first philosophy is for the disclosing of nature and the abridgment of art, so that it is the mother not of the contemplation of differences that never move but rather of universal conquest for bodies that are always on the move. In fact, then, first philosophy is not common and primitive; it is, as he said in book 1, the highest science of the furthest end of knowledge,[25] and so as soon as Bacon has called it primitive, he calls it the "common and high state" of learning. Bacon says that first philosophy is a mother to which all philosophy governed by experience will return, a fact attested by Aristotle's remark at the very beginning of the *Physics* about children recognizing their mothers.

But Bacon surely knows that, in the context of the remark, Aristotle is concerned to distinguish between wholes that are clearer to us than to nature and parts that are clearer to nature than to us. And Bacon also surely knows that Aristotle follows this distinction by criticizing the philosophers who make it impossible to distinguish between wholes and parts or who fail to preserve wholes as they are experienced as the horizon for all questions about the causes of things.[26] The new first philosophy is nothing like Aristotle's because the new first philosophy abstracts from the essential differences of things as they are objects of experience. And because he knows the reasons for Aristotle's critique of the ancient philosophers, Bacon reminds us that the new first philosophy is *not* morally neutral. According to its principles, we must treat government, nature, and religion by reducing them to their elements, and the model for such treatment is Machiavelli's procedure in *Discourses* 3.1. Bacon has had much to say about this procedure in the course of his apology.

In the course of Bacon's earlier discussion, we learned that Machiavelli proposed the regeneration of corrupt mixed bodies by bringing them back periodically to the harshness of origins, by way of well-managed but extraordinary executions. For Machiavelli, such artful return to origins imitates nature's self-purging natural or simple bodies. Bacon disagreed with Machiavelli, because unlike Ma-

25. *Advancement* 292–94.   26. *Physics* 184b15–89a10.

chiavelli, who vainly insisted on the difference between simple and mixed bodies, Bacon described mixed or political bodies by identifying them with individual, simple natural bodies.[27] In *Discourses* 3.1, Machiavelli's subject was the regenerated kingdom of the clergy, and in the present context, Bacon confirms that the spontaneous regeneration of this mixed body necessitates a new science that goes beyond Machiavelli's science insofar as that science only imitates the motions of simple bodies. But Bacon knows that Machiavelli strained beyond differences he was correct to acknowledge. Therefore Bacon knows *why* Machiavelli's science of honor is not and can never be moral and fully just, however timely it might be. Following Machiavelli's unintended path, charitable first philosophy dissolves the differences between simple and mixed bodies and between the experienced differences of things, and in so doing it dissolves the moral distinctions between man and the brutes or, to be more precise, between the objects that serve needs and the objects whose needs are served. Bacon speaks as if the status of first philosophy were simply self-grounding, but he shows that this cannot be so, because at the same time that it orders the whole of knowledges from beginning to end, it cannot order the human things. For this reason, first philosophy does not exhaust the comprehensive and highest forms of knowing, for it always competes with the older knowledge that begins and ends in wonder, not conquest. Therefore Bacon speaks not just of first philosophy but also of metaphysics as the inquiry into formal and final cause.

Bacon divides physics from metaphysics by distinguishing between material and efficient cause, on the one hand, and formal and final cause, on the other hand. Physics accords well with the order imposed on the sciences by first philosophy because physics treats of matter in motion as opposed to matter or bodies at rest. This order demonstrates Bacon's departure from the "received and sound division of causes"; unlike Aristotle, Bacon says that he does not allow for formal cause in the account of physics. Aristotle's *Physics* is a metaphysics of physics, but Bacon's allows no room for the perplexing question of how the same entity can both be in motion and at the same time be formed by what does not move; that is, how the same thing can be at the same time matter in motion and substance.[28] But actually the real difference between Bacon's and Aristotle's division according to causes is that Bacon's makes its way to perplexity, or wonder, much more slowly.

27. See Chap. 4 at nn. 108–10 above.
28. *Physics* 192b8–94b15, 198a14–200b8, 243a3–45b2, 256a4–67b26.

To the extent that physics appears to be certain, independent, or separable from the study of formal cause, or to the extent that physics does not open up to the perplexity of metaphysics, it accords with first philosophy in abstracting from essential differences to the sameness of all things as the result of original motions. But Bacon's example of a phenomenon of physics is taken from Virgil's *Eclogues,* which speaks not simply of the properties of fire but also of love and the way that love can assimilate both the hard and the soft.[29] Bacon's example provides a clue to his intention. According to the example, physics divorced from formal cause is like love in being able to embrace apparently opposite or separate things. But of course as the motion of a soul toward something it is not, love can never be the affect of something formless, lest the soul be nothing in itself and so be unable ever to be drawn to something beyond it. To separate physics from formal cause can accomplish the task of first philosophy, the conquest of nature; but as it only serves human purpose, it points to the question of what the formal limits of the human things are and how human experience can be intelligible and yet finite and always in motion.

In speaking of intelligible nature, or nature as it "supposeth . . . a reason, understanding and platform," Bacon separates metaphysics into the study of formal and final cause, to the same purpose as the separation of physics from metaphysics. Declaring his agreement with Plato in book 10 of the *Republic* that the forms of things are "the true objects of knowledge," Bacon disagrees with those who say that essential forms are not knowable. He says that he differs from Plato, however, in arguing that the forms should not be absolutely abstracted from matter, as they were by Plato. Plato of course never says that the *eide* can be completely abstracted from the particulars of which they are eide, and in the account of the realms of being in the *Republic,* the journey of the mind is never simply to wholly abstracted eide, but from appearances to the eide and *back* to appearances again.[30] The account of completely abstracted eide in book 10 of the *Republic* lives up to Glaucon's exaggerated demand for certainty (Socrates warns earlier that he might be giving a less than accurate account of the eide), and the absurdity of the account in book 10 is that the bed as such is one on which it would never be possible to lie.[31] In the *Republic* precisely the connection of the eide to the particulars of which they are eide is the ground of the intelligibility

29. *Eclogues* 8.17–109.   30. *Republic* 509d1–11c2; cf. 532a1–34a8.
31. Ibid., 358b1–62c8, 596a5–98c4.

of the visible world. The awesome mystery of this connection provokes the wonder with which the certainty-loving Glaucon is so impatient.

Bacon accounts for formal cause not so much to reconnect form and visible particulars as to dissolve the wonder to which such connection must lead. Therefore he speaks not of the forms of substances, or visible particulars as they occur in experience, but rather of the forms of nature abstracted from substances. Such an abstraction will serve those who keep a "continual, watchful and severe eye upon action, operation, and the use of knowledge,"[32] that is, those who are more like Glaucon than Socrates. Plato links formal and final cause because for Plato, form can never be abstracted from the entity as it is possibly self-sufficient or as it is part of a self-sufficient whole, that is, from things as they are encountered in moral experience. Bacon makes such a separation for the sake of use, and so again for the sake of use first philosophy draws the account of nature to itself. Final and formal cause must be separated for the sake of use, so that the study of formal differences is for the sake of mastering the common utility of all things. And this requires that such forms be abstracted from things as substances: as they are experienced or as they are possibly self-sufficient or final wholes. With regard to the intelligibility of experience, then, both the new physics and the new understanding of formal cause serve first philosophy, which looks past the knowledge of differences that is essential for any moral knowledge. In this sense, first philosophy draws the account of the whole to itself, for according to Bacon, the forms of substances, or things as they are experienced, are too "perplexed" to permit an investigation that considers how formal cause might open up to or be final cause. That is, they are too perplexed to consider their forms as the kind of thing that might instill awe or ground the possibility of moral restraint before a self-sufficient being. But in truth, first philosophy cannot completely draw the account of the whole to itself, because there is one substance of which the form can be discerned, and this substance is man himself.

Man is the only substance whose *substantial* form is discernible. Therefore man is the only substance whose form is the possible object of the inquiry into final cause. With respect to man, in other words, final and formal cause can be considered together, or the formal cause of man can be oriented toward a perfected or self-sufficient whole. The only being who inquires into final cause is himself the locus of

32. *Advancement* 355.

final cause. Metaphysics concerning final cause consists in man's self-knowing, which must itself consist at least in the knowledge of the difference between what is and what might be final cause but is not. When he says that the wisdom of God is "more admirable when nature intendeth one thing and providence draweth forth another," Bacon makes it clear that man, who is the object of the knowledge of final cause, in fact orders the final causes of the several beings in nature.[33] Created in God's image, man is the conqueror of divine providence, so man is himself the final cause of nature. But as evidence for the knowability of man as formal-final cause, Bacon refers to the difference between creation by divine power and creation by divine wisdom.

We have seen this distinction before, in the apology, where it turned out that, although a distinction between creating power and restful wisdom is intelligible, the two characteristics cannot be found together in any creator god. Reflection on this difference disclosed the true source of divine revenge, showing that as man is the master mover of nature, man exists as the image of divine jealousy or folly, which was unable to abide the existence of a motionless, restful, self-contemplating being superior to himself. When man comes to light as both the final cause of nature and the object of the inquiry into final cause, the difference between free self-sufficient perfection and the constrained need to make comes to light, as does the difference between rest and contemplation and needful, dependent artifice. If man is the final cause as he masters divine making, then such final cause lights the way to the end that making serves. But this end is not clear; rather it is the hidden cause of a perplexity, for it is not clear how the same being can be both at rest, or perfected, and yet animated by needy motions. It is not clear how the same being, man, can at the same time be the repository for final and efficient cause.

Man's mastery of first philosophy, physics, and the science of formal causes points not to wisdom but to wonder. But in the present charitable age, the age governed by the most robust divine revenge, wonder does not govern the order, or the pretenses, of the rational arts and sciences; it does not deter first philosophy from its pursuit of divine mastery. On the contrary, in the present age the part of

33. In the *De augmentis*, Bacon follows this remark with an argument that Aristotle, who mixed nature and final causes, did not have any need for God, but Democritus and Epicurus, who taught a doctrine of atoms and said that the universe came together by accident, were ridiculed universally. Searching for physical causes, in short, has no effect without resort to God (*De aug.* 570–71). In the present context it is clear that Bacon agrees with Aristotle and that Democritean or Epicurean atomism is more akin to Christian cosmogony than meets the eye.

metaphysics that serves first philosophy is the "most excellent," because it serves the duty so to unify the arts and sciences as to master
God's work from beginning to end. Even so, Bacon teaches that no
creator god can be the image of wisdom as such, so that such a god
is himself in need of a proper understanding of the difference between the human and the divine. And man understood as the equal
of the creator god is no less in need of a proper wonder before the
divine than the creator god himself. The desire for perfect mastery
is the perfect occurrence of divine jealousy, and Bacon makes it clear
that this is a political fact when he remarks that the operation of final cause as the intention of a creator god can be likened to the way
of a "deeper politique" in "civil actions."[34]

The love for a creator god's love is the result of the promise for
perfected moral certainty when it is no longer restrained by the conventional virtues. These virtues can be separated from their artful
origins only when they are measured by the restful perfection of
contemplation that is never so bold as to aspire to divine wisdom, let
alone to insist that such wisdom account wholly for the origins and
the motions of every formed, natural body. But of course in the best
of times virtue and moderation are rare, for political philosophy
cannot easily persuade on their behalf, because its refusal to practice
or to praise the productive arts must always appear to be unjust. And
as we have seen in great detail, in the present age there can be no
open speech on behalf of moderation and moral virtue or the philosopher's proper place in political life. It is not, however, useless to
know what this proper place is, for which reason Bacon's account of
natural philosophy and metaphysics does not simply reject the contemplative tradition of the ancients, and in particular of Plato and
Aristotle. Rather it demonstrates that the whole of man and nature
is rent by the tension between a stance toward nature oriented solely
by the origins of matter in motion and the contemplative perfection
that must but cannot moderate that stance.

The dangerous hopes of first philosophy serve the political hopes
of the productive arts, but in the age of Christian love there is no
alternative to abetting them. There is no alternative to letting the
spurious mother of the sciences embrace the philosophies of Empedocles, Pythagoras, Democritus, Parmenides, and the rest that would
reduce the appearances of things to a single demonstrable source or
origin, whether it be motion, as the physicists would have it, or wholly
divorced from any possible motion, as the mathematicians would have

34. *Advancement* 359.

it. Bacon criticizes all these views, referring to Aristotle's remark about mothers to remind us of Aristotle's intention always to begin with the wholes of things before descending to causal parts.[35] As human knowledge is governed by Christian providence, Bacon's, not Aristotle's, first philosophy draws the activities of the mind to itself. But Bacon knows this development exaggerates rather than overcomes the tension between the various claimants to the motherhood of the arts and sciences. And it is essential to know that it does so because, in the dangerous age to come, at least some must be free enough from dogma to be able to seek new political ways when old ones fail. Despite all of its complexity, the rest of the division of the sciences treats the human good when the activities of the mind are ruled by the productive arts and when the art of rhetoric is taken to be the art of arts. The political rule of the art of rhetoric is the real effect of the triumph of first philosophy. We are now prepared to learn what it means for rhetoric to eclipse the political claims of the philosopher, the statesman, and the poets, all of whom are capable of awe before the divine.

35. See n. 21 above.

[9]

Human Philosophy: Charitable Knowledge
and the Possibility of Political Science

To this point we know that as the arts and sciences are unified by the new first philosophy, the philosopher and the statesman are denied their proper places in political life. In the light of their absence, we must consider the places of the poets who animate the gods, the rhetoricians who persuade people regarding the laws, and the sophists who claim to teach an art of moral virtue. Bacon treats these subjects in their appropriate place—in the course of discussing the knowledge that is at once a part of and the "end and term" of natural philosophy, "the knowledge of ourselves." Knowledge of ourselves, or human philosophy, is broadly divided into knowledge of man "segregate or distributively" and knowledge of man "congregate or in society." The former is divided into the knowledge of body and mind, and the knowledge of mind is divided into knowledge respecting the understanding and reason, which produces "position or decree," and knowledge respecting will, appetite, and affection, which produces "action or execution." But before Bacon turns to the two major parts of the study of mind, he comments that it is true that the imagination is an "agent" in both of the mind's parts, the "judicial and the ministerial." Bacon accounts for the poets, the rhetoricians, and the sophists in the course of discussing first the imagination's agency between the two parts and then the first part, understanding and reason, by itself. To attend to the law of brevity, we turn now to these sections of the division.

[382–408]

In commenting on the imagination's agency between the two parts of the mind, Bacon says that because sense proceeds through the

260

imagination to judgment and reason proceeds through imagination to decree, imagination always precedes voluntary motion. Imagination consists of sister faces,[1] one printed with truth and the other printed with the good. But imagination also has its own authority, for as Aristotle says, the mind rules the body as a master, but reason rules the imagination as a "magistrate" rules a free citizen, who might also come to rule in his term.[2] In matters of faith and religion, the imagination is raised above reason, so that religion uses things like visions and dreams. Likewise, in eloquent persuasions, which "paint and disguise the true appearance of things," the imagination rules. Nevertheless, there is no separate science pertaining to the imagination, so he will not alter the former division, for poesy is the play of the imagination, not its work or duty. If poesy is a work, Bacon speaks now not of "such parts of learning as imagination produceth" but of the sciences that consider the imagination. He will not speak here about the knowledge produced by reason, including all of philosophy, but only about those knowledges inquiring into the faculty of reason. Therefore poesy "had his true place." Regarding the power of the imagination, he has treated this in the prior discussion of the soul, and finally, "for imaginative or insinuative reason," which is the "subject of rhetoric," this is best referred to the ensuing division of the arts of reason. Therefore Bacon is content with the former division of the mind's faculties into two parts: rational and moral.

The part of human philosophy treating reason itself is for most wits the least delightful and the most difficult. If knowledge is *pabulum animi*, then this is the most "celestial" and therefore to most men it is like manna, from which the Israelites turned to their fleshpots.[3] This *lumen siccum* offends most men, who are more at home in matters about which their affections, praise, and fortunes "turn and are conversant," such as civil history, morality, and policy. But Bacon says these "rational knowledges" are the "keys of all the other arts." Just as Aristotle said that the hand is the tool of tools and the mind is the form of forms,[4] so these can truly be said to be the "art of arts." They not only direct all of the arts but also confirm and strengthen them. There are four parts of rational knowledge, determined by four objects: invention or inquiry of what is sought or propounded, or the art of invention or inquiry; judgment of what is inquired, or the art of examination or judgment; retaining of what is judged, or the art of custody or memory; and the delivery of what is retained, or the

1. Ovid, *Metamorphoses* 2.14.   2. *Politics* 1254a17–55a3.
3. Cicero, *Lucullus*, rec. Plasberg (Leipzig: Teubner, 1961), 41.127; Num. 11.4–6.
4. *De anima* 431b20–32a14.

art of elocution or tradition. Invention is divided into two parts, one regarding the arts and sciences and the other regarding speech and arguments.

The former part of invention is deficient first because logic does not pretend to invent sciences or their axioms, so that *cuique in sua arte credendum.*[5] Celsus acknowledges, too, that in medicine cures are discovered before causes. Likewise Plato, in the *Theaetetus,* as Bacon says, maintains that the heart of the sciences is the middle proposition, derived from tradition and experience, and that generalities give no "sufficient direction" because particulars are infinite. Therefore those who discuss about the origins of things trace them to chance rather than to art and to various creatures rather than to men, as Bacon shows by referring to Virgil.[6] If one prefers the Greek account of Prometheus, still he discovered fire by chance rather than by knowledge of cause, and the same thing can be said of the "West-Indian Prometheus." Men trust to chance rather than logic, and Virgil's praise of practice and meditation is not better,[7] being no different from what the brutes can do, governed as they are by an "absolute necessity of conservation of being." Cicero's and Virgil's praises of practice and labor can also be applied to beasts such as birds, bees, and ants.[8]

The second reason for the deficiency is that the received induction, which seemed to be known by Plato and which generates the principles supplying middle propositions, is incompetent and vicious. Art should exalt and perfect nature, but it has wronged nature, for whoever observes how this knowledge is distilled from natural and artificial particulars, like Virgil's celestial honey,[9] will see that the unaided mind does better than the received induction describes it, for not to consider contradictions is the same as if Samuel had failed to see David.[10] It is so gross that it just makes way for theory and dogma, and just as regarding divine truth men cannot abide becoming like children,[11] so too men consider Bacon's kind of induction to be a kind of infancy.

The third reason the former art of invention is deficient is that

5. "The knowledge that pertains to each art must be taken on trust from those that profess it" (Spedding's trans.), *Advancement* 384; Aristotle, *Prior Analytics* 46a3–31; *Magna moralia,* trans. Armstrong, Loeb Classical Library (London: Heinemann, 1969), 1182b23–32.

6. *Aeneid* 12.412, 8.698; Plutarch, *De sollertia animalum* 974.

7. *Nic. Eth.* 1125b1–25; Virgil, *Georgics* 1.133.

8. Cicero, *Pro Balbo* 20; Virgil, *Georgics* 1.145; Persius, *Saturae,* text and trans. J. Jenkinson (Warminster: Aris and Philips, 1980), *Prologus* 8.

9. *Georgics* 4.1.   10. 1 Sam. 16.   11. Matt. 18.3.

even when principles or axioms are rightly induced, middle propositions cannot be deduced from them by way of syllogism. In popular sciences like morality, law, and divinity, this form has use; and in natural philosophy it suits argument or satisfactory reason, but it does not suffice for use, because nature is more subtle than these forms can ever be. If the words of propositions are the marks of ill-collected notions, then no attention to the forms of the syllogism can correct them. No wonder, says Bacon, that so many philosophers have been skeptics. It is true that Socrates' skepticism was ironic, used to enhance his knowledge, like that of Tiberius "in his beginnings." Likewise it is true that the acatalepsia of the later academy, embraced by Cicero, was not "held sincerely," because this sect, being made up of those who spoke well, gave glory to eloquence and variable discourse.[12] But many in these sects did believe it, their error being to blame the senses, which do not deserve it. The blame rests upon the weakness of the "intellectual powers" and the ill use of the senses.

Invention in speech or argument is the proper recalling and use of what we already know. This invention is divided into preparation and suggestion. Regarding the former, Aristotle's chiding of the sophists as ones who taught shoemaking not to show how to make a shoe but only to exhibit them is witty but hurtful because "if a shoemaker should have no shoes in his shop, but only work as he is bespoken, he should be weakly customed."[13] Jesus said that the kingdom of heaven is like a "good householder," bringing forth "old and new store,"[14] and the ancient teachers of rhetoric said that one must have set speeches for and against various matters. Cicero himself said that men should have premade speeches about whatever they will treat, so that all that must be done is to put in the appropriate names, times, and places.[15] Demosthenes agreed with Cicero about premade speeches, and the two of them may "overweigh Aristotle's opinion that would have us change a rich wardrobe for a pair of shears." Although the nature of this matter belongs both to logic and to rhetoric, Bacon will treat it further when he speaks of rhetoric. Suggestion is the way in which we return our mind to formerly collected knowledge. It is useful for disputing with others and for deliberating well with ourselves, and the "places" or mnemonic devices used in it direct as well as prompt our inquiry. Bacon mentions Plato's account of Meno's paradox,[16] concluding that the larger one's anticipation is, the better is the search. Mnemonic devices help us to question ex-

12. Cicero, *Lucullus* 5.15; *Brutus* 85.292–93; Tacitus, *Annales* 1.7; *Lucullus* 6.18.
13. *De sophisticis elenchis* 183b39–84b9.   14. Matt. 13.52.
15. *Orator* 14.45, 15.50.   16. *Meno* 80d5–e5.

perienced men, books and authors, and Bacon says that the study of them is not deficient. But Bacon does accept particular topics, as well as general ones, as mixtures of logic with the "matter of the science," so that every step in the particular sciences is a light to what follows.

From invention Bacon turns to the art of judgment. Regarding induction, this subject has coincidence with invention, for in all induction the same act of invention is also an act of judgment, "all one as in the sense." Such is not the case with syllogism, where "the invention of the mean is one thing and the judgment of the consequence is another." The syllogism is very pleasing to the mind. Therefore it has been well labored because, just as Aristotle tried to prove that there is a "quiescent" part in all motion and just as he described the fable of Atlas as the "axle-tree of heaven," so men wish to have an axle tree that might keep "them from fluctuation."[17] This art is the reduction of propositions to "principles in a middle term," with the principles agreed on by all, the middle term to be decided by free invention, and the reduction of the proposition either to the principle or to absurdity. There are two ways of accomplishing this end, by way of either direction or caution, that is, by analytics or by the discovery of sophisms, or "that which is termed Elenches." As Seneca says, we see gross sorts of fallacies,[18] but, says Bacon, the subtler ones are harder to see and so more likely to abuse the judgment. The study of elenchi is well done by Aristotle and more excellently by Plato in the example of the sophists and even Socrates who, claiming to affirm nothing, confounded what others affirmed by all the possible forms of objection and refutation. The degenerate use of this faculty is for "caption and contradiction," which is taken to be a great faculty and is advantageous. But Bacon approves the difference "made between orators and sophisters," that the one is like the greyhound, having its advantage in the race, and that the other is like the hare, which, being the "weaker creature," has its "advantage in the turn." The use of elenchi is wider than has been appreciated, however, and Bacon says that the knowledge of the "common adjuncts of essences," often taken to be metaphysics, is but the study of "wise cautions" against ambiguity, which is the "sophism of sophisms," and the same can be said for the "distribution of things into certain tribes."

There is another seducement that works by the overpowering of reason by the imagination, and Bacon will speak of it when he treats

17. *De motu animalium*, trans. Forster, Loeb Classical Library (London: Heinemann, 1937), 698b9–99b11.
18. *Ep.* 45.8–9.

rhetoric. After describing the causes of fallacy that are located in the defects of the mind itself and in our "nature and condition of life," Bacon turns to the third and fourth rational knowledges, custody and tradition of knowledge. Custody is divided into writing and memory, and tradition is divided into organ, method, and illustration. After describing the various methods of the tradition of knowledge, Bacon turns to an extended discussion of the illustration of tradition, which he says is comprehended in the science called rhetoric or "art of eloquence."

We begin by noting that, although by Bacon's argument the imagination is the nexus between the true and the good, he says that there is no science corresponding to the imagination because poetry is play rather than work. But despite this remark, Bacon says that the imagination does have its own work or power and that reason does not rule it despotically and that poetry may be work. Bacon's remarks are not perplexing if we think about what he has learned about reason and imagination or phantasia from the later ancients. We know that the imagination is like reason because it discloses connections between things. Because truth is never as immediate or self-evident as the "evidence" of the senses, reason cannot function apart from the imagination that acts upon the images of the senses. But of course the imagination's virtue is not the same as reason's, for as compared to reason's perfection by its submission to the truth, the imagination is free. As judged by reason's virtue, the imagination is the locus of error, but even so, the imagination's power actually to make what it will can have its own independent charm. Of course the poetic art depends upon such charm, so it is no wonder that Bacon says there is no science of the imagination. And this lack does not conflict with the imagination's having some work: the imagination's work is its service to science or reason.

But the poets do not claim to serve reason or to serve at all. Nevertheless, they do claim to disclose the truth about human and divine affairs, which can never wait for the slow and never-finished work of reason. The poets can disclose the truth without hard work because they are moved, not by reason or artful practice, but by effortless inspiration, as if their truth-disclosing art were seized by the perfectly free imagination or the imagination when it serves no truth at all. Bacon's opening remarks describe the poets to a tee: in claiming to be inspired, their art is effortless play, but insofar as their inspiration tells some urgent truth about man and god, their play is serious work. The imagination serves reason, although for mere human beings it has another serious work—it stands in for reason in practical affairs, for reason is never as timely as practice is urgent. Since poets are the imagination's custodians in this latter work, they cannot order stable practical affairs without being less free than they think they are. Rather they speak to practical affairs within the hidden boundary of knowledgelike opinion, over which they do not have complete control.

According to the later ancients, there was a quarrel between the political philosopher and the poets, both of whom claim to know something of divine and human affairs.[19] The ground of their difference was the philosopher's superior self-knowledge: the philosopher makes nothing, and most certainly not the truth to which he is attracted, but the poets do not know how comical it is for them to claim to be both truth tellers and inspired, effortless, imaginative makers. Now, from what we know so far it should be clear that the political philosopher must understand the poets and the poetic and must see their importance for practical excellence. The philosopher's freedom is indeed a model for practical virtue, but it cannot be a direct model, because the philosopher does not make, while virtuous deeds do make goods for public life. The poets, however, are makers whose imaginative charm is like the political philosopher's freedom from any making. The poets' freedom is just like the poets' art in being imitative, and the poets' claim to knowledge is always betrayed by their submission to some common opinion: poetic inspiration is approved or condemned by those who listen. Likewise, the virtuous person is one whose deeds make public goods and yet who does so in such a way as to be oblivious of his making and his thralldom to opinion. Virtuous deeds are imitative and self-forgetting; that is, they are poetic. The poets, then, are the intermediate model between the philosopher and the virtuous. And as such the poets depend upon the statesman's fashioning of right opinions—opinions that let the poets appear in their being poetic and yet deny the poets their perfect freedom while hiding their being the useful makers of gods who reward virtue. The poets are always given to the statesman as a model to be trimmed, for as long as men need both urgent art and untimely truth, there will always be poets. Virtuous men and deeds, however, are uncommon. The poets are the common, intermediate model for moral virtue, rather than vice versa, but not without the statesman's art.

19. Plato, *Republic* 377b5–83b7, 595a1–608b10.

Bacon understands the subtle nature of poetry, because he tells us that it is both play and work. Bacon understands the later ancients' teaching about the poets. But he has said that in the modern age the freedom of poetry is realized in the serious, practical scientific conquest of nature. And moreover, poetry serves magnanimity, morality, and delectation. We wonder just how in the modern age poetry serves these moral phenomena. If poetry is perfected in serious conquest, can it serve morality as a mixture of work and play? The answer to this question is important if the poets are to have an independent place in the political order of productive arts. The answer can be discerned from Bacon's subtle reference to Aristotle, whose remark from the *Politics* is said to illuminate the authority of the imagination.

Bacon says that Aristotle claims a separate authority for the imagination because he says that it is ruled by reason as a magistrate rules a free citizen who can himself come to rule. But in the context to which Bacon refers, Aristotle speaks not of reason's rule over the imagination but rather of reason's rule over the appetites being like either political or royal rule.[20] And moreover, Aristotle argues that reason cannot rule the appetites "politically," because for reason and the passions to be equal or in reversed positions is harmful in every respect. As Bacon appeals to Aristotle, the imagination stands for the appetites, and the autonomy of the imagination reflects the rule of the appetites over the rational and moral faculties. The imagination's work is untrammeled service to the appetites, which service cannot be effortless or playful because an appetite is always a need.

In the new age poetry is no longer poetic, because it is nothing but serious work; it exposes the truth that explodes poetic charm—the truth that poetry is *the* technological activity, having nothing to do with the appearance or image of effortless play. These poets claim to be more self-knowing than the older poets; eschewing inspiration, they make the explicit scientific claim that imagination and reason can be identical. The result is that imagination serves every need so that the imagination's charm, and with it the poets' charming imitation of philosophic freedom, disappears. Therefore the poets cannot be the intermediate model between the political philosopher and the morally virtuous actor. From all that we know so far, the new poets would seem to be no different from the sophists. In knowing the artful source of every opinion, the sophists understood the poets more clearly than the poets knew themselves, and in claiming to be the masters of every

20. *Politics* 1254a17–55a3.

productive art, the sophists claimed to be the real poets. This conclusion should be no surprise, for it was suggested at the very outset of Bacon's apology. Therefore it is perplexing for Bacon to say that imaginative reason, and not sophistry, is the subject of rhetoric. Could it be that even the sophists are somehow ejected from the modern order of the productive arts? If so, we need to know what such a fact means in the light of the later ancients' teaching. For Bacon's new project, rhetoric treats the connection between the true and the good. The next question concerns just what rhetoricians are in the political order where there are no political philosophers, statesmen, and poets and perhaps not even any sophists.

To discover the new power of rhetoric, Bacon does not continue to discuss the link between reason and will; rather he turns simply to the sciences of reason and understanding, as if reason and the power of the imagination to serve the appetites, or the will determined by the appetites, were the same. The question is how rhetoric emerges as the art of the whole of mind and nature, or as a comprehensive, all-knowing art of arts. Therefore Bacon begins by describing the "rational knowledges" as the keys to all the other arts. To make his point, Bacon refers to Aristotle's remark about the hand and the mind in the *De anima*. To the extent that Bacon speaks of the arts of arts, Aristotle's remark about the hand being the tool of tools is appropriate. But if the arts make, then the reference to Aristotle's "forms of forms" is not appropriate, because it occurs in Aristotle's description of "passive mind," the mind that makes nothing but only receives self-subsisting forms. Rather, in seeing the art of arts as the key rational knowledge, the art of arts is the same as what Aristotle calls "active mind," the mind that makes everything and that would be by itself perfectly free as to its procedure because it is not itself constrained or formed by any forms or differences.[21] Bacon of course knows that for Aristotle active and passive mind always occur together in man. By referring to Aristotle, he reminds us of the tension between the freedom to which the art of arts pretends and the form that cannot be made, the form of man himself. But since the new art of arts ignores this tension, Bacon criticizes the received inquiry and invention in the arts and sciences for its failure to provide useful axioms and for its trust in those who profess each separate art. However, to make his point against the tradition that respects the separateness of the arts and sciences, Bacon refers to the evi-

21. *De anima* 429a10–32a14.

dence of the later ancients who began it, Aristotle in the *Prior Analytics* and *Magna moralia* and Plato in the *Theaetetus*.

In the *Prior Analytics*, Aristotle argues that the principles of each science are different from the principles of the syllogism. The source of the principles of each science is experience, not trust simply, although, as Plato argues, experience is impossible without trust.[22] In the *Magna moralia*, Aristotle argues that no science or power is able to establish the goodness of its own end, for this is the subject of a different, separate power. He says also that, although there can be a science of the human good, there can be no general science of a general or universal good.[23] Aristotle means that the political art orders the arts as they serve a whole but, considered by itself, each art is the judge of its ways and means. It is precisely because of this limited self-sufficiency with respect to ways and means that there is a need for the political art, for each art can attend to its own business as the business of the whole of all the arts. Regarding the reference to Plato, contrary to Bacon's claim, Plato does not remark in the *Theaetetus* that particulars are infinite, that generalities give no direction, and that the arts differ from artlessness by way of propositions derived from experience. Rather in the *Theaetetus* Socrates blames Theaetetus' answers that give several particulars rather than an account of knowledge as such or in itself.[24] Again, in the *Philebus*, to which Bacon's reference can also be taken to refer, Socrates says that the way all inventions are brought to light shows how all things spring from one and many and have the finite and the infinite inherent in them, so that all inquiry must pursue the one eidos that is in every case the true object of knowledge. From the discovery of the one eidos, then, one proceeds to the discovery of many eide, but Socrates criticizes those who see nothing between one and many, unity and infinity.[25]

As Bacon presents his argument by way of Plato and Aristotle, the new art of arts is related at once to the erroneous claim that there is either only one being or infinite beings, none of which can be sorted according to kinds, and also to Theaetetus' erroneous claim that knowledge is the free work of sense as it serves the imagination. The new art of arts treats the many beings as if they cannot be sorted into

22. *Prior Analytics* 46a3–31; cf. Plato, *Republic* 511d6–e5.
23. *Magna moralia* 1182b23–83a8.
24. *Apology* 21e3–22e5; *Theaetetus* 146c7–48d7.
25. *Philebus* 16c5–17a5. In the *De augmentis*, Bacon omits the reference to the *Theaetetus* (*De aug.* 617). The following argument could not be discerned from the *De augmentis*.

kinds, because it claims to be the only way of art. But only the productive arts are tempted to such claims because to serve a part of the body's perfection—to get on its way—each art must be blind to its own dependence and must see itself in every other art as it thinks of itself as a unified, free whole. Which artists actually make this bold claim to artful but perfect freedom? Not the poets as represented by Virgil, for they attribute the origin of things to chance rather than to art, as if the gods themselves have no origins in some poetic art. For Virgil, art reveals man's dependence on harsh necessity and the amelioration of this necessity by divine gift.[26] But no art is godlike, so that there are always several different arts that depend upon each other. And surely no useful philosopher, not even the Roman Cicero, would make such a claim, for in the *Pro Balbo,* to which Bacon refers, Cicero distinguishes between art and mere practice and use, as if the cognitive superiority of art to mere practice separates art somehow from the utility that separates art from pure knowledge.[27]

The sophists and the rhetoricians are the ones who claim to order the whole of the arts by making the most extravagant claim latent in every art, the claim to be a perfectly unconstrained art of arts that makes the whole of the city. But when these practitioners shared their place with the political philosopher and the statesman, they presented their craft as an *imitative* art.[28] And because they combined claims to be imitative with claims to serious truth, both the sophists and rhetoricians could be restrained by a truthful political science. Therefore we wonder whether both the sophists and the rhetoricians are changed when they stand not for imitative art but for the method that actually subjects the whole of nature to the free imagination. Bacon shows just how they are changed when he discusses their immediate province, the recalling and use of those things discovered by the new method of invention.

In discussing the proper use and recalling of discovery, Bacon praises the sophists against Aristotle's critique of them. According to Aristotle, the sophists' art differs from practically all other arts, including the art of rhetoric, in regard to the discovery of first principles, or beginnings. Unlike the other arts, the sophists develop no first principles, which means that they cannot teach and only display the results of the other arts. In this respect Aristotle says they resemble the procedure of Gorgias, who taught both rhetoric and questioning and answering. Because they present no first principles, the

---

26. *Aeneid* 12.412, 8.698; *Georgics* 1.118–75.   27. *Pro Balbo* 20.45–46.
28. Plato, *Republic* 595a1–99a4; *Sophist* 235b8–36d3.

sophists are like but not the same as the rhetoricians, who, though they too profess all other arts without being able to teach them, do have first principles of their own arts.[29] In the present context, Bacon simply identifies the rhetoricians and the sophists, ignoring Aristotle's subtle distinction and forcing us to wonder about the difference between the rhetoricians and the sophists, a difference to which Aristotle points but which he does not describe. Whatever the difference, Bacon says that the common practice of the rhetoricians and the sophists, the mere display of the products of the other arts, is as useful as is the knowledge of the secrets of the kingdom of heaven, which "bringeth forth both new and old store." Bacon speaks in his next breath not of the sophists but of the rhetoricians, whose use of places is said by Cicero to be helpful not just for whatever occasion arises, as Bacon says, but for upholding any side of any argument.[30] The collection of these places belongs to rhetoric, to be treated later, but the other part, suggestion, deals with marks or places also, and insofar as these places also direct our inquiry, they help to overcome the argument against discovery voiced by Meno against Socrates.

In the *Meno,* Socrates discusses Meno's argument that nothing unknown can be inquired about and shows that it reflects the viewpoint of Gorgias, whose art of rhetoric differs from the sophists' art because, unlike the sophistical art, it makes no claim to be able to teach virtue.[31] This is the crucial difference between the sophists and the rhetoricians: because the sophists claim to teach virtue, they are actually in closer accord with conventional opinion than are the rhetoricians, who, though they serve conventional opinion, make no such claim. For this reason Socrates remarks in the *Republic* that the city itself is the greatest sophist.[32] But the sophists' bold claim to know the source of every opinion exposes the very problem of conventional opinion: if virtue cannot be taught, then the city's claim to punish injustice justly is unfounded; but if one grants the sophists' claim that it can be taught, then the basis of the city's justice, the defense of the law by the justice of the gods, is revealed to be merely the product of the poets' and the rhetoricians' arts.

Anytus could have been as disgusted with the rhetoricians as he was with the sophists, but he knew all too well that, while the sophistical city might do without sophists, even though their possibility is coeval with that of the city, it cannot do without the rhetoricians. The rhetoricians, unlike the sophists, know what art really is because they

---

29. *De sophisticis elenchis* 183a36–84b9; cf. *Rhetoric* 1355a3–b34.
30. Cicero, *Orator* 14.45–46.   31. *Meno* 70a1–71e1, 80d5–81a2, 95a2–96e5.
32. Ibid., 90e10–93a4; *Republic* 492a1–d1.

can teach the first principles of their own, and for this reason, though they must boast of universal facility like the sophists, they hold their tongues about the fact that virtue is the product of their own art. However alike the sophists and the rhetoricians may be, the later ancients considered their difference essential for the well-being of any city. The rhetoricians can be the efficient cause of a city because their genuine art is self-effacing, which preserves the city's justice. But the sophists cannot be such a cause because their spurious art and their candor undermine the city's conventional authority. For this reason it is possible for the rhetoricians, but not the sophists, to be the efficient cause of a moderate city; for this reason the rhetoricians can be adjuncts to the statesman's art and a truthful political science.

Bacon knows of the later ancient distinction between the sophists and the rhetoricians, and he calls the sophists the "weaker creature." But in discussing the new method for discovery he treats them as if they were the same. In the new world, not only are the rhetoricians the masters of the art of arts, having custody over memory and use of invention, but they do not hesitate to admit the object of their art, every opinion that could determine any virtue. When the rhetoricians and the sophists are the same, the rhetoricians are the efficient cause of a political order with no grounds for conventional authority. Such a city is possible when the rhetoricians are custodians for the actual conquest of nature. In such a city there are no honest sophistical mountebanks because the sophist-rhetoricians can actually accomplish their work. But as Bacon shows in the sequel, although the sophists' impossible city is now possible, it cannot be moderated by the older rhetoricians who could serve the statesman by telling knowing lies.

[408–417]

Turning to rhetoric, Bacon says that it is "a science excellent, and excellently well labored." It is true that rhetoric is inferior to wisdom, as is demonstrated by Moses' godlike superiority to Aaron; still, "with the people it is more mighty," as is shown by Solomon's proverb, which says that "profoundness of wisdom will help a man to a name or admiration but that it is eloquence that prevaileth in an active life."[33] Aristotle's and Cicero's emulation of rhetoricians led them to exceed themselves in their own works of rhetoric, and the examples of Demosthenes and Cicero "hath doubled the progression in this art." The deficiencies that Bacon notes pertain not to the art it-

33. Exod. 4.16; Prov. 16.21.

self but rather to its "handmaidens." The duty of rhetoric is to "apply reason to the imagination for the better moving of the will." Reason is disturbed by sophism, pertaining to logic, by imagination or impression, pertaining to rhetoric, and by passion and affection, pertaining to morality. Just as men are importuned in negotiation by cunning, importunity, or vehemency, so too in "negotiation within ourselves" men are "undermined by inconsequences, solicited and importuned by impression or observations, and transported by passions." But men are not so ill constructed that these powers cannot be used to help reason. The end of logic is to "secure reason," not "to entrap it"; the end of morality is to bend the affections to reason, not "to invade it"; and the end of rhetoric is to "fill the imagination to second reason, and not to oppress it." Therefore, Bacon concludes that Plato was unjust in blaming rhetoric as a "voluptuary art" like cookery.[34]

Bacon says that speech is better at adorning the good than the bad, for "there is no man but speaketh more honestly than he can do or think." In this vein Thucydides remarked that because Cleon took the bad side in "matters of estate," he was always "inveighing against eloquence and good speech; knowing that no man can speak fair of courses sordid and base."[35] Therefore, as Plato said that, if virtue could be seen, she would "move great love and affection,"[36] so if she cannot be seen by the corporeal sense, then the next best thing is to show her to the "imagination in lively representation." To show virtue only to reason was derided in Chrysippus and many of the Stoics because "sharp disputations and conclusions have no sympathy with the will of man."[37] If the affections were obedient to reason, then persuasion would not be needed, but because of the "continual mutinies and seditions of the affections," *video meliora, proboque; deteriora sequor*,[38] reason would become their captive if the "eloquence of persuasion" did not win the imagination from affection to reason against the affections. The reason is that, although the "affections themselves carry even an appetite to good, as reason doth," the difference is that the "affection beholdeth merely the present" and "reason beholdeth the future and sum of time."

The present fills the imagination more, so that "reason is com-

34. Plato, *Gorgias* 462b3–64b1.   35. Thucydides 3.42.

36. Plato, *Phaedrus* 250d3–e1; cf. Cicero, *De officiis* 1.5.15; *De finibus* 2.16.52; Rabelais, *Gargantua and Pantagruel*, trans. J. LeClercq (New York: Modern Library, 1936), 2.18.

37. Cicero, *De finibus* 4.18, 19; *Tusculanae disputationes* 2.18.42.

38. "Whereby they who not only see the better course, but approve it also, nevertheless follow the worse" (Spedding's trans.), *Advancement* 410; Ovid, *Metamorphoses* 7.20.

monly vanquished," but after eloquence has made the remote and future things appear present, then reason will be able to prevail over the "revolt of the imagination." Therefore rhetoric can no more be accused of "coloring the worse part than logic with sophistry or morality with vice." The doctrines of contraries are the same, "though the use be opposite," and logic differs from rhetoric not only as the "fist from the palm, the one close the other at large," but more because logic treats reason "exact and in the truth" and rhetoric treats reason as it is "planted in popular observations and manners."[39] Aristotle was wise to place rhetoric between logic and moral or civil knowledge because the proofs of logic are the same for all men and the "proofs and persuasions of rhetoric differ according to the auditors: *Orpheus in sylvis, inter delphinas Arion.*"[40] This application ought to extend to the point where, if someone says the same thing to several persons, he ought to speak to them "all respectively and several ways." It is possible for the greatest orators to lack this part of eloquence in private speech, so it is not "amiss" to recommend further study in this area, although Bacon is "not curious" whether it be placed here or "in that part which concerneth policy."

Bacon then "descends" to the "deficiencies," which he says are "but attendances." First, he does not find Aristotle's account of the "popular signs and colors of good and evil, both simple and comparative," well pursued.[41] These are the sophisms of rhetoric, which he has discussed before and of which he gives a present example. Aristotle's work is poor because it gives too few examples, because the elenchi are not attached, and because he intended them "only in probation" rather than "much more in impression," for many forms are "equal in signification which are differing in impression."[42] The second deficiency has to do with what he mentioned before "touching Provision or Preparatory store for the furniture of speech and readiness of invention," which is made of two kinds, *antitheta*, or arguments for and against, and formulae, which are "decent and apt passages or conveyances of speech, which may serve indifferently for differing subjects." Two remaining appendixes concerning the tradition of knowledge are critical and pedantical. Knowledge is either delivered by teachers or attained by proper endeavor; as the main part of tradition concerns the writing of books, so the "relative part" concerns the reading of them. Regarding pedantical knowledge, it

39. Aristotle, *Rhetoric* 1355b7–21; Cicero, *Orator* 32.113–15; *De finibus* 2.6.17.
40. Aristotle, *Rhetoric* 1356a20–35; "to be in the woods an Orpheus, among the dolphins an Arion" (Spedding's trans.); Virgil, *Eclogues* 8.56.
41. Aristotle, *Rhetoric* 1362a15–63b4.   42. Ibid.

"containeth that difference of tradition which is proper for youth," and to this category pertains several important considerations. First, it is important to know about the "timing and seasoning of knowledge," second, it is important to know how to proceed from the easy to the hard, third, it is important to know how to apply learning according to "the variety of wits," for no intellectual defect seems to be without cure, and various wits have various sympathies for different sciences, and fourth, the ordering of exercises is important. As the good or ill nurture of seedlings is most important for their thriving, and "as it was noted" that the first six kings, being the tutors of Rome in her infancy, were the "principal cause of the immense greatness of the state which followed,"[43] so the culture of youthful minds is so important and powerful that hardly any time and labor "can countervail it afterwards."

Bacon concludes the discussion of rhetoric by noting that he has now "come to a period in rational knowledges." If his division is new, he should not be thought to "disallow" all the divisions he does not use. Two necessities force his alteration of the division. One is "because it differeth in end and purpose, to sort together those things which are next in nature, and those which are next in use." Just as a "secretary of state" would distinguish between the nature and the use of his papers, in one case separating according to nature and in another case mixing different natures according to use, so in his "general cabinet of knowledge" Bacon has divided according to the "nature of things," while had he handled any "particular knowledge," he would have "respected the divisions fittest for use." The other necessity is that the bringing in of deficiencies "did by consequence alter the partitions of the rest." If the extant knowledges be fifteen and with the deficiencies they be twenty, the factors of fifteen and twenty are different, and so "these things are without contradiction, and could not otherwise be."

As we should expect, Bacon begins by boldly asserting the superiority of rhetoric to wisdom in "an active life." Since the aim of the new knowledge is the unity of theory and practice, such an assertion really means that wisdom and rhetoric are the same. Again, rhetoric aspires to be an all-knowing art of arts. This intent is borne out by the references to Moses, Aaron, and Solomon. Solomon actually says that wisdom leads to eloquence, not that they are somehow separate, and in an earlier context, Bacon made the apparent "mistake" of referring to the serpent that devoured the enchanters' serpents as having belonged to Moses rather than to Aaron.[44] According to the references, Moses' stammer could be seen as a form of hidden eloquence, using Aaron to speak for policies that had their source in Moses' will. But if wisdom and rhetoric come together in such hidden eloquence, then the wisest rhetoric can be compatible with the use of public goods for private or hidden purpose. And the full implication of this fact is demonstrated in the objection to Plato's condemnation of rhetoric and the example of Cleon's hostility to eloquence.

In the *Gorgias,* to which Bacon refers, Plato does not complain that rhetoric is "voluptuary"; instead he has Socrates argue that rhetoric is not an art but rather a kind of experience—a kind of flattery that can be no model of the mind's perfection.[45] Plato's point is not that the arts are the model of the mind's perfection but that they reflect something of this perfection insofar as they differ from mere experience. The point is not that rhetoric is simply not an art. Rather the point is that there is nothing elegant about rhetoric, so that of all the arts, rhetoric has the least of the uselessness that causes art to differ from mere experience. Rhetoric, like flattery, is the most constrained of all the arts, grappling as it does with every art only with regard to its usefulness rather than to its likeness to the knowledge of the being of things. But while Plato condemns rhetoric from the point of view of the love of wisdom and free contemplation, demonstrating the principles that subordinate rhetoric to political science, Bacon shows that in the charitable age, the rhetoricians attempt to identify the constrained usefulness of their art with the freedom of the love of wisdom. Charitable rhetoric merely repeats the partisan delusion latent in every productive art. The very reason why Plato's criticism of rhetoric is unjust demonstrates the potential injustice of charitable rhetoric. The injustice of charitable rhetoric springs from its attempt to identify freedom and harsh necessity, for in so doing, rhetoric cannot distinguish between goods that can be acquired freely by oneself and those that can be had only by recognizing one's need for others.

Actually, then, the problem of charitable rhetoric is demonstrated by Plato's "mistake," well noted in the example of Cleon's speech in the debate about the Mytileneans.[46] Cleon did not just blame wit and eloquence in his plea for harshness; rather he condemned them

44. *Advancement* 322–29; Exod. 4.16, 7.1, 12; Prov. 16.21.
45. *Gorgias* 462b3–64b1.
46. *Thucydides* 3.37–49; Strauss, *The City and Man,* 212–18.

in contrast to his praise of the laws, the laws of a free, imperial democracy that behaves as a tyrant toward the cities it rules. Bacon's argument that eloquence better adorns the good than the bad abstracts from the same thing that the harsh Cleon forgets: that the laws can be stubborn, unpitying, and possibly unjust despite their dependence on the softer wiles of persuasion. Cleon, the enemy of glory and the noble and the friend of merely vulgar self-interest, is unaware of the dependence of the laws on prior speeches about interest more generally understood. Therefore Cleon is ignorant of the difference between harsh unpitying laws directed to ignoble ends and the same laws directed to graceful or noble ends, that is, the difference between Sparta and the Athens he defends. But the function of rhetoric according to Bacon is not to bend the imagination and affections to the noble as opposed to the base; rather it aims to govern the way in which reason can make remote interest present or immediate.

In defending the rhetoric condemned by Cleon, Bacon grounds his argument on Cleon's grounds, which would have the laws be indifferent to interest construed broadly enough to include graceful softness in addition to the harshness of immediate advantage. Moreover, reminding us of the link between the graceful and the harsh, Bacon does not mention that, despite the success of eloquent speech, the Mytileneans escaped death only narrowly, nor does he mention the dreadful fate of the Melians. Cleon does not like deceptive, fine speeches, and Bacon disagrees because such speeches are incompatible with sordid and base policy. In this Bacon reminds us of the dialogue resulting in the horrible slaughter of the Melians, a collection of short, unadorned speeches that Cleon must have preferred.[47] At this point, whether charitable rhetoric is open to Cleon's terrible policy depends upon whether it practices short and unadorned speeches as well as long fine ones; in the sequel, Bacon shows that it does.

Bacon argues that rhetoric can be no more charged with "coloring the worse part" than logic can be charged with sophistry or morality with vice. We know, he says, that the "doctrine of contraries is the same, though the use be opposite," by which he refers to Aristotle's discussion of dialectic, sophistry, and rhetoric in the *Rhetoric*.[48] But closer examination shows that Bacon reverses Aristotle's argument. Aristotle says that dialectic and rhetoric are akin because they both concern all of the arts rather than any single art in particular. Rhetoric differs from sophistry because, unlike sophistry, while rhetoric

47. Thucydides 5.84–116.   48. *Rhetoric* 1355a19–b24.

studies the means to persuasion, it does not itself always persuade, just as every single art seeks the means to its end but does not simply and always produce it. Rhetoric is like dialectic because both seek the difference between the real and the apparent regarding either persuasion or syllogisms. The difference between dialectic and sophistry is determined by the intention or choice, but in rhetoric, those who act according to knowledge and those who act according to some intention or choice are both called rhetoricians.

According to Aristotle, then, rhetoric differs from sophistry because sophistry is sophistry solely by reference to the intention to deceive in order to claim what no particular art claims. But while rhetoric *is* an art, unlike sophistry, it can and does include the choice or intentions that make the sophist the sophist. Rhetoric, then, is like sophistry except for the sophists' delusional claim to be able always to produce by imitation what no particular art could always produce. The sophists are deluded, but the rhetoricians are knowing liars, and only as such liars can the rhetoricians serve the statesman in ordering a moderate whole of the productive arts. The sophists' short speeches are different from the rhetoricians' fine speeches because the sophists do not know that they deceive and the rhetoricians do. But Bacon argues that sophistry and rhetoric differ only in regard to use, with regard to the intention to deceive, *not* with regard to deluded and knowing deception. But if so, then the lying rhetoricians become no more self-knowing than deceiving sophists, and there can be no real differences between short and long speeches that deceive.

Aristotle's *Rhetoric* outlines a general art of arts, employed by the statesman in forming right opinion and enabling the statesman to use the poets and to restrain the sophists. For Bacon, this art is not tenable in the present charitable age. If rhetoric and logic differ only in regard to reason as it is planted in "popular opinions and manners" and to reason "exact and in truth," then charitable rhetoric can be like dialectic in its use of short speeches as well as long, fine ones. The new rhetoric serving the conquest of nature is as effective with short speeches as it is with long, fine ones. But if so, then the new rhetoric is no less open to the harshness of immediate interest or advantage than the short speeches that condemned the Melians. Likewise it is open to the fine speeches that are the source of the laws as understood and preferred by the harsh, violent Cleon.[49]

Bacon demonstrates this conclusion in the second reference to Pla-

---

49. In the *De augmentis* Bacon notes that Xenophon spoke so as to censure Cleon, a remark that does not appear in the *Advancement*. Again, the *Advancement* is closer to Bacon's intention, which can be discerned as well from the *De augmentis* (*De aug.* 672).

to's remark in the *Phaedrus*. According to Bacon, Plato said "eloquently" that if virtue could be seen, it would "move great love and affection." From this statement Bacon concludes that, since it cannot be shown "to the sense by corporeal shape," the next best thing is to show virtue to the imagination rather than just to reason as the Stoics did, for which they were derided. But Plato does not say that if virtue could be seen it would "move great love and affection." Rather he says that, in contrast to the beautiful, if the image of prudence (*phronēsis*) were given to sight, the sharpest of the physical senses, it would arouse a terrible or clever love.[50] Insofar as the imagination is linked to the body's organs, it cannot behold the image of prudence without inciting mad longing, mad because it is for something that no body could ever be. For Plato, prudence works only as it turns the mind to the intelligible things that limit the possibilities open to any individual, sighted body.[51] But the new first philosophy aims to free the imagination, and the mind, from the limits of the intelligible things, for the sake of practice and the body's desires. When Bacon says that rhetoric must show prudence to the imagination, he does so in order to remove the immediate objects of the body's affections from the way to the far ones, and Plato's wisdom shows that this course of action ultimately incites a terrible love.

Bacon knows that, for Plato and Aristotle, moral virtue is a fabricated image of the mind's rest, produced by the artful molding of the affections or passions. This art, presented by Aristotle, is itself productive and accommodates the body's needs by merely *likening* practical activity to the poetic images of the mind's possibilities. This art is rhetoric, which Bacon says Aristotle wisely placed between "logic" and moral and civil knowledge. But Aristotle did no such thing. Rather, as Bacon knows perfectly well, for Aristotle rhetoric stands ambiguously between dialectic and the statesman's moral and civil knowledge. Bacon knows that Aristotle so placed rhetoric wisely because, to preserve the distinction between mind and body, or to preserve the darkened, imagelike qualities of the virtues, it cannot reveal the true status of the moral virtues. It is clear now what happens when rhetoric does not mind its place: it promises to the body the perfection that philosophy promises to the mind, which leads to blindness to the limits of the human body, both with respect to itself and with respect to its needy dependence on other bodies. And as long as rhetoric serves the ambition of first philosophy, its intermediate status is untenable.

50. *Phaedrus* 250d3–e1.   51. See *Republic* 518c4–19a6.

Aristotle insists that since rhetoric treats the passions and the nature, origin, and manner of producing the virtues, it appears to be related to dialectic and to politics. But he says that rhetoric claims to be politics only because of the ignorance, boastfulness, or other weakness of those who possess it, because rhetoric is like dialectic in not being associated with one particular thing, as are the other sciences, including politics or ethics.[52] But in the *Ethics*, politics is described not as a science but as "sort of a science," and one reason is that like dialectic and rhetoric, politics treats a general good that orders a whole of specific goods.[53] Rhetoric's concern with origins requires that Aristotle make the link between politics or ethics and rhetoric, or the location of rhetoric between dialectic and politics, much more difficult to see than Bacon implies. Aristotle's caution accords with the truth of rhetoric's intermediate status. But Bacon says that rhetoric might easily be placed with policy as well as "here."

Bacon's greater boldness disguises his agreement with Aristotle, which does not differ from his earlier dissembling in the division of first philosophy and metaphysics, where a disagreement disguised a deeper agreement.[54] Likewise, from this point in the division, Bacon criticizes Aristotle's rhetoric as not well pursued; and in the upcoming discussion of the good he criticizes that rhetoric for including a brief and inadequate account of the "knowledge touching the several characters of natures and dispositions," which is part of a greater neglect of the origins, or the "roots and strings," of the virtues.[55] But immediately after this latter criticism, Bacon agrees with Aristotle's argument about teaching ethics to the young, an argument at the bottom of Aristotle's caution about the teaching of such "roots and strings." Nevertheless, in the present context, Bacon complains that Aristotle did not attend well enough to the popular signs of good and evil, which he says "are as the sophisms of rhetoric." He criticizes Aristotle for not showing that the common opinions of many are sophisms that can be reduced to dishonest maxims. We can see, then, that, however much Bacon agrees with Aristotle's caution about the nature and status of rhetoric, he recommends that the sophistical nature of every city be brought to light. Such a move brings the origins and production of the virtues more clearly to light than Aristotle judged to be prudent, for if common opinion is in fact sophistical, then the ultimate source of the virtues can be traced to unjust necessity. Bacon's greater boldness is accompanied by his careful attention

---

52. *Rhetoric* 1356a20–35.   53. *Nic. Eth.* 1094a1–b11.
54. See Chap. 8 above.   55. *Advancement* 420, 436–40.

to the reasons why such boldness might be unjust. We know that Bacon is not himself unjust, however, because he is constrained by a more compelling historical necessity than was experienced by Plato or Aristotle and because he offers the new conquest of nature to tame those who are unjust. By the light of the new charitable rhetoric, if the noble can be exposed as but a form of harsh advantage all harshness can be overcome. Bacon knows this is not really possible. The likes of Cleon can never be eliminated because interest and advantage will always be harsh. And if the necessary pursuit of advantage is to be moderated, the difference between noble and ignoble harshness must at least be understood if it cannot be preserved.

According to Aristotle, it is important for some to listen and not to know, a point appearing most explicitly in his argument that the young are not fit students of politics.[56] The young must listen to the speeches that form them to the virtues, but they are not, as young, proper inquirers into the nature and the origins of the virtues themselves. Aristotle does not take youth to mean simply a given number of years, because there are men of age who must be said to be youthful as regards morality. Therefore the teaching of the nature and the origins of virtue is a problem for man's life in the city, comprised as the city is of the young and the old as such. In the apology, Bacon taught that, for charitable learning, lawless, youthful knowing is superior to aged listening.[57] The new project for taming the sects treats all, the many and the few, as if they were all young knowers rather than listeners and to do so abandons Aristotle's caution regarding rhetoric, which can no longer serve political science in forming the practical virtues after the poetic imitation of restful contemplation.

In the charitable age rhetoric will no longer lie, but because of this new-found integrity, it can no longer serve the truth. In the light of this integrity, the statesman appears to be an imposter with no tools, and the traditional political science seems otiose. And for practical purposes it is, for in the new charitable age the necessary claims to knowledge about political life—the claims of the sophists, the poets, the statesman, and the political philosopher—are so transformed or silenced that they can no longer be properly ordered by political science. The new rhetoric means the end of rhetoric, without which political science cannot be practiced. And without this science the most extreme political partisanship is freed from moderating restraint. This development would not be troubling if one could rely on the new conquest of nature to absorb the unjust force of such partisanship.

56. *Nic. Eth.* 1094b29–95a10.   57. Chap. 4 at nn. 15–21 above.

To some extent it can. But in fact, the unjust partisan rule by truth-telling rhetoric is caused by the new modern project that transforms, and does not escape, the need for political knowledge. The old political science discloses the true limits and dangers of the new charitable learning. And the old political science must not be forgotten, because while the new age will need Machiavelli's political science, that immoral science is itself in need of the new, more powerful project for mastering immediate interest and advantage.

[10]

Appetite, the Body, and the Private and Public Good: The True Roots of Good and Evil

Having shown in great detail how the new project for learning subverts the old political science without adequately replacing it, Bacon must still show on what moral grounds the new learning can be dangerous and must demonstrate that any concrete political teaching contained in it is in fact the deficient and immoral political science of Machiavelli. In the present chapter we therefore consider Bacon's thematic account of the human good, and in the next chapter we consider his explicit remarks about civil knowledge. The account of the good is contained in the division of the second part of the knowledge pertaining to the mind, the knowledge concerning man's appetite and will.

[417–445]

Bacon begins by arguing that teachers have given fine pictures of good, virtue, duty, and felicity, showing them to be the "true objects and scopes" of will and desire, but they have ignored the matter of how "to attain these excellent marks" and how to "frame and subdue" man's will to them. This neglect cannot be excused by casual remarks such as those that Aristotle made about the role of nature and habit or the distinction between those "generous spirits" who are won by "doctrines and persuasions" and the vulgar who are won by reward and punishment.[1] The reason for the neglect is the contempt men have for ordinary and common matters, for men have made sciences consist of merely resplendent matter for the sake of the glory

---

1. *Nic. Eth.* 1103a14–b25, 1179a34–81b24.

of subtlety or eloquence. To remedy this difficulty, Bacon divides moral knowledge into two main parts: the "exemplar or platform of good" and the "regiment or culture of the mind."

Bacon begins the division of moral knowledge with the "platform or nature of good," which is either simple, concerning the kinds of good, or is compared, concerning the degrees of good. Regarding the degrees of good, the Christian faith has "discharged" the infinite disputes about the highest good, felicity or beatitude, "the doctrines concerning which were as the heathen divinity." As Aristotle said that young men can be happy but only by hope,[2] so "we must all acknowledge our minority" and embrace the happiness "which is by hope of the future world." Freed from the philosopher's heaven, which took man's nature to be higher than it was, as is demonstrated by Seneca's praise of human divinity,[3] we can profitably receive the "inquiries and labors" of the ancients. The ancients have excellently described the "nature of good positive or simple," distributing it into "kinds, provinces, actions and administration," and they have commended them well and fortified them as much as possible against "corrupt and popular opinions." They have also excellently handled the "degrees and comparative nature of good" in their "triplicity of good," comparison of action and contemplation, distinction between "virtue with reluctation" and "virtue secured," comparison of honesty and profit, and the balancing of virtue and virtue.[4] Nevertheless, if they had considered the "roots of good and evil, and the strings of those roots" before they came to "popular and received notions" of virtue, vice, pleasure, and pain, they would have given "great light to what followed," especially if they had "consulted with nature" and so made their doctrines "less prolix and more profound." Because they partly omitted and partly confused these matters of roots, strings and nature, Bacon "will endeavor to resume and open in a more clear manner."

There is formed in everything, says Bacon, "a double nature of good," first as a thing is "a total or substantive in itself" and second as a thing is "a part or member of a greater body." The latter is the better because "it tendeth to the conservation of a more general form." This double nature and "the comparative thereof" is even more "engraven upon man" if he is not degenerate. For man, duty to the public is superior to the conservation of life and being, and it must be af-

---

2. Ibid., 1099b33–100a9; *Rhetoric* 1388b31–89b13.   3. *Ep.* 53.12.
4. *Nic. Eth.* 1098b12–20, 1176a30–79a32, bk. 7; *Rhetoric* 1362a15–63b4; *Nic. Eth.* 1113a15–b2.

firmed that no religion or other discipline has so exalted the "communicative" good and depressed the private and particular good as "the Holy Faith." This "being set down and strongly planted" decides most of the controversies of moral philosophy. It decides against Aristotle's preference for the contemplative over the active life because his arguments for the contemplative life are all private, not unlike Pythagoras' likening of philosophy and contemplation to those who come to the games only to look on.[5] "But men must know," says Bacon, that only God and angels can be lookers on in "this theatre of man's life." Divinity knows nothing of socially useless contemplation. The exaltation of communicative over private good also decides the controversy between Zeno and Socrates and their schools, who "placed felicity in virtue simply or attended," which "chiefly" embraced society, on the one hand, and, on the other hand, the Cyrenaics and Epicureans, who made virtue the servant of pleasure, the reformed Epicureans, who made felicity consist in serenity and freedom from perturbation, as if Jupiter were deposed and Saturn restored along with "the first age,"[6] and Herillus, who made felicity the end of "disputes of the mind," saying that there was no fixed good and that things are good according to the "clearness of the desires" (a view adopted by the Anabaptists), all of which prefer repose and contentment over society.[7]

After discussing the superiority of the communicative to the private good, Bacon "resumes" the division of the private good. The private good is divided into active and passive, a difference also formed in all things. It is best disclosed in the "two several appetites" in creatures, self-preservation or continuance and the desire to "dilate or multiply themselves." The latter seems to be "the worthier," for in nature, the superior heaven is the agent while the inferior earth is the patient, the pleasure of generation is greater than the pleasure of eating, in divine doctrine it is more blessed to give than to receive,[8] and "in life" no one's spirit is so soft "but esteemeth the effecting of somewhat he hath fixed in his desire more than sensuality." The priority of the active good is upheld by our mortality and exposure to fortune. If we could have certain perpetuity of our pleasures, then the mere "state of them would advance their price." But when we see with Seneca that it is but *magni aestimamus mori tardius* and, from Proverbs, *ne glorieris de crastino, nescis partum diei*, we wish

---

5. Cicero, *Tusculanae disputationes* 5.3.   6. Ovid. *Metamorphoses* 1.107.
7. Cicero, *De finibus* 4.14.   8. Acts. 20.35.

to have something secure from time, which can only be our "deeds and works," as it is said, *opera eorum sequuntur eos.*[9] The priority of the active good is also upheld by man's natural affection for "variety and proceeding," of which the pleasures of the senses, the "principal part of the passive good," participate but little, as is shown by Seneca's remark in his tenth epistle. In "enterprises, pursuits and purposes of life" there is much variety, "whereof men are sensible with pleasure" in the various motions to their ends, as was well said by Seneca, *vita sine proposito languida et vaga est.*[10] The active good of which Bacon speaks is not the "good of society," or the communicative good, even though they are at some times similar or coincident. Rather the active good has to do with a man's own or private "power, glory, amplification or continuance." The "gigantine state of mind" that possessed the troublers of the world, such as Lucius Sulla and other "smaller models" who wanted all men happy or unhappy as they were their friends and enemies and would form the world after their own humors ("the true Theomachy"), pretended to the active good, even though "it recedeth furthest from the good of society, which we have determined to be the greater."[11]

Bacon next resumes the passive good, dividing it into conservative and perfective. At this point he asks that we take a "brief review" of what has been said. He first spoke of the good of society, which "embraces the form of human nature," of which we are parts, but not our "own proper and individual form." Then he spoke of the active good as a part of private good, because all things have a triple desire proceeding from self-love: a first of preserving and continuing their form, a second of advancing and perfecting their form, and a third of "multiplying and extending their form upon other things," with the latter being called the active good. There "remaineth the conserving of it, and perfecting or raising of it," of which the latter is "the highest degree of passive good" because to preserve "in state" is not as good as to preserve "with advancement," and so in man, *igneus est ollis vigor, et caelestis origo.*[12] Man's approach or assumption to "divine or angelical nature" is the perfection of his form, and the

9. "We think it is a great matter to be a little longer in dying" (Spedding's trans.), Seneca, *Naturales quaestiones* 2.59.7; "boast not thyself of tomorrow, thou knowest not what the day may bring forth" (Spedding's trans.), Prov. 27.1; "their works follow them" (Spedding's trans.), Rev. 14.13; *Advancement* 424.
10. *Ep.* 77.6; "life without an object to pursue is a languid and tiresome thing" (Spedding's trans.), *Ep.* 95.46.
11. Plutarch, *Sulla* 38.
12. "The living fire that glows those seeds within remembers its celestial origin" (Spedding's trans.), *Advancement* 426; Virgil, *Aeneid* 6.730.

false imitation of this good is "the tempest of human life, while man, upon the instinct of an advancement formal and essential, is carried to seek advancement local." Just as the sick toss as if in changing position they will also change internally, so ambitious men who fail "of the mean to exalt their nature" are in a "perpetual estuation to exalt their place." For these reasons, then, the passive good is divided into conservative and perfective.

Taking up the conservative good, or the "good of comfort," Bacon says that it consists in the "fruition of that which is most agreeable to our natures." This seems to be the purest and most natural of the pleasures, but in fact it is the softest and the lowest. Fruition is good either in the "sincereness" of it or in the "quickness and vigor" of it. The one is "superinduced by the equality" and has less mixture of evil, and the other is superinduced "by vicissitude" and has "more impression of good." Two questions regarding this difference, which has not been inquired, concern first which of the two is better and second whether man's nature might not be capable of both. The first question is "controverted," and the second question has not been inquired. The first question was debated between Socrates, who placed happiness in "an equal and constant state of mind," and a sophist, who placed it in much desiring and enjoying.[13] According to Bacon, both opinions do not want for support. Socrates is supported by the "Epicures," who think that virtue is important in felicity and that virtue has more to do with "clearing perturbations than in compassing desires." The sophist's opinion is favored by the superiority of "advancement" over "simple preservation," because every "obtaining a desire hath a show of advancement," as circular motion appears to be progression. However, properly decided, the second question obviates the first, and surely there are some who enjoy pleasures more than others and yet who are less troubled by their loss. The philosophers have tried to make men's minds too "uniform and harmonical," not training them for "contrary motions," because the philosophers were dedicated to a "private, free and unapplied course of life." Men should imitate jewelers, who will grind a stone if it can be done without taking too much, and so men should procure serenity without destroying magnanimity.

Having "deduced" the private human good as far as seems fit, Bacon turns to the part of the knowledge of man particular that is communicative, or to the good of man "which respecteth and beholdeth society," which he terms "duty." Duty is toward others, and virtue

---

13. Plato, *Gorgias* 462–94.

concerns a well-formed mind in itself, although neither duty nor virtue can be understood without some relation to the other. Duty may be thought to concern political or civil science, but this is not so, because it concerns the government of every man over himself, not over others. As in architecture and "mechanicals" the direction for framing a whole differs from the manner of joining the parts, but in expressing one you express an aptness for the other, so the doctrine of conjoining men to society differs from men's conformity to society. Duty is divided into two parts, the common duty of every man as a member of a state, and the special or "respective duty" of each man to profession, vocation, and place. The first is extant and well labored, as he has said; the second more dispersed than deficient. Such dispersion is proper, for no one could write an account of duty for every vocation, profession, or place.

Corresponding to the duties of professions is the handling of the frauds, impostures, and vices of the professions, which has been done more satirically than seriously. Men have more tried to blame what is good in professions than to root out what is bad, and the fable of the basilisk shows that if evil arts be spied, they die, but if not "they endanger," so that we are indebted to Machiavelli and others "who write what men do and not what they ought to do." One cannot combine "serpentine wisdom with columbine innocency" unless one knows the serpent, or "all forms and natures of evil." Otherwise virtue is unprotected. In fact, the honest cannot reform the wicked without knowing evil. The corrupt think that honesty springs from "simplicity of manners," listening to preachers and schoolmasters, and "exterior language," so unless one can get them to see that one can see the "utmost reaches" of their corrupt opinions, they despise all morality.

The knowledge concerning "good respecting society" is also handled "comparatively." This treats the weighing of duties between persons, cases, and publics as can be seen in the case of Lucius Brutus against his sons, which was extolled and yet about which was said *infelix, utcunque ferent ea facta minores.*[14] The most common such question concerns great goods won from small injustices. Jason of Thessalia decided that some injustice is necessary for justice in many things,[15] but it is good to reply to this that one can control a justice

14. Livy 2.5; "Unhappy man! Whatever judgment posterity shall pass upon that deed," etc. (Spedding's trans.), *Advancement* 431; Virgil, *Aeneid* 6.822.
15. Plutarch, *De tuenda sanitate praecepta* 24 (135); *Praecepta gerendae reipublicae* 24 (818).

to be done but one cannot control what is to follow. Men must pursue present justice and leave the future to divine providence, and with this pious remark, Bacon "passes on" from his general remarks about "the exemplar and description of good" to the question of ways and means, or to the "husbandry" of the good.

Without considering the "husbandry" of the good, the account of the good is but a lifeless statue. To establish this point Bacon refers to Aristotle's argument that the object of ethics is to become good, to know the how of the good, not just to know what virtue is.[16] Cicero commended Cato for wanting to live like a philosopher, and although Bacon also agrees with Seneca that in the present age men do not care about the whole of life, which makes husbandry seem useless, he concludes with Hippocrates that the medicine is important even to make men aware of the disease.[17] The care of the mind belongs to divinity, but moral philosophy is the handmaid. As excellent as it is, it is strange that this part of knowledge is not reduced to written inquiry, especially since it is so common and wiser in common speech than in books. It is important to propound it "in the more particularity" because it is worthy, deficient, and is "otherwise conceived" by those who have written about it.

It is important to know what can and cannot be done so as to distinguish between alteration and application. Two things are beyond command, nature and fortune, and since our work is tied by them, we have to proceed with them "by application" or by wise and industrious suffering.[18] We do so by accommodating or applying, which tries to derive use and advantage from what seems adverse. The first part of application is to list the "several characters and tempers" of men's dispositions, especially those that are basic or most frequent. It is not enough to describe them just to show the virtues as means, as Aristotle described magnanimity; it is also important to describe the usefulness of various kinds of minds and dispositions.[19] Bacon is surprised that this aspect is omitted in morality and policy. Plautus said it was wondrous to find an old man beneficent, Paul thought Cretans needed harsh discipline, Sallust thought kings desired contradictories, Tacitus noted how fortune mends the disposition, and Pindar believed that great and sudden fortune defeats men, as the

16. *Nic. Eth.* 1103b26–31; *Magna moralia* 1182a1–8.
17. Cicero, *Pro Murena* 30.62; *Ep.* 71.2; Hippocrates, *Aphorisms*, trans. Jones, Loeb Classical Library (London: Heinemann, 1923–31), 2.6.
18. *Aeneid* 5.710.   19. *Nic. Eth.* 1123a34–25b25, 1126b11–27a12.

psalm shows.[20] But this kind of observation is only treated cursorily by Aristotle, in the *Rhetoric* and elsewhere, and not in "moral philosophy," where it belongs.

Another "article of this knowledge" concerns the affections, for after knowing the different characters it is necessary to know the mind's diseases, which are the perturbations of the affections. Again, Bacon finds it strange that Aristotle considered the affections only cursorily, as they are moved by speech, but that he omitted them from where they belong. It is not enough just to consider pleasure and pain, and the Stoics treated the subject in their typically oversubtle way. Bacon finds some elegant writings on the affections, but the "poets and writers of history are the best doctors of this knowledge," considering as they do the ways of the affections, the most important of which for moral and civil matters is the way the passions "do fight and encounter one with another." The use of reward (hope) and punishment (fear), upon which states rest, is the use of one affection against another, without which "we could not so easily recover."

Regarding the things within our command, Bacon will consider custom and habit because it would take too long to "prosecute all." Aristotle is wrong to say that custom cannot change nature.[21] He is correct where nature is "peremptory," which Bacon cannot here explain, but not where nature allows latitude. Those natural things that are perfected or altered by use are more akin to manners, but granting to Aristotle that the virtues and vices consist in habit, he should have done more to teach ways of forming these habits, which are as numerous as the exercises of the body. Bacon will recite a few of these precepts. One is not to take too high a strain or too weak, another is to practice both when the mind is best disposed and worst disposed, another, noted by Aristotle, is to bend against the natural inclination,[22] and another is to dissemble the real intention of an exercise because of the mind's natural hatred of "necessity and constraint." There are many such axioms that can "prove another nature," but when left to chance they ape nature and produce what is "lame and counterfeit."

Regarding books and studies, there are diverse precepts of caution and direction. One of the fathers called poetry *vinum daemonum*, and Aristotle opined "worthily" that the young are not fit auditors of moral

20. Plautus, *Miles gloriosus* 3.1.640; Titus 1.12–13; Sallust, *Bellum Iugurthinum*, ed. L. Watkiss (London: University Tutorial Press), 113; Tacitus, *Historiae*, 1.50; Pindar, *Olympian Odes*, trans. Sandys, Loeb Classical Library (London: Heinemann, 1961), 1.55; Ps. 62.10.

21. *Nic. Eth.* 1103a14–26.   22. Ibid., 1109a20–b26.

philosophy because they are too moved by the affections and are not well experienced.[23] It is because they are taught to the young rather than to the mature that the ancient books and discourses, which praise virtue and scorn popular opinion, have been of so "little effect towards honesty of life." But it is also true that young men are not fit auditors until they have been "thoroughly seasoned in religion and morality" lest they become corrupted and recognize no differences between things other than "utility and fortune," as Seneca's and Juvenal's verses teach "satirically and in indignation on virtue's behalf" but as books of policy "do speak seriously and positively," which can be seen in Machiavelli's comment about Caesar and Cataline, "as if there had been no difference but in fortune, between a very fury of lust and blood, and the most excellent spirit (his ambition reserved) of the world."[24] There is even a caution against moralities themselves, "lest they make men too precise, arrogant, and incompatible," as Cicero said about Cato.[25] There are many such axioms about studies, as well the other points, "of company, fame, laws, and the rest," which Bacon recited at the beginning. But there is an even more accurate culture of the mind, grounded in the fact that all men's minds are at some times more perfect and at other times more depraved. The good times may be fixed by vows or resolutions and the evil times obliterated by redemption and expiation of the past and resolution for the future. This procedure is sacred and religious, which is just, because morality is religion's handmaid.

Bacon will conclude with the "most compendious and summary point" and the most "noble and effectual" for reducing the mind to "virtue and good estate." This is a man's propounding to himself as "good and virtuous ends of his life" as can be attained. If one propounds such ends and is resolute and constant, he will "mould himself into virtue at once." This is like nature's work; the other is like the hand's. The hand can only work on parts of an image, but nature forms all the parts of creatures at once. Obtaining virtue by habit is partial, but in application to good ends, one is disposed to whatever virtues spring from the "pursuit and passage toward those ends." Aristotle called this state of mind divine, not virtuous, just as brutality is different from vice, and we can see what "celsitude of honor" Pliny the Second attributed to Trajan when he said that men should

23. St. Augustine, *Confessions,* trans. E. Pusey (New York: Dutton, 1926), 1.16; *Nic. Eth.* 1094b27–95a11.

24. Seneca, *Hercules furens,* trans. Miller, Loeb Classical Library (London: Heinemann, 1917), 251–52; Juvenal, *Satirae* 13.100–18; *D* 1.10.

25. *Pro Murena* 29.61.

pray that the gods should continue to be to them as Trajan had been, as if Trajan had been not an imitation but a pattern of divinity.[26] These heathen passages are but a shadow of the "divine state of mind": the Holy Faith directs men to divinity by imprinting charity on their souls. Charity is well called the bond of perfection because it comprehends all of the virtues.[27] A mind inflamed with charity will be formed more perfectly than any doctrine of morality, which is, by comparison, sophistry.[28] Xenophon noted that the affections raise the mind only by distorting excess, but love both exalts and settles the mind,[29] and the same, says Bacon, is true of all other excellences except charity, which "admitteth no excess." In aspiring to divine power the angels fell, as Isaiah says, and by aspiring to divine knowledge man fell, as Genesis says, but no man or angel ever transgressed and fell aspiring to divine love.[30] We are called to that imitation by Matthew and Luke, so in describing the "first platform of the divine nature itself" the heathen say "best and greatest" but the holy scriptures refer to "God's mercy over all of his works."[31] Bacon then concludes his discussion of the culture and regiment of the mind, commenting that those who say that he has collected what others have omitted as mere common sense will have judged well.

We begin by noting that Bacon presents his account of the human good as a reconsideration and reformation of the ancients', and particularly Aristotle's, teaching. Bacon is clear about why the ancients must be reconsidered and why their teachings must be amended, but he is not immediately clear as to whether their deficiency, an insufficient concern for the virtues' "roots and strings," detracts from their true excellence. On the one hand, the ancient teaching, and especially Aristotle's, is deficient regarding the question of ways and means, causing Bacon to distinguish between the "platform or exemplar" of the good and the "regiment or cultivation of the mind." But on the other hand, when he divides the platform of the good into the good simple and the good compared, with the Christian faith discharging the latter so as to free men from the philosopher's heaven, he says that the ancients' teachings were excellent with respect to both. Because the ancients are excellent when freed from the doctrine of the philosopher's heaven, Bacon still praises Aristotle's comparative

26. *Nic. Eth.* 1145a15–33; Pliny (the Younger), *Panegyricus,* trans. Radice, Loeb Classical Library (London: Heinemann, 1972), 74.
27. Col. 3.14.   28. *De aug.* 742.   29. Xenophon, *Symposium* 1.9–10.
30. Isa. 14.14; Gen. 3.5.   31. Matt. 5.44; Luke 6.27–28; Ps. 145.9.

treatment of action and contemplation. And no wonder, for Aristotle's view of the superiority of contemplation to action is true, apart from the doctrine of the "philosopher's heaven." Actually, Aristotle professed no such doctrine. Rather, for Aristotle, contemplation is not wholly private and action is not wholly communicative, because, as Bacon has demonstrated that he knows, the contemplative political philosopher does have a place in the city as the ultimate model for moderate practical virtues. To put this point differently, the distinction between contemplation and action corresponds precisely to a distinction between private and communicative only if no conventional moral virtues stand between them. Even divorced from the philosopher's heaven (to which it was never really married), Aristotle's preference for contemplation over action is incompatible with the charitable, Christian preference for the communicative over the private good.

Bacon knows that Christianity decides against contemplation and for action because Christianity erroneously takes the former to aspire to a "philosopher's heaven"; Christianity cannot abide contemplation because it cannot abide the practical moral virtues for which contemplation can be a model. We recall that at the outset of his treatise Bacon had to ignore the distinction between public and private propriety in order to address the Christian prince. By this point, and in the present context, we understand why: because Christianity cannot see the proper link between theory and practice, and hopes rather for their charitable unity, it does not understand the proper relationship—and tension—between public and private life. In the Christian age, the extreme hope for the perfect harmony of the public and private good produces immoderate contempt for the difference between what is and what is not properly one's own. And the new charitable knowledge is fashioned on just this Christian model. As we see in this chapter, Christianity and charitable knowledge force our attention on the virtues' "roots and strings," but in the light of the ancient wisdom, this focus reflects a distorted understanding of contemplation as the philosopher's heaven. Such a focus is the ultimate source of the confusion of public and private goods, a confusion that is greatest in the worst of all partisan claims.

Bacon confirms these points by the subtle, opening references to Aristotle purporting to show how insufficiently the ancients attended to ways and means. Considered together, the two references—the first to Aristotle's discussion of nature and habit and the second to the discussion of the difference between those who can be moved by theoretical speech and the many whose poor nature can be moved only by pleasure and pain—do not show that Aristotle failed to consider the virtues' "roots and strings." Rather they disclose the reasons why Aristotle thought them to be unfit for anything but a partial discussion. In the first discussion, Aristotle concludes that the moral virtues, unlike the intellectual virtues, are formed by habit and that, although they presuppose a natural disposition, they are not formed by nature.[32] The virtues are instructed habits: they are like the arts because one acquires them by practice and because they always gild one or another art, but in fact they differ from the arts because, for the possession of the virtues, knowledge is of little importance.[33] In the second discussion, Aristotle describes the transition from the account of ethics to the account of politics, where he explains that in discussing politics one shows how moral habit is generated and supported by the force of the laws.[34]

These two references demonstrate that the virtues are not simply teachable, but at the same time, they show that the virtues are not solely produced by pleasure and pain, as would be the case for the brutes and, perhaps, for the many who could never be virtuous. Bacon knows these facts perfectly well, because the list of the ancients' satisfactory teachings about the comparative good is a perfect scheme of the problems of the intermediate status of the virtues as disclosed in the *Nicomachean Ethics*. The distinction between external goods, the goods of the body, and the goods of the soul points to the possibility that, requiring some good fortune, the virtues may not be simply teachable; and it discloses that although the distinction between the body and the soul may be clear, as is the distinction between man and the brutes, the distinction between the parts of the soul, and between reason and the passions, is not so clear.[35] The difference between virtue with reluctation and virtue secured shows that, while moral virtue is different from intellectual virtue, it cannot be wholly different: although mere self-restraint is more philosophic in wondering about the substance of happiness than is moral virtue springing from unknowing habit, both differ from vice and neither is brutish and irrational.[36] In fact, while the virtues are not simply teachable and are the product of the legislative art that both persuades and coerces, they derive their worth from their likeness to the self-sufficient activity of the mind's contemplation of the teachable and the knowable. And for precisely this reason the subject of ways

32. *Nic. Eth.* 1103a14–b25.   33. Ibid., 1105a17–b18.
34. Ibid., 1179a34–81b24.
35. Ibid., 1098b9–20, 1099a31–b25, 1102a26–3a3.
36. Ibid., 1145a15–b1, 1151b32–52a6.

and means is not suitable for full public discussion; more to the point, for precisely this reason politics should not be taught to the young, a judgment with which Bacon agrees.

Aristotle does not focus upon the ways and means of the virtues, because to do so would expose the mere likeness of moral virtue to self-sufficient activity of the mind's contemplation of the teachable and the knowable. And for precisely this reason the subject of ways *Rhetoric*, Aristotle argues that the young are hopeful of the future rather than rememberers of the past, because they are guided more by courage and the noble than by calculation. Preoccupied with bodies that lack many days, the old are cowardly and calculating.[37] But the young can forget their bodies and can therefore be formed to noble virtue. But the very openness to the noble becomes vicious when it knows the true link between the noble and the objects of calculation or when it knows the truth of the virtues' service to the arts that serve the objects of calculation. When Aristotle says that the young are unfit students of politics, he does not mean that they are merely calculating or that politics is to be taught only to the old. Rather it can be taught only to those who are already virtuous, a paradox showing that the origins of the virtues must always be mostly in the dark.

When the link between contemplation and action is properly understood, ethics should not focus on the question of roots and strings. Bacon knows this, just as he also knows that Christian charity makes all men like the young, not those whose love of the noble makes them virtuously eschew the safety of the body, but rather those whose love of the noble can be calculating and overreaching because, knowing too much, it only serves the body's needs. Because Christian charity cannot abide virtuous practical freedom modeled on the political philosopher, it can only see a perfect freedom of the mind, the philosopher's heaven, and the absolute neediness of the body. In the present world it rejects the first for the second, but it promises their perfect harmony in the world to come. And the result of such a view is that in the present world, where no merely practical freedom is respected, the young and the old alike become sinful young old men, the very kind whose calculating love of self-sufficiency is always the most partisan and the most dangerous of all.

To sum up, it should be obvious that Bacon has begun the account of the human good by recalling the arguments developed so far. The ancient teaching about the good is now obscured by Christian char-

37. *Rhetoric* 1388b31–90b13; *Nic. Eth.* 1099b33–100a9.

ity, but the new knowlege that tames the violence of the charitable sects is no less charitable in its focus upon the satisfaction of the body's needs. The new power to serve such needs moderates the harsh violence of the sects. However, the new project cannot overcome the dangers of political partisanship precisely to the extent that it is modeled after the hopes of Christian charity. Charitable science replaces the old political science, but only that science can explain why the new project for knowledge is grounded upon a faulty understanding of the true human good.

According to Bacon's account, the good is divided into the private good, or the good of a thing as it is in itself, and the communicative good, or the good of a thing as it is part of some greater whole. By this division, Christian charity exalts the latter over the former. Likewise the private good is divided into the passive good that preserves and continues the self and the active good that dilates or multiplies the self, and again, Christian charity demonstrates the superiority of the latter to the former. But with regard to the private good, the priority of the active good is upheld also by a consideration of man's mortality and by the natural desire for variety: since men always wish to live longer, they attempt to live on through deeds and works. Only if the pleasures could be made perpetual could the passive good be equally important.

Now, lest we think that these latter arguments do not accord with Christian charity, Bacon says that the fact that words and deeds live on is attested by the saying from Revelation: *opera eorum sequuntur eos.* But in the context to which Bacon refers, it is not promised that words and deeds will last beyond the body, as is implied in the references to Seneca, who counsels noble death. Rather it is promised that those who die for Jesus will be resurrected and will have nothing to fear from a second death.[38] But as this promise is fulfilled, either by divine grace or the artful conquest of nature, it is not simply the case that the certain perpetuity of the pleasures would satisfy our desire, because as Seneca makes clear in the passages to which Bacon refers, the satisfaction of the pleasures is stultified by satiety long before their perpetual certainty might occur.[39] For a perpetual creature, the love of the new could never be conquered: in a world of actual or virtual plenty, the only way to escape such satiety would be to procure what is not properly one's own. By Bacon's division, and by the testimony of heathen philosophy and Christian scripture,

38. *Naturales quaestiones* 2.59; Rev. 14.13, 20.4–6.
39. *Ep.* 77.6–20, 95.30–46.

as the good is governed by charitable hope the private good can aspire to the very scope of the communicative good. By this combined testimony, the likes of Sulla would be no mere *pretenders* to the active private good. Bacon admits that the active private good and the communicative or social good can be coincident. Moreover, under the influence of charitable knowledge, this coincidence can occur by way of the troublers of the world who are moved by a "gigantine state of mind" and who, in ignoring the difference between private and public propriety, give form to the world according to their own humors. The charitable project can result in the "true theomachy," the ultimate ground for which is explained in the ensuing division of the passive private good.

Like the active private good, the passive private good is twofold, with perfective passive good being superior to the conservative passive good. Bacon does not tell us what this perfective passive good is, except to say that it has to do with the "preservation with advancement," that its perfection is an approach to divinity, and that it likens the soul to the ones seen by Aeneas in his descent to the river Lethe. But he does tell us what happens when it is mistaken, that is, when it is taken to be a good that can be perfected by way of the body's motions. Such a mistake springs from ambition, causing men to lie like fevered bodies that change place in a fruitless attempt to alter or improve their internal natures. Now, the divine souls seen by Aeneas were the very few who became truly free from their bodies,[40] but according to the hopes of charitable faith and charitable knowledge, the perfected soul is just the one that can hope for a body whose pleasures are perpetually secure. And when this occurs, we remember, the active and passive private goods become coequal. The human body has limits that cannot be overcome, not the least of which are that the body is always one's own and is always in need. The greatest love of honor can be traced to the artful belief that the human body might be perfected, which is why such honor is excessive ambition, or what Bacon has called a "gigantine state of mind": one strives to perfect the body only by distorting the soul, and such striving is the true root of good and evil.

"Gigantine" ambition is the fruit of the Holy Faith's preoccupation with the body it professes to hate. Again, such ambition is the consequence of the new project for knowledge that makes Jesus' secret orientation by carnal desire explicit. Nothing could be farther from Socrates' understanding of the private good; for him it consisted in

40. *Aeneid* 6.703–51.

"equal and constant peace of mind," and he differed from the sophists about whom Bacon speaks. Actually, in the *Gorgias,* to which Bacon refers in discussing the conservative passive (private) good, Socrates argues this point not with a sophist, but with Callicles, a harsh, nonplayful version of Thrasymachos, in the context of an argument about whether rhetoric, not sophistry, is an art. Bacon's "mistake" is not unreasonable, though, because Socrates argues that while the sophists and the rhetoricians really differ, they are so mixed together as to be considered the same.[41] Bacon knows that when they are, rhetoric is no art, and certainly not one that has a rightful place in the city. We know now that charitable knowledge does consider them simply to be the same. Bacon says that Socrates' view and the sophistical view that private perfection is much desiring and enjoying might be harmonized. But in the present charitable age, such harmony is possible only as a unity that distorts both theory and practice, and the perfect unity of mind and desire is therefore taken to be the perfected human body. With such a unity, the troublers of the world include the likes of Callicles along with Sulla, who both had dangerous contempt for the difference between public and private goods.

When Bacon turns briefly to the communicative good, he reminds us that it is not to be confused with "science civil and politic," which is a part of the broad division of the knowledges concerned with man "congregate and in society." Bacon reminds us, in effect, that the communicative good is a part of man considered separately and not in society. But nevertheless, when earlier comparing the active private and communicative goods, he said explicitly that the latter is social. In fact, the whole division according to the charitable reformation of the ancient wisdom confuses private and public goods. To drive this point home, Bacon argues that the communicative good and political science differ as "putting to work" and the "form of a whole" differ in architecture. But if so, then the putting to work of the political whole is ultimately a private matter, and it is obvious that the question of moral ways and means is a part of man considered separately. The conclusion Bacon forces is that, in the charitable age, private men, not the statesman, must be expected to consider the ways and means of the common whole. Any modern account of government and ways and means must take this point into account. In the modern world, men are governed by the ways of private acquisition,

41. Plato, *Gorgias* 465b6–c7, 482c3–86d1, 494a6–97d8.

and despite the promise of the new project for acquiring the whole of nature, this truth will always obtain.

For this reason moral philosophy must focus on the "husbandry of the good," even though for Aristotle such husbandry was the same as teaching politics to the young and even though when he discusses the husbandry of the good, Bacon asserts as strongly as did Aristotle that the young are not fit auditors of moral philosophy.[42] If so, then despite its demonstrable dangers, in the present age all must be instructed in opinions that were condemned on behalf of virtue by heathen poets but are taught "seriously and positively" by Machiavelli. Bacon says that Machiavelli's wisdom exposed the likeness of the excellent Caesar to the bloodthirsty Cataline. Actually, in the context to which Bacon refers, Machiavelli shows that the difference between Caesar and Cataline was that Caesar, who seized Rome with one stroke, was more able than Cataline to compel writers to praise him. We and Bacon know that Machiavelli discusses Caesar's likeness to Cataline (and Romulus) in order to justify the tyrant and the new writer's use of the tyrant.[43] Bacon knows that this justice is merely urgent, but there is no alternative to it as long as moral philosophy is the handmaid to religion or, to be more precise, as long as Christian charity is the bond of perfection that claims to give perfect clarity and unity to the several aspects of human goodness.

In this vein Bacon says that the dedication and application of a man to good practical ends rather than to habitual virtue makes a man like God. To demonstrate his point he refers to Aristotle, who distinguished between habitual virtue and human divinity and between vice and brutality. As Bacon knows, Aristotle teaches that man is open both to the divine and to the bestial. But he also knows that, for Aristotle, while divine possibilities *are* greater than virtue, actual divinity is impossible for man, and only virtue's opaque likeness to divinelike philosophy prevents the kind of deliberation that tempts men more to bestial possibilities than it ever could to true godliness.[44] And this deliberation is just the charitable kind that looks beyond the partial perfections of the habitual virtues to a perfect, heavenly practical whole that mirrors a merely sinful world. Bacon says that the heathen passages he quotes are but a shadow of the charitable divine state of mind. This statement is certainly true, because when the mind is inflamed with charity, moral virtue becomes nothing but sophistry, as

42. See Spedding's note, *Advancement* 440.   43. *D* 1.10.
44. *Nic. Eth.* 1145a15–33, 1178b7–79a32.

he says. When unrestrained by virtue, charity knows no bounds, and it confuses the goods of the body and the goods of the soul, not knowing just how they differ and how they can come together in moral virtue. Consequently, when Bacon praises Xenophon's praise of love's boundless moderation, he is silent about Xenophon's distinction between the heavenly love of souls, which knows no limiting surfeit, and the vulgar love of bodies, which does.[45]

For these reasons, the new charitable knowledge cannot but turn from the moral virtues to the realistic management of the affections, which in the present age must be the rocks upon which civil society rests. When Bacon complains that Aristotle had not considered well what can and cannot be changed by custom, he simply restates Aristotle's own distinction between those instances in which nature is "peremptory," to use Bacon's term, and those instances in which nature allows latitude.[46] Bacon knows that Aristotle understood well the subtle distinctions between nature and convention. Therefore, it is from Aristotle that we can learn about the natural measure and limit of art. However, Aristotle mistook the vigor with which men might strive against those things that are not within their power. He did so not because he had no knowledge of charity but rather because he did not anticipate the Christian doctrine of sin and the power with which such a doctrine would elevate charity as the corrosive bond of all the virtues. For this reason Aristotle did not include in his *Ethics* the very substance of rhetoric and did not include the harsh maxims that are actually contained in the sayings of Plautus, Sallust, Tacitus, and Pindar, to whom Bacon refers.[47] The harsh urgency of charity does not permit the leisurely separation of ethics and rhetoric. And the corresponding focus upon the roots and strings of virtue leads not to Aristotle's caution but to a new attempt to justify the tyrant.[48]

With the new project for conquering nature, the tyrant is excused not by an impossible solitude but by freeing his desire against nature, which then serves the unbridled desires of all. Bacon's use of the tyrant appears to be a more successful justification of the tyrant than Machiavelli's, because whereas for Machiavelli the tyrant is forced to serve a perfect, solitary writer, for Bacon knowledge serves the tyrant as it accommodates every human need. Actually, it is Bacon, not Machiavelli, who focuses unflinchingly on what men actually do as opposed to what they ought to do. But also, again unlike Machia-

45. *Symposium* 8.6–42.   46. *Advancement* 438–39; *Nic. Eth.* 1103a14–b25.
47. *Advancement* 436; see n. 20 above; Titus 1.5–16; Pindar, *Olympian Odes* 1.23–93; Ps. 62.5–10.
48. Cf. Aristotle, *Politics* 1313a34–16a.

velli, he focuses equally on both, so that his knowledge of the old and new accounts of the good discloses the limits of the modern project, which subverts the essential and ineradicable differences between private and public propriety. The new universal project will never be free from politics and the harshest effects of partisan and tyrannical ambition. The new project may tame the ferocious Christian sects, but it will always be forced back upon Machiavelli's unjust and immoral political science.

[11]

Civil Knowledge and Government:
Naive and Realistic Utopia

In the chapter on the human good, Bacon made it clear that in the charitable age and with the new charitable knowledge, the political whole is governed not by virtuous statesmanship but by the managing of the private acquisitive passions. But his primary purpose was to demonstrate the grounds upon which such rule can be judged to be dangerous and unjust, even if by the standard of urgent necessity it must be pursued as if it were just. It now remains only for Bacon to show concretely that, for the new knowledge, government is simply Machiavellian, and he does so in the division of civil knowledge.

[445–476]
Bacon says that, of all knowledge, civil knowledge is the "most immersed in matter," but even so, as Cato remarked that it is easier to move a flock of Romans than to move just one,[1] moral philosophy is more difficult than policy. Since moral philosophy requires internal goodness but policy requires only external goodness, it is possible to have "evil times" in good governments. However, whereas persons often change quickly, matters of states move more slowly, for which reason the great difficulty of civil knowledge is somewhat qualified.

As the aims of society are comfort, use, and protection, there are three corresponding kinds of wisdom, the wisdom of behavior or conversation, the wisdom of business or negotiation, and the wisdom of state or government. The wisdom of conversation is honorable in itself and is also useful for business and government, and Bacon says

1. Plutarch, *Cato* 8.

that it is well handled and not deficient. But the wisdom of negotiation or business has not been collected into writing, which explains why this section is the longest separate section in the entire treatise. This knowledge is "less infinite" than government, so it can be reduced to precept. The ancient Romans taught it in "the saddest and wisest times" as Cicero reports of the senators who would give advice about private business.[2] Therefore, there is a wisdom of private counsel arising from "an universal insight into the affairs of the world," not a few excellent principles of which can be seen in the aphorisms of Solomon, whose heart was like the sands of the sea, "encompassing the world and all worldly matters."[3] Bacon thinks these principles important enough to give twenty-four examples, spending more time on Solomon's "politic sentences" than is appropriate for an example because he wishes to give authority to this important but deficient part of knowledge. Bacon claims that his brief interpretations do not offend the texts, even though he knows they may be applied to "more divine use." But even in divinity, some interpretations have "more of the eagle" than others. As examples for life, the sayings would have "received large discourse" if he had broken and illustrated them. It is not only the Hebrews who give us such wisdom, for it can be found also in ancient parables, aphorisms, and fables. But now that "the times abound with history," fables, serving when examples are not available, are not necessary. The best form of writing for matters of negotiation is discourse on histories or examples, which "Machiavel chose wisely and aptly for government." But history of times is best for government, as "Machiavel handleth," and the history of lives is the best discourse of business "as more conversant in private actions." Even better than them both is discourse on letters, as we see in Cicero's letters to Atticus.

Bacon says that he has concluded the matter and form of the civil knowledge touching negotiation. But another part of this knowledge differs from the one just discussed as being wise for oneself, which moves toward the center, differs from being wise, which moves "to the circumference." For just as there is wisdom of counsel, so too there is a "wisdom of pressing a man's own fortune." This knowledge is also deficient, not because it is not practiced, but because it has not been reduced to writing. And lest it be thought that it cannot be reduced to axioms, it is required, as with the former part, to give some "heads or passages of it." This knowledge may seem new and unwonted, and everyone would seem to want to be a disciple until they

2. *De oratore* 3.33.   3. 1 Kings 4.29.

see that fortune lays as many impositions as virtue and that it is as hard to be a true politique as it is to be truly moral. But this knowledge is important because it is important for practical men to know that learning is like the lark and also the hawk. Nothing in being and action should not be drawn into "contemplation and doctrine." Moreover, learning should not consider this part inferior, because although a man's fortune is not "an end worthy of his being," yet it deserves consideration as "an organ of virtue and merit."

The most general precept of the "architecture of fortune" is to get the window "Momus required," which is to discover persons' natures, desires, ends, advantages, weaknesses, moods, and times, and the like. One should know this about actions as well as persons, because men change with actions. This knowledge is possible, and although the knowledge is always particular, pertaining to particular persons, the instructions for acquiring it can fall under precept.

The first of these precepts is that the "sinews of wisdom are slowness of belief and distrust," that one should believe faces and deeds more than words, and that one should believe surprised words more than purposeful ones. One ought not fear *fronti nulla fides*,[4] as we see in Tacitus' examples of Gallus and Tiberius, Tiberius' different speeches commending Germanicus and Drusus, and Tiberius' different way of speaking when discussing things "gracious and popular."[5] No master of dissimulation can completely hide a feigned tale. Even deeds are not sure enough not to warrant careful consideration of their "magnitude and nature." As Livy says, treachery first shows fidelity in small things for the sake of greater deceit.[6] Small favors lull men to sleep and are, according to Demosthenes, *alimenta socordiae*.[7] How false some deeds are can be seen from Mutianus' false reconcilement with Antonius Primus.[8] Words are not to be despised, as Tacitus shows from the way Agrippina's stinging words brought Tiberius "a step forth of his dissimulation," so that the poet (Horace) well calls the passions torturers that urge men to confess their secrets.[9] Few men cannot be moved to "open themselves," especially if they are "put to it with a counter-dissimulation." In knowing men by reports, one should doubt reports by men's equals or superiors, in front of whom men are "more masked," and the best disclosing of men is "by their natures and ends," the weakest men by their natures and the wisest by their ends.

4. "No trusting to the face" (Spedding's trans.), Juvenal, *Satirae* 2.8–15.
5. *Annales* 1.12, 1.52, 4.31.   6. Livy 28.42.
7. Demosthenes, *Third Olynthiac* 33.   8. Tacitus, *Historiae* 4.39.
9. Tacitus, *Annales* 4.52; Suetonius, *Tiberius* 53; Horace, *Epistulae* 1.18.38.

However, princes are best disclosed by their natures because, having "for the most part no particular ends whereunto they aspire," their hearts are inscrutable. Regarding ends and natures, we need only heed the "predominancy" as we see demonstrated by Tigellinus' use of Nero's fears to break Petronius Turpilianus' neck.[10] The most "compendious" way of this inquiry rests in three things: first, to have at least one friend who has "general acquaintance" with and looks most into the world, and especially into diversity of business and persons; second, to "keep a good mediocrity in liberty of speech and secrecy"; and finally to keep the habit of making "account and purpose, in every conference and action, as well to observe as to act." Just as Epictetus recommends that a philosopher act and yet keep his course, so too a politic man must do things and learn from them.[11] Bacon has "stayed longer" upon the matter of obtaining information because it "answereth to all the rest." But above all, men must not let this knowledge "draw on much meddling"; we should pursue it only to choose better and freer those actions that concern us and to conduct them well.

Another important precept of pressing one's own fortune is for men to know themselves. Men ought to weigh their strengths and weaknesses for the following purposes: they should see how these qualities fit with the times, so as to be either open or retiring, a difference Tacitus observed between Tiberius and Augustus. They should see how they sort with profession and course of life, so as to stay or change, as was done by Duke Valentine. They should see how they sort with competitors, so as to be most alone in their endeavors, as Julius Caesar did when he chose between rhetoric, or civil and popular greatness, and martial greatness. They should choose compatible friends and dependences, as Caesar did, and they should know what kinds of examples to emulate, as we see in the solemn, majestic Pompey's mistaken and less effective emulation of the fierce and violent Sulla.[12]

Another precept is the proper "opening and revealing" of a man's self. It is good to show one's virtues and to hide one's weaknesses, as Tacitus said Mutianus did, who was "the greatest politique of his time."[13] This requires some art, but ostentation is a vice more in manners than in policy, for as Plutarch says, one should slander boldly because something always sticks. Therefore unless one is simply ri-

---

10. Tacitus, *Annales* 14.57.   11. *Encheiridion* 9.

12. Tacitus, *Annales* 1.54; Guicciardini, *Storia d'Italia* 6.3; Plutarch, *Caesar* 3; Cicero, *Att.* 9.10.2.

13. Tacitus, *Historiae* 2.80.

diculous, one ought to put oneself forward boldly because something will always stick, at least with the ignorant many, whose favor countervails the disdain of the few.[14] If it be done well, in proper circumstances, it greatly adds to reputation, and too much moderation in this regard is not profitable. Still, it is important not to debase virtue below "the just price" by obtruding oneself, by doing too much, and by being pleased with too little, which Cicero warns against by saying that easy satisfaction shows unfamiliarity with great things.[15]

After discussing the ways to cover up defects, Bacon says that another precept is obedience to occasions. As Cicero says, nothing impedes fortune like continuing the same when it is no longer fit. In the same vein Livy called Cato, who was an "architect of fortune," one who had *versatile ingenium*.[16] Some are hard to turn and some think that what has been good ought not be changed, as Machiavelli noted wisely about Fabius Maximus.[17] Some have bad judgment and do not know when things have ended, as Demosthenes said of the Athenians, who were like slow fencers.[18] Some are slow to "leese labors past" and think they can ply occasions only to be too late to change, and nothing is more politic than to make the wheels of the mind move with fortune. A related precept is to run with occasions and choose the most passable way, which will appear moderate and successful and will add to reputation, and yet another precept, which seems opposed to the former two but is not in fact, is what Demosthenes said in "high terms," that as captains ought to lead armies, the wise should lead affairs and not be forced to follow events.[19]

Another precept is mediocrity in declaring or not declaring a man's self. Although secrecy and proceeding like a ship through water is sometimes good, since it often breeds mistakes the "greatest politiques" have openly "professed their desires." Sulla wished men happy or unhappy as they were his friends or enemies, and Caesar admitted his ambition when he first went into Gaul, so that, Cicero noted, he demanded to be called a tyrant.[20] In a letter to Atticus, Cicero describes how Augustus revealed his tyrannical ambition to the people, except that he did it so as to make men wonder if it were true.[21] Unlike these men, Pompey, whom Tacitus and Sallust said concealed his intention, hid a shameless mind with an honest tongue, secretly

14. *Quomodo adulator ab amico internoscatur* 65c–d.
15. *Ad Herennium*, ed. Marx (Leipzig: Teubner, 1964), 4.4.
16. Cicero, *Brutus* 95; Livy 39.40. 17. *D* 3.9. 18. *First Philippic* 51.40.
19. Ibid., 51.39. 20. Plutarch, *Sulla* 38, *Caesar* 11; *Att.* 10.4.2.
21. *Att.* 16.15.3.

designed to seize the state unseen, and failed because he had to return to the "beaten track" of putting arms in his hands "by color of the doubt of Caesar's designs."[22] Tacitus says such deep dissimulation is an inferior cunning compared with true policy. He attributed one to Augustus and the other to Tiberius when he said that Livia combined her husband's art with her son's dissimulation.[23]

Still another precept is to value things properly in relation to our ends. Men fall in love with things that are but matters of envy, peril, and impediment, and some measure things only by the labor they require, as Cato the Second was said to do by Caesar.[24] As for the true marshaling of pursuits toward fortune, Bacon says they are first the amendment of one's mind and second wealth and means. He condemns those who would place the second first with the same reason that Machiavelli condemns the view that money is the sinew of war. Machiavelli said the true sinews were men's arms, a military population, which he demonstrated by referring to Solon's remark to Croesus upon seeing the king's gold.[25] Likewise, the sinews and steel of the wit, courage, audacity, resolution, temper, industry, and the like are the sinews of fortune. In third place Bacon puts reputation, and in last place he puts honor, better won by the first three than vice versa. To conclude this precept, Bacon notes that men must watch time and not fly to an end when they should make a beginning, not observing the precept noted by Virgil, *quod nunc instat agamus*.[26]

Another precept is not to do things taking too much time but to heed Virgil's remark that, while one makes ready, the time for doing is gone.[27] Still another precept is to imitate nature, which does nothing in vain, by being ready to take whatever degree of success can be achieved. Not to be flexible is to lose "intervening occasions," and men must heed the rule *haec oportet facere, et illa non omittere*.[28] Another precept is always to have some means of escape from any situation, and the final precept of the architecture of fortune is the ancient one

22. Tacitus, *Historiae* 2.38; Suetonius, *De grammaticis*, coll. Brugnoli (Leipzig: Teubner, 1963), 15.

23. *Annales* 5.1.   24. Caesar, *De bello civili* 1.30.

25. *D* 2.10; Cicero, *Philippicae*, ed. Fedeli (Leipzig: Teubner, 1982), 5.5; Diogenes Laertius 4.48; Plutarch, *Cleomenes* 27; Lucian, *Charon* 10–12; *Essays* 446.

26. "Despatch we now what stands us now upon" (Spedding's trans.), *Advancement* 469; *Eclogues* 9.66.

27. *Georgics* 3.284.

28. "These things ought ye to do, and not to leave the other undone" (Spedding's trans.), *Advancement* 470; Matt. 23.23; Luke 11.42.

of Bias, who warns that we ought to love our friends as if they may some day be our enemies and vice versa.[29]

Bacon gives so many examples because he does not want it thought that this matter is light, even though he knows that fortune merely tumbles into some men's laps. Just as Cicero did not think that every pleader must be his perfect orator, and just as princes and courtiers are described according to the perfection of the art and not common practice, Bacon knows this ought to be done likewise for the politic man for his own fortune. He wants it known that the precepts he has listed may be considered to be honest arts. As for evil arts, if one sets down for oneself such precepts as Machiavelli gives regarding the appearance of virtue and the utility of fear, or the principle cited by Cicero *(cadant amici, dummodo inimici intercidant)* in describing the Triumvirs, or the "protestation" of L. Catalina that if his fortunes be fired he will put them out with ruin rather than water, or Lysander's remark that children are to be "deceived with comfits" and men with oaths, such dispensation from charity and integrity may be hastier and more compendious, but the shortest way is commonly foulest, and the fairer way "is not much about."[30]

In condemning Machiavelli's evil arts, Bacon notes that if men are not carried away with ambition, they will remember that all is vanity and vexation of the spirit, that being without well-being is a curse, and that virtue is rewarded in itself and wickedness punished in itself, as the poet Virgil says well in his verse about the reward for noble deeds.[31] Even if we refrain from evil, the "sabbathless" pursuit of fortune does not leave the tribute of time we owe to God; it is to little purpose to have one's face toward heaven but one's spirit groveling on the earth like the serpent who fixes to earth the divine spark. If anyone thinks that one will use his fortune well even though it is ill obtained—as was said, first of Augustus Caesar and then of Septimius Severus, that the man ought never to have been born or else never should have died—it should be remembered that these "compensations and satisfactions" are well used but never well purposed.[32] Finally men should remember Charles V's comment that fortune is like a woman: the harder wooed, the farther off. This ad-

29. Aristotle, *Rhetoric* 1389b13–90a1; Cicero, *De amicitia*, trans. Falconer, Loeb Classical Library (London: Heinemann, 1923), 16.

30. *P* 17–18; "down with friends so enemies go down with them" (Spedding's trans.), *Advancement* 471; Cicero, *Pro rege Deiotaro* 9.25; Cicero, *Pro Murena* 25.51; Plutarch, *Lysander* 8.

31. Eccles. 2.2; Virgil, *Aeneid* 9.252–54.

32. Sextus Aurelius Victor, *Epitome de Caesaribus* 1.28; *Historiae*, Spartianus, *Severus* 18.

vice is for the corrupt, however, so men should build on the foundation of the advice of philosophy and divinity that concur about what should be sought first. Divinity advises us first to seek the kingdom of God, and philosophy advises us to seek the things of the mind, and both say that from their respective quests will follow all else.[33] Although human foundations have "somewhat of the sand," as M. Brutus showed when he said that he took virtue for a reality but found it to be an empty name, "yet the divine foundation is upon the rock."[34] With this remark Bacon concludes his discussion of the deficient architecture of one's own fortune and all but the last part of civil knowledge.

Regarding government, Bacon says that it is a subject "secret and retired," in the same two respects in which "things are deemed secret." Some things are secret because they are hard and others because they are not fit to utter. And all governments are as obscure and invisible as Virgil describes the diffusion of the great mind in the mass of the great body to be.[35] God's government of the world is hidden because it contains irregularity and confusion. For the same reason the soul's rule over the body is also hidden. As befits a secret subject, the poetic wisdom of antiquity compares the "offence of futility" to rebellion in the shadowy fables about Sisyphus and Tantalus.[36] It is true that this subtle wisdom refers to particulars, but even the general rules and discourses of policy and government require "reverent and reserved handling." However, "contrariwise in the governors toward the governed" all should be manifest and revealed as far as the "frailty of man permitteth." The scriptures say that to the government of God the world is like a crystal, while it is dark and shady to man. Therefore "the natures and dispositions of the people, their conditions and necessities, their factions and combinations, their animosities and discontents" ought to be clear and transparent to princes, states, wise senates, and councils. But he will pass over this matter in silence because he addresses one who is the mas-

---

33. Matt. 6.33.
34. Ibid., 7.24–27; Dio Cassius 47.49; Plutarch, *De superstitione* 165a.
35. *Aeneid* 6.726. In the *De augmentis,* Bacon omits the argument that government is a secret science, but he says that the account he will give should demonstrate the art of silence, which he has forgotten to mention. It is of course not true that he has forgotten to mention this art, because in book 6, chapter 2, he mentions the acroamatic method of tradition used by the ancients. In the *De augmentis,* then, Bacon presents government as a secret science because of his own intention to be silent. Bacon's reason for silence is different from that of the ancients (*De aug.* 745–46, 664–65).
36. Homer, *Odyssey* 11.582; Cicero, *Tusculanae disputationes* 1.5.10, 4.16.35; Virgil, *Aeneid* 6.616.

ter of this science, desiring as he does to gain the certificate of the philosopher who knew how to hold his peace.[37]

Regarding the more public part of government, concerning laws, the deficiency is that philosophers have given imaginary laws for imaginary commonwealths, which are like stars that give little light because they are too high. The lawyers write only about received law and not about what the law ought to be, so that the wisdoms of the lawyer and the lawmaker differ. The lawmaker must consider the fountains of justice, the means whereby the laws might be made certain, and what causes them to be uncertain, along with many other considerations of administration and animation. Bacon will insist upon this less because he has begun such a work in the form of aphorisms, but it can be noted in the meantime to be deficient. As for the king's laws, Bacon could say much of their dignity and something of their defect, but they must surely excel the civil law, because the latter was "not made for countries which it governeth." But with this Bacon ceases to speak, for it is not his wish to mix matters of action with matters of general learning.

To comply again with the law of brevity, we begin by noting that, when he turns to the last part of civil knowledge, the knowledge of government, Bacon says that it is secret and retired. Bacon explains this point in the light of the two reasons why any subject should be secret, its being too hard or its being unfit to be uttered. It is certain that all government is obscure, he says, and it is possible that it is unfit to be uttered. But if the former is true, it is unclear why, at the very beginning of the division of civil knowledge, he says that policy is easier than moral philosophy because policy requires only the production of external goodness. In order to see Bacon's argument in this section, we have to explain the contradiction. As usual, we do so by attending to the examples Bacon uses.

To demonstrate that civil philosophy is easier than moral philosophy, Bacon refers to Cato's remark about the Romans, which occurs in Plutarch's record of his famous sayings. Now, according to Plutarch, Cato's remark shows that men's characters are revealed more

---

37. Rev. 4.5–6; Plutarch, *De garrulitate* 503e–504c; Diogenes Laertius 7.24. In the *De augmentis* Bacon refers to this story in commenting on his silence about the whole of the art of government. There he does not distinguish between the part of government that is simply secret and the part that might be treated as far as man's frailty permits. Of course both parts are treated secretly, just as both are presented openly as far as frailty permits (*De aug.* 745–46).

by their words than by their looks, as most believe. Cato did liken the Romans to sheep whose bellies are more subtle than their ears. But in the context to which Bacon refers, Cato also repeated a saying of Themistocles to the effect that, while the Romans ruled all men, they were themselves governed by their wives.[38] According to the example, then, the government of imperial peoples is like the rule of women over men, a similarity that has less to do with the difference between internal and external goodness than with the private, indirect management of the internal affections for some external end. And the difference between internal and external goodness, on the one hand, and political rule by the private management of private passions, on the other hand, has less to do with the difference between ease and difficulty than with the difference between a proper and an improper stance toward public and private propriety. If producing external goodness makes policy easy, and vice versa, then according to Bacon's example government must in fact be hard because it produces good ends by way of internal means. This consequence casts doubt on the goodness of its results, as is suggested by the other reason for secrecy and as we see from closer inspection of Bacon's reference to Virgil.

Virgil is said to account for all governments' obscurity, but actually Virgil does not speak explicitly about government in the context to which Bacon refers. Rather the quotation refers to Anchesis' explanation to Aeneas of how the soul pervades the body so as to give rise to the passions that disturb the soul.[39] The obscurity of government has to do with the relations of the soul, the body, and the passions. But Bacon says that government is obscure only for the governed toward those who govern, *not* for the governors toward the governed. In other words, anyone who would rule wisely should know the ways of the ensouled body's passions; but this knowledge cannot be public. Such an argument expresses succinctly the problem that Aristotle ascribed to all free government: because politics is an unfit subject for the young, the ultimate source of the virtues in the passion-forming laws must be obscure to the virtuous, who are themselves the most competent to rule. By this argument, good and free government must be hard because its principles must remain in the dark. But if so, then such government will be rare. This is precisely the teaching of the political philosophers who treated the "public part of government" in imaginary commonwealths, a teaching with which Bacon has agreed throughout his treatise. As we noted in the intro-

38. Plutarch, *Cato* 8.   39. *Aeneid* 6.719–51.

duction, the ancient utopians' "star that gives little light because it is too high" is the model for Bacon's own utopia, and only after the course of the present treatise do we know why, although the utopian teaching of the ancients is correct, Bacon turns from it to the project of modern scientific conquest. And in this light we can explain the apparent contradiction between government's secrecy and the assertion that civil philosophy is easier than moral philosophy.

The ancient utopians taught that the more public part of government, the part dealing with "external goodness," was difficult because there is no decisive separation between politics and moral virtue. For them knowledge of the sources of the virtues in the habit- and passion-forming legislative art was unfit to be uttered, because it could accord with both justice and moral virtue only when it remained at least partly in the dark. Only with the advent of Christian sin and charitable knowledge is such knowledge made public, and when it is, the link between moral virtue and justice is broken, so that justice becomes wholly subservient to urgent necessity. The problem is, then, that, when such knowledge is easy because its external end is served by open and public internal means, it is no longer fit to be uttered because it is not necessarily moral or just. The reason is grounded in the truth of government's difficult obscurity as it is explained by the heathen poet Virgil.

As Virgil reports, government is obscure because the relations between body, the soul, and the passions are obscure. In the light of the ancient teaching, the justice-loving attempt to overcome such obscurity by reducing one element to the other, by realizing their perfect unity, is both unjust and immoderate. Such a spurious harmony always seems to be easy, because it is the hope that ultimately sets every productive art upon its way. We recognize this harmony as the explicit promise of the new charitable knowledge. In any age, civil states rest upon the balancing of the affections. But in the modern age, such balance cannot be for the sake of moral virtue because the harsh candor of sin illuminates what should remain in the dark. Therefore, such balance cannot preserve the just difference between public and private affairs. Precisely because of Christian charity, Bacon knows that no writer can be virtually alone or completely hidden by the dark—Machiavelli's solitary conspiracy against the tyrant is impossible, and so it cannot be just. In the modern age, a reason will always be found why the boldest things said by Machiavelli in secret will be taken to be commonplace or even charitable. Had Machiavelli understood the modern age, he would not have written as he did, or else he would have done for different reasons, in which case he would not have presented the teaching that he did. But for all the reasons that we know, Machiavelli would have included what he said because the new learning can supplement but never outstrip his political science.

In the modern age, then, any practical civil philosophy consists in the wisest courses for private ambition, the maxims of which are demonstrated in the "wisdom of negotiation" or business, which is in fact sufficient because it is presented in Machiavelli's art of apparent reputation. Bacon makes this point clear when he divides the wisdom of negotiation into two parts: the wisdom of private counsel and fortune and the "wisdom of pressing a man's own fortune," or the "architecture of fortune," which differs from the former in being directed to "the circumference." Both are modes of civil knowledge; consequently, the difference between them cannot be determined by the difference between private and public matters. Rather, as we see by examining Bacon's examples, the distinction between them merely demonstrates that dark and disingenuous private ambition is appropriate for both private and public affairs. Bacon demonstrates this fact "enigmatically," but not, as Machiavelli did, because he thought he could be perfectly just. Rather he writes enigmatically because he was moderate and therefore as just as necessity would allow.

In saying that there is a wisdom of private fortune, Bacon argues that the Romans knew it in the saddest and the wisest of times. His evidence for this contention is Cicero's remark about the senators as reported in *De oratore*. But the remark to which Bacon refers in *De oratore* is Crassus' complaint about the disunity of the sciences and his praise of Cato's successful attention to both public and private business.[40] The modern age remedies Crassus' complaint, and it follows Cato, for whom private business enhanced the power of public means. And any management of public affairs must assume that these means are not fashioned after moral virtue. Brevity prohibits our recounting and commenting on Bacon's long list of Solomon's wise maxims concerning private uses of public counsel. But the reader will find that none is interpreted to show the worth of virtue in itself and that all are interpreted so as to be compatible with Machiavelli's preference for the appearance of virtue to the actual possession of it.[41] However, we can notice that Bacon claims to have done no violence to the sense of Solomon's precepts and that his defense for this is as

40. *De oratore* 3.33.
41. *Advancement* 448–53; see, e.g., Eccles. 7.21; Plutarch, *Pompeius* 20; *Sertorius* 27; Eccles. 4.15; Plutarch, *Pompeius* 14; Tacitus, *Annales* 6.46; Prov. 27.19; Ovid, *Ars amatoria* 1.760.

extraordinary as the claim itself. He says his interpretation is appropriate, because among divine writings some have more of the eagle than others. Of course the opposite companion of the skyward eagle is the earth-bound snake, which means that by Bacon's argument, Solomon's wisdom, being more tied to mundane affairs than other divine writings, has more of the serpent than they do. And in the discussion of the architecture of fortune, Bacon says that the serpent's wisdom consists of the evil arts as they are taught by Machiavelli and are blamed by Cicero and Plutarch.[42]

Likewise, when Bacon turns to the long account of the architecture of fortune, the examples show its precepts to betray contempt for moral virtue. According to the first precept of pressing one's fortune, we should be slow to believe or to trust, but we should not believe with Juvenal that there is no trusting the face, as Bacon says is proved by Tacitus' three remarks about Tiberius. But according to the context in Tacitus to which Bacon refers, Gellius' knowledge did not help to quench the emperor's anger, and the same could be said for Germanicus. Although Tiberius spoke more clearly when dealing with charity, Tacitus' point is that people wondered at Tiberius' fluent pardon of Gaius Cominus when he usually took a more inscrutable path and that it is difficult to say when the act of an emperor is applauded sincerely or feignedly.[43] In the first two cases, the knowledge did no good, and in the latter case the perspicuousness of Tiberius' speech was matched by the inscrutability of praise of that speech.

According to Juvenal, the face is not simply inscrutable but also just like the sleek buttocks from which the grinning doctor cuts the piles.[44] With such an obscene joke, Bacon informs us that, according to the architecture of fortune, all men must be taken to be unfaithful liars and that one should take small favors to be a strategem of treachery. But according to the context in Tacitus' *Historiae* to which Bacon refers, treacherous deeds can be used to produce peace in a city, as Mutianus' treachery toward Antonius did.[45] From Livy we learn that, knowing what he knew about treachery, Scipio successfully invaded Africa with his own volunteers, not expecting any African friends

42. *Advancement* 471–72. In the *De augmentis*, Bacon interprets ten more proverbs. He mentions the Romans in six and Machiavelli in two, in one the *Prince* and in another the *Discourses*. In the *De augmentis*, the compatibility of Solomon's wisdom with Machiavelli's is much more apparent. This instance, regarding the Old Testament and not the New, is the only instance in which the *De augmentis* is both fuller and bolder than the *Advancement*. Even so, there is no difference between the teachings of the two accounts. See *De aug.* 751–68.

43. See n. 5 above.   44. *Satirae* 2.8–15.   45. Tacitus, *Historiae* 4.39.

or trust.[46] Horace's teaching about how the passions cause men to betray secrets is in fact a warning never to reveal secrets to a friend.[47] If a well-used lie can indeed discover the truth or, rather, the secrets entrusted to a friend, then the architecture of fortune cannot be compatible with true friendship, because, for safety's sake, one ought to tell lies to one's friends, especially since others will think that one's virtues, abilities, conceits, and opinions are best discovered from one's friends.

According to this same first precept, weak men and princes are best known by their natures, and the wise and private men are best known by their ends, not because princes are at the top of human desires, but rather because of their likeness to the weak in pursuing no specific end. The inscrutability of the princely heart springs from its flexible application to all ends, from the fact that such a perfect prince would have a mind like a perfectly multipurpose tool. But in fact the same indeterminacy applies to strong, ambitious private men as well. In the passage from Tacitus to which Bacon refers,[48] Tigellinus did not probe Nero's fear in order to overcome Petronius, who actually killed himself, but simply to bind himself and his evil arts closer to the emperor. To do so he assisted in the murders of Sulla and Plautus, but his end was as broad as the emperor's. By the first precept, then, the true fortune of the wise is to serve the desires of peoples and princes, which means that they serve the unlimited desires of the weak, princes, and private persons who would be like princes with the aid of their evil arts. Bacon says that the politic man should observe as well as act so as to be in political matters what the philosopher was said by Epictetus to be in practical matters. But according to what Epictetus actually says, it is impossible to mix the freedom of philosophy with the pursuit of power and wealth.[49] The wise pursuit of fortune must assume the unfaithfulness of all men, because for the new knowledge, there is no shared pursuit like philosophy in which a gain is not another's loss.

Regarding the precept that men ought to know themselves, Bacon says that one ought to know one's compatibility with the general state of the times. The examples Bacon gives of this principle are taken from Tacitus' description of Augustus and Tiberius, of whom Tacitus actually says that the former thought it proper to indulge in the pleasures of the crowd and that the latter lacked the courage to lead a softened nation to the ways of austerity.[50] The example of the wise

46. Livy 28.42.   47. See n. 9 above.   48. Tacitus, *Annales* 14.57.
49. *Encheiridion* 1, 9.   50. *Annales* 1.54.

choice of profession is Cesare Borgia, who followed his inclination toward the exercise of arms. Bacon speaks as if Cesare's choice would have been as bad for a prince as it would for a priest, and he is silent about the rejoicing that Guicciardini reports occurred in Rome at the death of Cesare's father, Alexander VI, who was rumored to have been poisoned by accident in an attempt to murder the cardinal of Corneto.[51] The example of one who correctly judged his competition is Caesar, who chose arms and rule so as not to compete with many fine orators but who we know made this choice for the sake of determining just what praises and blames such speakers might be free to bestow. The imprudent judge of examples is Pompey, who according to Bacon was ineffectual because he was solemn rather than fierce and violent and who, according to Cicero, would have made war against both Caesar and Rome.[52] But as we know, Pompey's defeat sealed the fate of the free republic.

Regarding the precept counseling the proper opening and revealing of a man's self, Bacon says that Mutianus, the greatest politique of his time, knew how to do this well. But actually, according to Tacitus it was not about himself that Mutianus spoke to the crowd at Antioch. Rather he lied to them about Vitellius' intention to transfer them to the cold of Germany, so that they would pledge their service to Vespasian.[53] As a model for self-disclosure, Bacon offers the saying of Medius, the leader and master of the flatterers, whose slander condemned his rivals such as Callisthenes, Parmenio, and Philotas. According to Bacon's source Plutarch, Medius' calumnies led Alexander to be brought low by allowing himself to be worshiped like a barbarian idol. But Bacon advises against Plutarch, who warns against undeserved conceit that puts one at the mercy of the insignificant and the mean.[54] Bacon advises that if ostentation be used well with the many it will "countervail" the disdain of the few, adding to reputation, the lack of which inhibits the rise of "solid natures." According to the precepts of politic self-revealing, virtue must be taken to have a price that can be calculated according to gain and loss. The example of such calculation is taken from Cicero's remark about those who know the art and not just the labor of rhetoric.[55] The artful rhetorician knows just how to measure small delights for the sake of the large, and according to the present context, this means not that one must disdain the applause of the many but rather that one must

51. *Storia d'Italia* 6.4.   52. See n. 12 above.   53. Tacitus, *Historiae* 2.80.
54. *Quomodo adulator ab amico internoscatur* 65c–e.   55. See n. 15 above.

use it rightly in order to receive the full value for the coin of one's virtue.

To this point, Bacon's argument and examples show that *Machiavelli* has written "wisely and aptly" about government. But the examples of the precept about obedience to occasions demonstrate Bacon's ultimate difference from Machiavelli. Bacon says that nothing hinders fortunes more than being able to be pliant and obedient to occasions, commenting that Fabius Maximus is one who could not change because he had found former courses to be good, which Machiavelli noticed wisely. Now, according to Bacon's earlier reference to Livy, used to illustrate the precept of slow belief and trust, Scipio applied the knowledge of treachery in using faithful arms against Africa when Fabius Maximus would not.[56] And in *Discourses* 3.9, Machiavelli presents Fabius and Scipio together to show that republics are superior to principalities because republics contain the diverse natures that can respond to changing circumstances. According to Machiavelli, Fabius' senatorial caution served Rome well when Hannibal's impetuous youth had faded. In each case, the players acted according to nature, not choice, and after hinting at this lack of flexibility in individuals, Machiavelli then offers a modern example suggesting that in fact republics are not any safer than principalities from changing fortunes. This statement reminds us of *Discourses* 1.53, where impetuous appeals to peoples are said to be dangerous for republics.[57] Republics respond well to the external changes of times, but they are slow to change themselves.

As it turns out, for Machiavelli the changing of a republic requires an individual who can change with the times, even though such an individual is not sufficient to guide an unchanged republic. To change with the times means to overcome human nature, which is satisfied with what has come before, so that nature as it appears in changing occasions can be overcome by changing and varying human ways of proceeding. For Machiavelli, such mastery of times and fortune is in the hands of one who changes political ways and means by changing his own. Fortune can be mastered as long as man's lubricious flexibility is understood to move within the bounds of political possibilities. Machiavelli, a lone writer free from every constraint of moral virtue, can change every political body in order to bend fortune to human will. But in the present context, Bacon speaks only of the individual who could change, even though at the beginning of this sec-

56. Livy 28.40–45.   57. Mansfield, 347–50.

tion he says that the quick changeability of individuals moderates the sluggishness of states. Bacon mentions no state, whether republic or principality, that could change. Quite the contrary: in the speech of Demosthenes to which Bacon refers, we are told that the Athenians were unable to anticipate events because they would not let a statesman guide circumstances.[58] But more important, where Machiavelli points to the single variable mind that can reorder republics, Bacon praises Cato, whose flexibility confronted fortune only by submitting to nature's urgent necessity and who could not have tried to change the ways of Rome because her ways in fact reflected his own. Bacon subscribes to Machiavelli's method, but he doubts its certain efficacy.

For Machiavelli, fortune can be mastered because by accommodating necessity man can learn to manage perfectly his tractable political nature. But for Bacon, man's intractable political nature is at once the reason why he must challenge fortune and the limit to any real mastery. Bacon shows clearly that for the new charitable knowledge political science must be Machiavellian. But for Bacon, this fact signifies the limits to the broader assault against the whole of nature that was missing from Machiavelli's political science. Machiavelli's political science is far more optimistic than Bacon's new project for learning and the productive arts. For Bacon, the force of charity so affects political life that the only effective management of political affairs is to direct the hope for local freedom to a hope for more universal freedom. This hope is put to work for the sake of the body in order to tame the charitable sects, but it cannot conquer the possibility of sects, however much it might for a time manage them. It cannot conquer them because, for charitable knowledge, there is no principle for managing the passions that does not always serve the private use of public orders. The universal body for the sake of which the new science conquers is always someone's own, especially for the most ambitious lovers of universal mastery. Therefore, the Machiavellian management of the passions will be the foundation of states even in the modern age. But unlike Machiavelli, Bacon has no naive illusions about the power of such new principles. In fact, one can grasp the modern age only by understanding how its deepest problems are reflected in Machiavelli's subconscious, naive dream to overcome the very limits of political life. Machiavelli is a naive utopian, reflecting as he does the dream of every productive art. Bacon, on the other hand, is a realistic utopian after the manner of the later ancients. Ba-

58. Demosthenes, *First Philippic* 51.39–40.

con fashioned his account of the perfect society after Plato's because like Plato, Bacon knew that utopia means "no place."

In the immediate context, Bacon demonstrates his knowledge of the difference between naive and realistic utopianism in two ways. First, after discussing the precept about obedience to occasions, Bacon returns to precepts whose examples show them to be harshly Machiavellian. He concludes with the very worst one, the ancient precept of Bias that we love our friends as if they will some day be our enemies. But Bacon is silent about what he doubtless knows: as reported by both Cicero and Aristotle, this maxim either destroys friendship altogether or is a vice of suspicious old age.[59] Bacon knows that in the charitable age, the true architect of fortune is an old man made young. And by Aristotle's measure, such a one could never be free or even happy.

Second, Bacon tells us subtly that his precepts are Machiavellian while telling us explicitly that Machiavelli's "arts" are evil. Bacon knows about Machiavelli's evil not from the common opinion of the charitable age, but from the only source from which it could be understood properly: the utopian teaching of the later ancients. Bacon says that he wants it known that his precepts may be considered to be honest arts, unlike the harsh maxims of *Prince* 17–18, the verse cited by Cicero about the Triumvirs, the saying attributed to Cataline by Cicero, and Lysander's artful use of oaths as reported by Plutarch. But of course his maxims are as bad or worse: Bacon has agreed with Machiavelli in praising apparent reputation, and he has recommended the precept about friends that Cicero is said to have attributed to the Triumvirate. Actually, Bacon attributes the vile use of friends to the Triumvirs, who (considering both the first and the second Triumvirates) included two of the three greatest politiques, not Cicero, who denies it.[60] However, by now we know that the honest arts of policy are the same as Machiavelli's evil arts, not because Machiavelli's are dispensations from charity and integrity, but because the honest arts are nothing but charitable. As the ancient utopians taught, it is when one dispenses with charity that the constraint on man's ambition can be said honestly to be the vanity of all things and that virtue and vice are rewarded and punished in themselves. This latter truth is demonstrated, Bacon says, by Virgil's verse about the reward for noble deeds.

Bacon refers to the threefold reward promised to the Trojan sol-

59. See n. 29 above.   60. Cicero, *Pro rege Deiotaro* 9.25.

diers by the aged Aletes. And as usual, the reference does not quite match Bacon's stated purpose. Virgil does not speak only about the self-sufficiency of noble deeds. Rather the soldiers will be rewarded not just by nobility but also by divine reward and by a third reward to be paid by Aeneas.[61] Bacon knows that nobility and divine rewards conflict; he knows that Aeneas' ambition did not even respect the bounds of the netherworld. But he also knows that the three kinds of rewards can never wholly be separated, because every political whole is open to the vengeful promise of divine reward. Therefore, virtue will always be rare because it is always tempted by charity. But for the same reason, with the triumph of charitable knowledge it is important to remember that, the more fortune is wooed, the more "she is the farther off." It is precisely when philosophy and divinity meet, or rather *because* they meet, that it is more important to remember that foundations have "somewhat of the sand." According to Bacon, Brutus discovered this fact when he found virtue to be nothing but a name. But according to Bacon's source Plutarch, such a bad view is the painful alternative to the painless but bad opinion that virtue is only a state of the body and not the soul. Plutarch says that the latter view is the analogue of painless atheism, whose freedom from the sabbath contrasts with the painful torments of the superstitious.[62]

Bacon could not conclude the architecture of fortune more artfully: insofar as Plutarch illustrates his argument, we are driven by the torments of superstition to the atheistical view that virtue is the body's good. But in our attempt to assure the truth of atheism, we are not thereby freed from bad opinions. Bacon knows that men can assure the truth of atheism only by imitating the way of the creator god. But this is just such a god's divine revenge, a revenge that is coeval with man's political life. In the modern age, as in every other age, the truly wise should never think it possible to be free from painful, and partisan, superstition. There will always be a need for Machiavelli's imperfect political science because scientific hope seeks conquest for the sake of the body's desires. These desires are the grounds of contentious opinion. Moreover, they fuel the tyrant's boundless ambition. Such tyranny would be benign if it could be harmonized with the desires of the many, but no such political economy is possible. In order to comport well toward charity and justice, Bacon's new science must point man's mind beyond Machiavelli's political horizons. But justice and charity are never wholly compatible

61. Virgil, *Aeneid* 9.246–80.   62. Plutarch, *De superstitione* 165a, 169d–e.

because together they actually require a turn to Machiavelli's harsh Machiavellianism. And for the last time, there is no truth to Machiavelli's naively utopian claim that wise management can justify the tyrant.[63]

63. Bacon has taught much about the true limits of empire, which in the *De augmentis* is presented in the chapter on the art of government (*De aug.* 793–802). See Chap. 4 at nn. 63–93 above. An example of a work touching universal ways and means is inserted in the *De augmentis* (*De aug.* 803–27). This collection of aphorisms is *not*, however, the realistic account of origins presented in the *Advancement* and in the *De augmentis* itself.

Conclusion: The Utopian Roots
of Modernity

[477–491]

In an appropriate step from the end of the last chapter, Bacon concludes the treatise by dividing the knowledge of "sacred and inspired" divinity. So far Bacon has divided divine learning according to the parts of the mind exercised, so that divine history was discussed along with history and divine poetry was discussed along with poetry. Accordingly divine precept is considered as an appendage to the knowledge concerning the human faculties of reason and appetite and will: because it concerns belief, divine precept concerns both reason and will. With this subject, Bacon returns to the problem of belief and to the deepest sources of mankind's political life.

Bacon says that God's prerogative extends to man's reason as well as to his will; even as we obey divine law against our will, so we believe God's word even though we find "a reluctation in our reason." If we believe only what is agreeable to our sense, we treat God as a suspect witness. It is worthier to believe than to know because belief springs from the spirit rather than the sense, but when men are glorified, then "we shall know as we are known," that is, as we are known by spirit, or God.[1] From this statement Bacon concludes that divinity is grounded on God's word, not on the light of nature, for while it is written that God's glory shows in heaven, it is not written that heaven declares God's will. Rather it is said, *ad legem et testimonium: si non fecerint secundum verbum istud, etc.*[2] This rule applies not only to the mys-

1. 1 Cor. 13.12.
2. "To the law and to the testimony: if they do not according to this world," etc. (Spedding's trans.), *Advancement* 478; Isa. 8.20; see Ps. 19.1.

322

teries of deity, creation, and redemption but also to "the law moral truly interpreted," such as the admonition to love one's enemies and to imitate God, who lets his rain fall on the just and the unjust.[3] To this it should "be applauded," *nec vox hominem sonat.*[4] We see that the "heathen poets" speak of law and morality as if they were "opposite and malignant to nature: *et quod natura remittit, invida jura negant.*"[5] Man does have natural notions of virtue and vice, justice and wrong, good and evil, but as far as the perfection of this knowledge is concerned, there is a kind of natural spiritual notions only in the experience of conscience, which checks vice but is not able "to inform law."

Despite the limits, the use of reason in spiritual matters is great, especially for Christianity, which is between the extremes of the heathen, who had no "constant belief or confession," and the "law of Mahumet," which altogether "interdicteth argument." In religion, reason conceives and apprehends God's revealed mysteries and infers and derives doctrine and direction from them. But in the former case reason works "by way of illustration" rather than by argument. Only in the latter case does reason work by way of "probation and argument." In the former we see God's mysteries as God "vouchsafeth to descend to our capacity," or as they may be "sensible to us" and open to human understanding. And in the latter we use argument, but we use reason only secondarily in reasoning from the *"placets* of God," unlike the investigation of nature. Bacon thinks that the question of reason's limits in divinity has not been sufficiently inquired, so that people err either as did Nicodemus or as did the questioning disciples.[6] He has insisted upon this point because to determine reason's proper limits in divinity would prevent the schools' curious speculations and the fury of controversies. Many controversies concern matters not revealed or positive, or else they spring from weak inferences or derivations. Regarding the latter, it would be better if men would speak as Paul did, *ego, non Dominus* and *secundum consilium meum,* "in opinions and counsels, and not in positions and oppositions." As it is, however, men usurp the style *non ego, sed Dominus.*[7]

Bacon divides divinity into matter revealed and nature of the rev-

---

3. Matt. 5.44–45.   4. "Not a voice with a human sound," Virgil, *Aeneid* 1.328.

5. "And what nature suffers envious laws forbid" (Spedding's trans.), *Advancement* 478; Ovid, *Metamorphoses* 10.330.

6. John 3.4, 16.17.

7. "I, not the Lord," 1 Cor. 7.12; "according to my counsel," 1 Cor. 7.40; "not I, but the Lord," 1 Cor. 7.10 (Spedding's trans.).

elation, beginning with the latter because it coheres most with what he has just "handled." The nature of revelation has three parts: limits of the information, the sufficiency of the information, and the acquiring or obtaining of the information. Limits concern how far persons continue to be inspired, how far the Church is inspired, and what the limits of reason are, a part already shown to be deficient. Sufficiency concerns what points of religion are fundamental and what points are to be further perfected, and how far the differences of light depending upon "the dispensation of times" bear on the sufficiency of belief. It is important to distinguish between fundamental and perfective points because it will abate the fury of the controversies produced by failure to attend to the proper limits of reason. Obtaining information depends on the proper interpretation of the scriptures, which can be done either by method or "solutely" and "at large." The former is more likely to be corrupt because it seeks summary brevity, compacted strength, and complete perfection, which lead respectively to extreme dilation, merely apparent strength or overcomplicated weakness, and an impossible completeness. The latter kind of interpretation can be "curious and unsafe," but it must be confessed that scripture is not like profane books because the author of scripture knew four things unknown by man: the mysteries of the kingdom of glory, the perfection of the laws of nature, the secrets of the heart of man, and the "future succession of all ages."

Regarding the kingdom of glory and the laws of nature, analogical and philosophical senses are drawn. Analogy gives men some liberty, although it cannot be pressed too far lest it intoxicate the mind. Regarding philosophical senses, to find all natural philosophy in the scripture, condemning all other philosophy, is to impute an enmity between God's word and his works that does not exist for God. Also, this debases the scripture, for though heaven and earth will pass, God's word will not.[8] God's purpose in the scripture is not to "express matters of nature" except "in passage." Regarding men's hearts and the succession of time, it is important to distinguish scripture from all other books, for Jesus and the scripture speak to and are written for men's thoughts and the succession of all ages, anticipating heresies, contradictions, and differing estates of the Church. For this reason the words of scripture are not limited to one time and place but rather extend to all. Therefore they are not to be interpreted as one would a profane book. Bacon thinks that the interpretation of the scriptures is not deficient, but he comments that the best observations on

8. Mark 13.31.

the scripture are to be found in Britain in the "space of these forty years and more."

Bacon divides the matter of divinity into two parts, belief and opinion of truth, and matter of service and adoration, with the former being the judge and the director of the latter, the soul of religion itself. The heathen religion was itself an idol because it had no soul, being doctored by poets and having gods that were not jealous. From these two parts come the four branches of divinity: faith, manners, liturgy, and government. The matter of divinity is handled either by instructing in truth or confuting falsehood or the declinations from religion, including, in addition to the privative, or atheism, heresy, idolatry, and witchcraft.

Bacon "passes over" the matter of divinity "so briefly" because he does not think it deficient, and he thereby concludes his "globe of the intellectual world." His intention has been to amend and make proficient, not to change and to differ. And as with anything "well set down," if a first reading moves an objection, he intends for a second reading to make an answer. His errors are his own, but he has not made them worse by litigious arguments. The good, if any, is due *tanquam adeps sacrificii*,[9] to be burned to honor first God and then the king, to whom on earth Bacon is "most bounden."[10]

Bacon concluded the last chapter by returning to the question of the gods. There we learned that the attempt to assure the truth of atheism—the displacing of the creator god by the artful mastery of nature—is actually the result of divine revenge as this phenomenon has been explained in the course of the treatise. The modern project, we understand, can never be free from the question of the gods. Therefore it is fitting that Bacon concludes the treatise by dividing the knowledge of sacred and inspired divinity. It is likewise fitting that his discussion concerns the difference and the relations between reason and belief.

Bacon divides divinity from the rational sciences of nature and man by distinguishing between knowledge and belief. This distinction depends upon the difference between the heavens that show God's glory and that are the objects of sense and God's word that shows his will and that is the object of the conscience. But this distinction provokes

9. "As the fat of the sacrifice" (Spedding's trans.), *Advancement* 491; Lev. 1.8.
10. As befits Bacon's practice in the *De augmentis*, this section is greatly shortened in the translation. He will keep silent, he says, because divinity is beyond reason and philosophy (*De aug.* 829).

an obvious question: does not one have to hear even a divine word, and is not hearing as much a sense as is sight, touch, taste, or smell? Bacon answers this question affirmatively in two different ways, one direct and the other indirect. First, at the very outset of the division of the sciences, Bacon provisionally divided divinity according to the categories of human learning because God's "oracle" is received by the human spirit, which he says is "the same" with regard to what is received even though the "revelation of oracle and sense be diverse."[11] Bacon means that, even though the source of revelation is different from the source of what is sensed, the human spirit can receive God's revealed oracle because the human spirit and divine oracle are akin. From this argument, Bacon concludes that divine learning is divided into divine history, divine poetry, and divine doctrine or precept. But this threefold division corresponds to the three faculties of the understanding—memory, imagination, and reason—each of which is inseparable from the senses, as Bacon makes clear in drawing his present distinction between sensory knowledge and belief. He says here that belief springs from the spirit rather than sense, but from the earlier context at the beginning of the division it is clear that the human spirit is nothing but memory, imagination, and reason, all of which operate in some fashion through the senses.

Second, the present distinction between the heavens that show divine glory and the word that shows God's will has been denied in the course of the treatise: we have learned that for the creator God, all divine phenomena, divine will and works, are merely workmanlike. And in the present context Bacon confirms this point by his subtle reference to the psalm that explains the heavens' glory. The psalm proclaims that divine glory is workmanlike and that this glory is expressed by the heavens rather than by divine law and testimony. But Bacon is silent about the second verse of the psalm he quotes, where it is said that, since the heavens declare God's glory, the night to night declares knowledge and the day to day issues speech.[12] Of course the day to day and night to night require and entail each other. If the psalm confirms Bacon's point, then what can be known about God and God's creation is inseparable from what can be believed of divine speech given to hearing, a sense that like the others reports to knowing reason.

The result of these affirmative answers is that reason and belief cannot be separated, precisely because divine work and divine law are really the same. But for the new knowledge, the artful conquest

11. *Advancement* 329.   12. Ps. 19.1–4.

of nature can therefore never wholly overcome the claims of conscience. This consequence is important in the light of Bacon's comment about the heathen opposition of nature to law and morality. Bacon argues that there is some conformity between them but that the knowledge of virtue, justice, and the good, and their opposites, is perfected in conscience, which cannot inform law. For the new knowledge, the opposition of nature to law and morality is replaced by the opposition of nature and law to conscience. But if in fact reason and belief are not really separate, then neither can the knowledge of nature, law, and conscience be separate. And conscience, we know, can always be harshly contentious and invisibly subversive.

By Bacon's argument, the new knowledge will always have to contend with the subversion of law by conscientious belief, especially as nature conquered is pressed to harmonize with the human good as opposed to the merely legal. In distinguishing between the heathen distinction and divine conscience, Bacon has Virgil speak for the latter and Ovid speak for the former. It is fitting that Bacon has two Romans speak for the full legacy of the Jews. Considered together, the two contexts to which Bacon refers show again, and finally, that the convergence of knowledge, law, and conscience is the very form of divine revenge.

Virgil may have thought Venus unjust when, fearing Jupiter's wrath, she diverted Aeneas from his quest and thrust him into Dido's arms.[13] But Virgil would have thought so because he did not understand that a creator god can never be just, so that Venus could well have been less unjust than Jupiter in opposing him as she did. This we know from Bacon's reference to Ovid as the spokesman for divine conscience: if for the new knowledge law, nature, the good, and conscience converge, then they must countenance Myrrha's natural but revolting desires, about which Bacon is silent but about which the context in Ovid speaks.[14] The real legacy of Aeneas' Roman quest is the charitable conquest of nature, and we know by now that Aeneas' success, not his temporary diversion by Venus, served Jupiter's jealous wrath. But we could not expect a creator god to do otherwise, because such unjust divine revenge is simply an essential fact of all human beginnings: it is simply the utopian dream that sets every order of productive arts on its way; it is coeval with man's political nature, which cannot be outstripped as long as there is a difference between the human, the bestial, and the truly divine. Bacon knows why

13. Virgil, *Aeneid* 1.305–417.
14. Ovid, *Metamorphoses* 10.229–55; Plutarch, *Alexander* 65.

there is no divine speech about justice in Plato's account of the best city. He knows why Plato's speeches tell us that perfect justice is impossible.

If we recall the discussion in the Introduction, the justice-loving Timaeus, Glaucon, and Thrasymachos were correct to imply that Zeus was the unjust source of the divine purge that destroyed the Athenians. But these three did not see that their necessary love of justice—their robust patriotism— is the reflection of such divine injustice. Because Bacon knows that it is, there is no divine speech about justice in his account of the best city. Although he serves the likes of Socrates' interlocutors by promising perfect new beginnings, Bacon imitates the form and agrees with the substance of Plato's speeches. Both Plato and Bacon know that the city is always the home of the creating, speaking Olympian gods who set the productive arts on their way. But both Plato and Bacon write to remind us that the city is also the home of the silent gods before whom the political philosopher is struck by wonder. They tell us that the tension between these gods can never be overcome and that the belief that it can is an ever-present danger of the human condition. As such a belief is pursued, by some at any time and by Socrates' three interlocutors, the worst of divine rancor occurs. In modern times, such a belief must be tempted and managed more boldly than Plato thought possible, but only because the creator god has found a subtler and more tempting voice with which to speak.

Therefore, in modern times it is important for some to remember that the mastery of nature will never overcome the harsh and contentious beliefs that interpret the unjust creator god's will. It is certainly true, then, as Bacon says in the present context, that to establish the boundaries of reason and belief would prevent the furious controversies that wrack the universal Christian empire. Bacon's example of a wise understanding of such boundaries is Paul, and no wonder, since above all else the new project for learning is charitable. But in the context to which Bacon refers, Paul did not speak for his own opinion only, separating it from God's wisdom; rather Paul spoke for himself and for God. And to speak as Paul spoke for himself could never answer the question of whether Paul or any other interpreter of revelation speaks for God or for himself.[15] According to Bacon's argument, there is always the need for new signs that attend our need for scripture, or the likes of scripture, so the limits of

15. 1Cor. 7.10, 12, 40.

information and revelation concern how far individuals and the Church, or their likes, are or continue to be inspired. Controversy about signs can be decided only by reference to the end of days, but because they are always farther off, they are themselves interpreted only by way of controversial signs.

In the modern age, human art aspires to know the end of days by masterfully causing them to occur. But this striving is always vain—it is never free from the need for signs and the conscience that interprets them. This modern project can quiet the doctrinal controversies of Christendom, but it cannot prevent, and in fact it assures, the reappearance of these controversies in other conscientious and universal guises. And it cannot always prevent their service to the harsh and unjust claims of the tyrant or of any partisan whose love of justice is the same as tyrannical ambition. No perfected human art can overcome the need for controversial political belief about the whole. The new project will always be threatened by the danger of the creator god's call: the desire for mastery and moral certainty can blind us to the knowledge that belief is always necessary, rather than true. But this is the knowledge that can inspire wonder, the only true ground for moderating every belief.

With these remarks we bring our study to a close. There is no need to summarize Bacon's argument, for that would be to turn too quickly from the riches that have been passed over or missed and to turn too quickly from the need to think again about what has so far been discovered. And to spin out the implications of his argument for our contemporary experience would require more than one additional book. But even so, we have to remember that we began by seeking some bearings for grasping the present problem of technology. And in the light of this beginning, some very general remarks can and should be made.

After coming to grips with Bacon, we should not be surprised to find that the technological project seems to tie the great, competing political systems of our day together, because this project is rooted not just in some accident of time and place but rather in the very fabric of political life as such. It should likewise be no surprise that the scientific promise of liberty seems belied by scientific determinism. The reason is that the liberty promised by the *new* technological project—the one begun by Bacon—is grounded upon the charitable understanding of need. And it should now be no surprise that our rush to human self-reliance is accompanied by a search for missing gods, because the gods can never really take leave of political life and

329

because the new technological project is rooted in man's necessary, political openness to the divine and is modeled upon the charitable promise of Christianity, whose defects it was designed to mitigate.

It is not surprising that in the modern age we are confused about the relations between means and ends. Modern idealism and materialism make these relations obscure because they ignore what it means for human life to be situated between the bestial experience of urgent need and the divine splendor of what never has to move. From Bacon we learn that the modern theoretical enthusiasms are but opposite sides of the same charitable coin: our crafty, Machiavellian realism is in fact naive utopian idealism, and our moralistic idealism is in fact naive utopian realism. These two faces of naive utopianism are simply inseparable from political life because human beings are never free from the need to order the productive arts. In this regard, then, the "problem of technology" is not wholly new. Our present times differ from others in regard to how this problem is to be managed, for in our age naive utopian hope is deflected to the mastery of nature's ways and means. But from Bacon we learn that the new project for managing naive utopianism is itself charitable and naively utopian, so that there is always the danger that the principles of the cure will obscure the true nature of the disease. Or to be more precise, there is always the danger that we will take the promise of modern cures too seriously.

It should by now be obvious that to understand the ever-present danger of naive utopianism, we must look to the doctrine of the realistic utopians. This is primarily located in the teachings of the later ancients, but for modern times especially, it is located in the treatise we have just examined. For understanding the modern age, the difference between naive and realisitic utopianism is decisive. Naive utopianism is the necessary hope of man the productive animal. Realistic utopianism is the teaching about naive utopianism that comes from thinking about man the political animal. The roots of the modern age are utopian in both senses: the modern scientific project is the most manageable form of Christian charity, the sinful version of naive utopianism, and the one who first recommended this project, Bacon, showed the way to the older, realistic utopian wisdom we will always need.

From Bacon's realistic utopian teaching we can take a clue as to how we might gauge the effects of the modern technological project on our own democratic liberties. These liberties rest on the doctrine that free government must represent the equal rights of individuals who are equal by way of these rights. As this rights tradition sprang from the Hobbesian interpretation of Machiavelli's political science, the right that all share equally and from which all other rights derive is the right to acquire for the sake of needs that can be separated from any controversial political claim. And as this doctrine of right was transformed into the tradition of liberal, representative democracy, it was from the start associated with the promise that such liberated acquisition would master nature and so would facilitate "long life and commodious living," the very conditions upon which it is thought that acquisition need not be politically controversial.

However powerful this teaching is, and however much it has secured a liberty heretofore unknown to mankind, Bacon teaches us that it is grounded upon the potentially dangerous, and charitable, delusion of the productive arts that takes political controversy to be caused by scarcity rather than vice versa. And the same could be said for the idealistic, or Kantian, interpretation of the doctrine of equal and abstract rights, which construes rights not as licenses for acquisition but as the hedges of equal "moral dignity." For just as the materialistic interpretation is grounded upon the delusional moral orientation by the body's needs alone, the "Kantian" interpretation is grounded upon the delusional orientation by the mind's freedom alone. Each is in fact the mirror image of the other, and both are at once the product of our technological nature and encouraged by the necessary, modern management of that nature. Both views are naively utopian. As such they tempt us to hope for a justice so perfect that it is no longer political. This is the real "problem of technology," and to the extent that liberal democracy depends upon the promise of technology, what Bacon taught following the later ancients is true in our time: the science of government will always be hard, and good government will be rare.

From Bacon we learn that the roots of the modern project are both realistically and naively utopian. Therefore, we should not think that our age is wholly new and not tied to a wisdom older than its founders. This latter fact is the source of possible moderate hope. But Bacon was not naive; he did not think a return to older times possible, especially since, though there may be an older wisdom, there can never be older times if there are really no wholly new ones. Bacon began the project that promised to mankind unfettered freedom, but when we read his work, we see that, to the extent that he did, the proximal beginnings of our once cheerful age were both sober and realistic after the manner of the later ancient utopians. Perhaps it

331

was because Bacon so deftly combined the new promise of freedom with the realistic knowledge of its limits that Rousseau, the only other modern realistic utopian, called him *le plus grand, peut-être, des Philosophes.*[16]

16. J. J. Rousseau, *Discours sur les sciences et sur les arts*, ed. François Bouchardy (Paris: Gallimard, 1964), 29.

INDEX

333

nus Pius, 204; and charity, 54, 57-58, 61-63, 67; on meat eating and vege-tarianism, 54-57, 59, 61-63; and Olympian gods, 67; warns against phi-losophy, 50-51, 62. *See also* Machiavelli
Peoples and princes, 82, 106-107, 117, 128, 161, 315
Pertinax, 106
Philo Judaeus, 63-66
Philosophy: and art, 184; difference from political philosophy, 30-32; and divinity, 308-309, 320; and power and wealth, 315; and scripture, 324; source of in political life, 24, 53. *See also* Classical utopian thought; Political philosophy
Phocion, 73, 83-87, 113. *See also* Socrates
Physicist's dream, 66-67, 258
Physics, 60-62. *See also* Cause; Material-ism; Metaphysics
Pindar, 289, 300
Pius V (pope), 72, 80
Plato, 20, 28, 70, 87, 101, 103, 149-150, 172, 262-264; on age of Cronos, 24; attacked by Machiavelli, 26; criticizes sophists, 47-48; and Dionysius, 177; on doubt and certainty, 189-190; on final cause, 249; and first philosophy, 252-253; on form, 247, 255-256; on the Good, 183-185; on knowledge and remembrance, 43-44, 47; on moral virtue, 279; and Parmenides on unity, 248; on persuasion and force, 111-114; on philosopher kings, 202, 208; *Republic*, 22-24, 26, 327-328; *Republic, Timaeus,* and *Critias,* 28-34; on rheto-ric, 273, 276-277; on separateness of the arts and sciences, 182-187, 268-269; on Socrates, 121; *Theaetetus,* 68-69; on virtue and knowledge, 161. *See also* Atlantis; Beginnings; Charity; Ci-cero; Cosmogony; *New Atlantis*
Plautus, 289, 300, 315
Pliny the Elder, 205, 218
Pliny the Younger, 203, 214, 291-292
Plutarch, 149-150, 252; on Alexander and Achilles, 218-219; on architecture of fortune, 305-311, 314, 316, 319-321; on Cato the Elder, 88; on Cato the Elder and Socrates, 75, 88; *De esu carnium,* 59-62; on Demosthenes, 85-86, 117-118; on myths and politics, 143-144; on Phocion and Cato the Younger, 83-84
Poetry, 239-243; and the imagination,

261, 265; and the mastery of nature, 241. *See also* Technology
Poets, 265-268; intermediate model of virtue, 266-267; knowledge of the city, 230; movers of God, 89, 242, 271; place in city, 239, 242-243, 259-260, 267-268, 281; quarrel with political philosopher, 266; superior to philoso-phers, 241-242. *See also* Affections; Moral virtue; *New Atlantis;* Political philosopher; Rhetorician; Sophists; Statesman
Political philosopher: difference from statesman, 113-114; model for states-man, 237, 243, 250-251; model for virtue, 25, 153-154, 186-187, 266, 281, 293-294; place in city, 53, 185-186, 239, 258-260, 268, 281, 293, 328; and poets, 266-268. *See also* Poets; Socrates
Political philosophy, 24, 30-31; and af-fairs of the world, 237; and art, 67, 258; and knowledge and belief, 162; and philosopher's heaven, 284, 292-295; and poetry, rhetoric, and sophis-try, 76, 243; and virtue and modera-tion, 258. *See also* Classical utopian thought; Philosophy; Political philoso-pher; Socrates
Political science: and the affections, 290-291; Bacon's description of, 34-35, 309-313; Machiavelli's, 92-94, 100-101, 282; new and old, 282-283, 296, 298, 301; not taught to young, 78-79, 94, 109-110, 280-281, 290-291, 295, 299, 311; and rhetoric, 281-282; sta-tus as a science, 280; taught to young, 108. *See also* Civil philosophy; Con-science; Progress
Polyphemus, 230, 235
Pompey the Great, 48, 124, 305-306, 316
Poverty, 95-101, 158-159
Praise and blame: Machiavelli's com-mand of, 107; tyrant's command of, 105-106
Pride, 78, 101; cause of fall, 53-54
Productive art: and convention, 69-70; and desire for certainty, 184-190; and human nature, 21-22; model for all the arts and sciences, 169, 259; model for just city, 28-34, 208-210; place in city, 243; and politics, 175-176, 258; and reason, 23, 27, 32, 62, 88, 113-114, 191; and rhetoric, 276; social or-

Library of Congress Cataloging in Publication Data

Weinberger, J.
    Science, faith, and politics.

    Includes index.
    1. Bacon, Francis, 1561–1626. Advancement of learning. 2.
Science—Methodology. 3. Logic,   Modern—17th   century.   I. Bacon,
Francis, 1561–1626. Advancement of learning.   II. Title.
B1194.W45   1985   121   85-47707
ISBN 0-8014-1817-8 (alk. paper)